Lecture Notes in Geoinformation and Cartography

Qiming Zhou · Brian Lees · Guo-an Tang (Eds.)

Advances in Digital Terrain Analysis

Springer

Editors

Prof. Qiming Zhou
Hong Kong Baptist University
Department of Geography
Kowlon Tong
Kowlooon
Hong Kong/PR China
qiming@hkbu.edu.hk

Prof. Brian Lees
The University of New South Wales at ADFA
ACT 2600
Australia
b.lees@adfa.edu.au

Prof. Guo-an Tang
Nanjing Normal University
Key Laboratory of Virtual Geographic Environment
210046 Nanjing
China, People's Republic
tangguoan@njnu.edu.cn

ISBN: 978-3-540-77799-1 e-ISBN: 978-3-540-77800-4

Lecture Notes in Geoinformation and Cartography ISSN: 1863-2246

Library of Congress Control Number: 2008921722

Cover design: deblik, Berlin

Printed on acid-free paper

9 8 7 6 5 4 3 2 1

springer.com

Preface

The chapters in this book mostly started as presentations at the Terrain Analysis and Digital Terrain Modelling conference hosted by Nanjing Normal University in November 2006. As far as I am aware this was the first international conference devoted specifically to this area of research, and since it was also my first visit to China it was an exciting and unique experience for me. The participants ranged from leaders in the field discussing visions and challenges for the future to students grappling with the possibilities and exploring new directions. These papers are a selection of the many presentations at the conference and give some indication of the breadth of research on show at the meeting.

Digital terrain analysis has moved beyond a research tool into routine application, such as determination of catchment areas and flow pathways in hydrological analysis, supporting soil mapping through spatial prediction and the definition of landform elements, and the use of slope and other attributes for land capability analysis. But there are still many areas of active research refining these methods or exploring new approaches, as this book shows.

One recent development explored in several of the papers in this book is the availability of global or near-global terrain data in several forms, GTOPO-30 and SRTM 3 second data being the most significant. Reliable global topographic data opens the doors for truly global analysis, consistent analysis on different continents and the generation of collective experience that is transforming the field of geomorphometry into a robust science.

Another theme reflected in these papers is the increasing sophistication in our understanding of issues related to scale, accuracy, uncertainty and error propagation in digital terrain analysis. As these methods are increasingly used to support important decisions, information on uncertainty becomes vital for the rational use of predictions. There is still some way to go before we have tools for estimating and representing uncertainties that meet the needs of our user community.

Other papers demonstrate the continued demand for improved methods to classify and segment the land surface into useful units for land management or mapping; showcase innovations in representing and characterising the land surface; highlight a growing focus on processes rather than statistical correlations for understanding the earth's surface; and exemplify the ongoing development and testing of new algorithms addressing deficiencies in quality and efficiency of existing methods.

At the Nanjing conference, I was astonished by the number of students from China and elsewhere training in this research area and by the variety and innovation of their work. I was also impressed by their probing questions and contributions to the discussions. The conference provided an opportunity to renew some old friendships, make new friends and meet for the first time some of the people whose names I knew from their published papers. I greatly enjoyed the interaction with so many disciples in thc ficld of terrain analysis and consider myself fortunate to have had the opportunity to participate in this meeting. I am hopeful of many more stimulating and rewarding meetings and discussions as part of the TADTM initiative in the coming years.

Dr John Gallant
CSIRO Land and Water, November 2007

Contents

List of Contributors **xi**

Introduction **1**

ZHOU Qiming, Brian G. LEES and TANG Guo-an 3
Advances in Digital Terrain Analysis: The TADTM Initiative

Section 1: Digital Representation for Terrain Analysis **11**

George Ch. MILIARESIS 13
Quantification of Terrain Processes

Peter A. SHARY 29
Models of Topography

LI Zhilin 59
Multi-Scale Digital Terrain Modelling and Analysis

ZHAO Xuesheng, BAI Jianjun and CHEN Jun 85
A Seamless and Adaptive LOD Model of the Global Terrain Based on the QTM

Section 2: Morphological Terrain Analysis **105**

TANG Guo-an and LI Fayuan 107
Landform Classification of the Loess Plateau Based on Slope Spectrum from Grid DEMs

Josef STROBL 125
Segmentation-based Terrain Classification

Lucian D. DRĂGUȚ and Thomas BLASCHKE 141
Terrain Segmentation and Classification using SRTM Data

LU Huaxing 159
Modelling Terrain Complexity

LIU Aili 177
DEM-based Analysis of Local Relief

YANG Qinke, David JUPP, LI Rui and LIANG Wei 193
Re-Scaling Lower Resolution Slope by Histogram Matching

Section 3: Hydrological Terrain Analysis **211**

John P. WILSON, Graeme AGGETT, DENG Yongxin and Christine S. LAM 213
Water in the Landscape: A Review of Contemporary Flow Routing Algorithms

Petter PILESJÖ 237
An Integrated Raster-TIN Surface Flow Algorithm

TIAN Yuan, WU Lun, GAO Yong, WANG Daming and ZHANG Yi 257
DEM-based Modelling and Simulation of Modern Landform Evolution of Loess

Section 4: Uncertainty in Terrain Analysis **277**

ZHOU Qiming and LIU Xuejun 279
Assessing Uncertainties in Derived Slope and Aspect from a Grid DEM

LIU Xuejun and BIAN Lu 307
Accuracy Assessment of DEM Slope Algorithms Related to Spatial Autocorrelation of DEM Errors

DENG Fengdong, WANG Lili, ZHUO Jing and LIU Anlin 323
Modelling Slope Field Uncertainty Derived From DEM in the Loess Plateau

ZHU A-Xing, James E. BURT, Michael SMITH, WANG Rongxun and GAO Jing 333
The Impact of Neighbourhood Size on Terrain Derivatives and Digital Soil Mapping

Brian G. LEES, HUANG Zhi, Kimberley VAN NIEL and Shawn W. LAFFAN 349
The Impact of DEM Error on Predictive Vegetation Mapping

Section 5: Applications of Terrain Analysis 363

Igor V. FLORINSKY 365
Global Lineaments: Application of Digital Terrain Modelling

John B. LINDSAY and James J. ROTHWELL 383
Modelling Channelling and Deflection of Wind by Topography

ZHANG Ting, LI Jun, WANG Chun and ZHAN Lei 407
Spatial Correlation of Topographic Attributes in Loess Plateau

YANG Xin and XIAO Chenchao 425
Terrain-based Revision of an Air Temperature Model in Mountain Areas

James R.F. BARRINGER, Allan E. HEWITT, Ian H. LYNN and Jochen SCHMIDT 443
National Mapping of Landform Elements in Support of S-Map, A New Zealand Soils Database

Concluding Remarks 459

Brian G. LEES 461
Progress in Digital Terrain Analysis

List of Contributors

Graeme **AGGETT**, Riverside Technology Inc., 2290 East Prospect Road, Suite 1, Fort Collins, Colorado CO 80525
E-mail: gra@riverside.com

BAI Jianjun, Department of Surveying, China University of Mining and Technology (Beijing), D11, Xueyuan Road, Beijing 100083, P.R. China

James R. F. **BARRINGER**, Landcare Research, PO Box 40, Lincoln 7640, New Zealand, Email: barringerj@landcareresearch.co.nz

BIAN Lu, Key Laboratory of Virtual Geographic Environment, Nanjing Normal University, Ministry of Education, Nanjing, Jiangsu 210046, P.R. China

Thomas **BLASCHKE**, Researchstudio iSPACE, ARC Austrian Research Centers, Leopoldskronstr. 30, 5020 Salzburg, Austria

James E. **BURT**, Department of Geography, University of Wisconsin-Madison, 550 N. Park St, Madison WI 53706, USA

CHEN Jun, National Geometric Centre of China, No.1 Baishengcun, Zizhuyuan, Beijing 10004, P.R. China
Email: chenjun@nsdi.gov.cn,

DENG Fengdong, Shaanxi Remote Sensing Information Centre for Agriculture, Email: phoenixlet@yahoo.com.cn

DENG Yongxin, Department of Geography, Western Illinois University, Macomb, IL 61455, E-mail: y-deng2@wiu.edu

Lucian D. **DRĂGUȚ**, GIS-Centre for Geoinformatics, Salzburg University, Schillerstr. 30, 5020 Salzburg, Austria
Email: lucian.dragut@sbg.ac.at

Igor V. **FLORINSKY**, Institute of Mathematical Problems of Biology, Russian Academy of Sciences, Pushchino, Moscow Region, 142290, Russia, Email: iflorinsky@yahoo.ca

GAO Jing, Department of Geography, University of Wisconsin-Madison, 550 N. Park St. Madison WI 53706, USA
Email: jgao3@wisc.edu

GAO Yong, Institute of RS and GIS, Peking University, Beijing, 100871, P.R. China

Allan E. **HEWITT**, Landcare Research, PO Box 40, Lincoln 7640, New Zealand

HUANG Zhi, The Australian Government Department of Environment and Water Resources, Email: zhi.huang@environment.gov.au

David **JUPP**, CSIRO Marine and Atmospheric Research, CS Christian Building, CSIRO Labs, Clunies Ross St., Black Mountain ACT, 2601, Australia

Shawn W. **LAFFAN**, School of Biological, Earth and Environmental Sciences, University of New South Wales, Australia
Email: shawn.laffan@unsw.edu.au

Christine S. **LAM**, GIS Research Laboratory, Department of Geography, University of Southern California, Los Angeles, CA 90089-0255, Email: csl@usc.edu

Brian G. **LEES**, The University of New South Wales at ADFA, Canberra, ACT 2600, Australia, E-mail: b.lees@adfa.edu.au

LI Fayuan, Key Laboratory of Virtual Geographic Environment, Nanjing Normal University, Ministry of Education, Nanjing, Jiangsu, 210046, P.R. China

LI Jun, Key Laboratory of Virtual Geographic Environment, Nanjing Normal University, Ministry of Education, Nanjing, 210046, P.R. China

LI Rui, Northwest University, No 229, Northern Taibai Road, Xi'an 710069, P.R. China

LI Wei, Northwest University, No 229, Northern Taibai Road, Xi'an 710069, P.R. China

LI Zhilin, Dept. of Land Surveying and Geo-Informatics, The Hong Kong Polytechnic University, Hong Kong, Email: lszlli@polyu.edu.hk

LIANG Wei, Northwest University, No 229, Northern Taibai Road, Xi'an 710069, P.R. China

John B. **LINDSAY**, Uplands Environments Research Unit (UpERU), School of Environment and Development, The University of Manchester, Oxford Road, Manchester, M13 9PL, UK,
Email:john.lindsay@manchester.ac.uk

LIU Aili, School of Remote Sensing, Nanjing University of Information Science & Technology, Street No.114 Pancheng New, Nanjing, Jiangsu 210044, P. R. China. Email: ailii66@126.com

LIU Anlin, Shaanxi Remote Sensing Information Centre for Agriculture, Email: phoenixlet@yahoo.com.cn

LIU Xuejun, Key Laboratory of Virtual Geographic Environment, Nanjing Normal University, Ministry of Education, Nanjing, 210046, P.R. China

LU Huaxing, Key Laboratory of Virtual Geographic Environment, Ministry of Education, Nanjing Normal University, No.1 WenYuan Road , Nanjing, Jiangsu, 210046, P.R. China
Email: huaxinglu@163.com

Ian H. **LYNN**, Landcare Research, PO Box 40, Lincoln 7640, New Zealand

George Ch. **MILIARESIS**, Department of Geology, University of Patras, Rion 26504, Greece, Email: gmiliar@upatras.gr

Petter **PILESJÖ**, Lund University GIS Centre, Lund University, Solvegatan 12, SE-223 62 Lund, Sweden
Email: Petter.Pilesjo@giscentrum.lu.se

James J. **ROTHWELL**, Department of Environmental & Geographical Sciences, Manchester Metropolitan University, Chester Street, Manchester, M1 5GD, UK

Jochen **SCHMIDT**, National Institute of Water and Atmosphere (NIWA), PO Box 8602, Christchurch, New Zealand

Peter A. **SHARY**, Institute of Physical, Chemical and Biological Problems of Soil Science, RAS, 142290 Institutskaya Street Bldg. 2, Poushchino, Moscow Region, Russia
Email: peter_shary@hotmail.com

Michael **SMITH**, Department of Geography, University of Wisconsin-Madison, 550 N. Park St., Madison WI, 53706, USA

Josef **STROBL**, Centre for Geoinformatics, Salzburg University, Hellbrunnerstrasse 34, 5020 Salzburg, Austria
Email: josef.strobl@sbg.ac.at

TANG Guo-an, Key Laboratory of Virtual Geographic Environment, Nanjing Normal University, Ministry of Education, Nanjing, 210046, P.R. China

TIAN Yuan, Institute of RS and GIS, Peking University, Beijing, 100871, P.R. China, Email: wulun@pku.edu.cn

Kimberley **VAN NIEL**, School of Earth and Geographical Sciences, University of Western Australia, 35 Stirling Hwy, Crawley WA 6009, Australia, Email: kvn@segs.uwa.edu.au

WANG Chun, Key Laboratory of Virtual Geographic Environment, Nanjing Normal University, Ministry of Education, Nanjing, 210046, P.R. China

WANG Daming, Institute of RS and GIS, Peking University, Beijing, 100871, P.R. China

WANG Lili, Shaanxi City and Country Planning & Design Research Institute

WANG Rongxun, Department of Geography, University of Wisconsin-Madison, 550 N. Park St, Madison WI, 53706, USA

John P. **WILSON**, GIS Research Laboratory, Department of Geography, University of Southern California, Los Angeles, CA 90089-0255, Email: jpwilson@usc.edu

WU Lun, Institute of RS and GIS, Peking University, Beijing, 100871, P.R. China

XIAO Chenchao, Institute of Remote Sensing and GIS, Peking University, Beijing 100871, P.R. China, Email: chenchaox@gmail.com

YANG Qinke, Northwest University, No 229, Northern Taibai Road, Xi'an 710069, P.R. China, Email: qkyang@nwu.edu.cn

YANG Xin, Key Laboratory of Virtual Geographic Environment, Nanjing Normal University, Ministry of Education, Street No.1 Wen Yuan, Nanjing, Jiangsu 210046, P. R. China, Email: xxinyang@163.com

ZHAN Lei, Key Laboratory of Virtual Geographic Environment, Nanjing Normal University, Ministry of Education, Nanjing, 210046, P.R. China

ZHANG Ting, Key Laboratory of Virtual Geographic Environment, Nanjing Normal University, Ministry of Education, 210046, Nanjing, P.R. China, Email: tting.zhang@gmail.com

ZHANG Yi, Institute of RS and GIS, Peking University, Beijing, 100871, P.R. China

ZHAO Xuesheng, Department of Surveying, China University of Mining and Technology (Beijing), D11, Xueyuan Road, Beijing 100083, P.R. China, Email: zxs@cumtb.edu.cn

ZHOU Qiming, Department of Geography, Hong Kong Baptist University, Kowloon Tong, Kowloon, Hong Kong Email: qiming@hkbu.edu.hk

ZHU A-Xing, State Key Laboratory of Resources and Environmental Information System, Institute of Geographical Sciences and Natural Resources Research, Chinese Academy of Sciences, Beijing 100101, P.R. China

ZHUO Jing, Shaanxi Remote Sensing Information Centre for Agriculture, Email: phoenixlet@yahoo.com.cn

Introduction

Advances in Digital Terrain Analysis: The TADTM Initiative

ZHOU Qiming, Brian G. LEES and TANG Guo-an

Background

Digital terrain modelling has been one of the most active research and application fields in geo-spatial information science and technology. Using the techniques of computer graphics, the land surface, or terrain surface, can be represented digitally using large volumes of regularly or irregularly distributed sample points, instead of solely relying on the traditional contours or other cartographic symbolism. The term digital terrain model (DTM) is now widely recognized as the digital representation of the terrain surface for a given geographical region.

Compared to traditional contour maps, Li *et al.* (2005) outlined the specific features of a DTM as:

1. A variety of representation forms,
2. No accuracy loss of data over time,
3. Greater feasibility of automation and real-time processing, and
4. Easier multi-scale representation.

Despite its obvious advantages as listed above, the effective use of DTMs, however, requires more effort than the interpretation of traditional paper maps. Just as terrain information extraction from a contour map requires the techniques of map reading, interpretation, and measurement, deriving terrain features and measurements from a DEM also demands information extraction methods and techniques based on digital representation of the terrain. This leads to the focus of this volume – digital terrain analysis (DTA).

If we use the term digital elevation model (DEM) to refer to terrain models with elevation information only, while the term digital terrain model (DTM) refers to a much broader concept of terrain representation, including terrain parameters such as slope and aspect, terrain features such as ridges and valleys and other geographical/environmental characteristics, DTA specifies the process that transforms DEMs to DTMs, using the principles and knowledge of geography, or other application fields (Figure 1). This process was also previously termed "DTM interpretation" (Hutchinson and Gallant 1999).

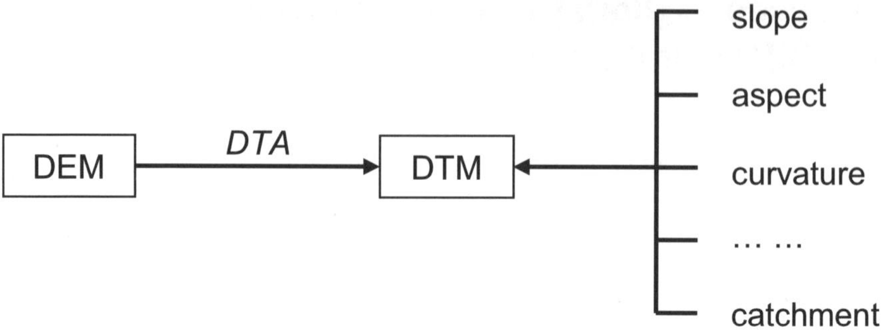

Figure 1. DEM, DTM and DTA.

Development of Computer-aided Terrain Analysis

Computer-aided terrain analysis is not a new concept. It has been an active field of study for some years and has attracted effort from many researchers including geographers, surveyors, engineers, and computer scientists. However, due to lack of communication across various disciplines, the efforts seem to be quite isolated and have mainly focused on problems within individual application fields. For example, in the field of geomorphometry, the research focuses on the extraction of morphological features of terrain and simulation of geomorphological processes (Miliaresis 2001). In the field of hydrology, researchers concentrate on deriving quantitative derivatives from DEMs for hydrological modelling (Band 1999), while in the field of geo-sciences the statistical features of the terrain in a geographical region attract more attention (Davis 2002).

The different application fields share the basic principle of terrain analysis as they fundamentally deal with the same thing – extracting quantitative derivatives from digital terrain data. Thus it is natural to expect the breakdown of the "communication barrier" between different interest groups so that the technology can further advance for problem solving, which will eventually deliver the benefits to everybody. In the past decades, there have been some efforts to bring people in the field of terrain analysis together. The pioneering works of Moore *et al.* (1991) and others are mostly focused on the hydrological applications (Beven and Moore 1991), but potential applications in the fields of geomorphology and biology were also reviewed. Wilson and Gallant (2000) presented a collection of techniques, methods, and applications of terrain analysis in the fields of hydrology, geomorphology, ecology, and soil sciences. Though the majority of the methods and applications were demonstrated through a dedicated software tool – TAPES (Terrain Analysis Programs for the Environmental Sciences), comprehensive and in-depth discussions on principles and algorithms of digital terrain analysis were well covered.

Although the methodology and technology of DTA have been developed well with the rapid advances in geo-spatial information technology, they remain for the most part in the research and development domain. Driven by commercial motivations, today's popular GIS software packages are more interested in DTM visualization and presentation, the quantitative analysis and information extraction from digital terrain data being poorly supported. Even though more advanced and accurate algorithms have already been developed in DTA for some years, their implementation in commercial GIS software packages has been slow and seldom supported.

The poor implementation of DTA technology is largely due to the following reasons:

- Most DTA research is narrowly focused with assumptions and limitations that only apply to the local conditions.
- Quantitative models in many application fields such as geography, geomorphology, ecology, and soil sciences are either poorly developed or too generous to be tested in real-world conditions.
- Lack of awareness of advanced technology in DTA has restricted broader DTA applications.
- Lack of communications and inter-discipline collaborations results in slower progress in the advance of DTA.

With the rapid growth of Geographical Information System (GIS) technology, particularly the establishment of high resolution digital elevation models (DEM) at national level, the challenge is now focused on delivering justifiable socio-economical and environmental benefits, i.e. extracting and presenting parameters and features inherent in the DEM for more direct use in applications. To make this possible, more collaborated inter-discipline effort is undoubtedly needed.

About the TADTM Initiative and the Symposium

In response to the above shortfalls, the Terrain Analysis and Digital Terrain Modelling (TADTM) initiative was proposed in early 2006 and overwhelmingly supported by workers in the DTA fields around the world. The initiative proposed actions such as:

- Identifying research groups that had been active in research fields related to digital terrain analysis.
- Establishing communication platforms (or "access points") such as websites and mailing lists, so that regular exchange of ideas could be conducted.
- Organizing an international workshop on digital terrain analysis.
- Publishing selected research papers in an edited volume or in journals.
- Establishing a workgroup on digital terrain analysis in related international communities.

One of the above actions was the organization of the International Symposium on Terrain Analysis and Digital Terrain Modelling, which was held in Nanjing Normal University, Nanjing, China during 23–25 November 2006. The Symposium attempted to create a platform for better communications and scholarly exchange among researchers in the fields of terrain analysis, geomorphometry, environmental modelling, and geographical information sciences. The Symposium comprises about 60 papers covering broad areas of terrain analysis and modelling, including:

- Feature extraction from DEM
- Terrain classification and spatial analysis
- Terrain modelling and DEM management
- Scales in digital terrain analysis
- Uncertainty in digital terrain analysis
- DTM and geomorphometry
- DTM and land cover modelling
- DTM-based soil-landscape modelling
- DTM-based environmental change modelling and simulation
- Urban DTM

Three keynote papers were presented by John WILSON on "From precipitation to streamflow: simulating the movement of water within landscapes", George MILIARESIS on "Terrain modelling for specific geomorphologic processing", and John GALLANT on "Multiscale methods in terrain analysis". Eight invited papers covering a wide range of topics were also presented in plenary sessions by CHEN Jun, Josef STROBL, Peter SHARY, TANG Guo-an, Petter PILESJÖ, ZHU A-xing, LI Zhilin and Igor FLORINSKY. A panel discussion session was also organized to stimulate free discussion between the audience and panellists. The Symposium was well received by over 100 participants including many young scholars and postgraduate research students, who we believe will form the backbone of the future DTA community.

About this Volume

This volume collects 23 contributions from papers presented in the TADTM 2006 Symposium. The selected papers were first revised by the contributors to meet the publication quality required for book chapters. The manuscripts were then peer-reviewed by an international expert panel and the authors were requested to revise and respond to the criticism and comments raised by the reviewers before the contributions were presented in their final form.

The contents of this volume are divided into five sections:

Section 1 "Digital Representation for Terrain Analysis" focuses on the terrain data model and representation for DTA. From the point of view of geomorphometry, Miliaresis provides a review of the research on quantitative models of terrain processes – the theoretical foundation of DTA. Similarly, Shary describes the mathematical foundations of the representation of digital terrain models and terrain derivatives. Li addresses the issues related to the multi-scale representation of the terrain, which may have great impact on the outcomes from DTA. For DTA at a global scale, Zhao *et al.* propose a seamless and adaptive LOD (level of details) model of the global terrain based on QTM (Quaternary Triangular Mesh).

Section 2 "Morphological Terrain Analysis" contains six chapters focusing on the extraction and interpretation of morphological features and measurement of terrain. Tang and Li propose a statistical approach that employs "slope spectrum" for landform classification in the Loess Plateau of China. Strobl presents a review of the basic conceptual foundations for segmentation in terrain classification. Also on terrain classification, Dragut and Blaschke report on research that segments and classifies Shuttle Radar Topography Mission (SRTM) data into specific landforms using object-oriented image analysis. As well as classification issues, new methods for terrain description are also addressed in this section. Lu proposes the compound terrain complexity index (CTCI), which is made of four traditional terrain morphological indices including total curvature, roughness, local relief, and local standard deviation as a quantitative measurement of terrain complexity. Liu proposes a method for extracting local relief from a 1-km resolution DEM of China to investigate the large-scale geomorphological features. Yang *et al.* describe an approach by slope histogram matching to re-scale a coarser resolution slope histogram into the slope histogram at a finer resolution, so that the statistical characteristics of the higher-resolution DEM derivatives will be maintained to minimize the "slope-reduction" effect.

Section 3 "Hydrological Terrain Analysis" addresses principles and algorithms of hydrological modelling with DTMs. Wilson *et al.* present a comprehensive review of the advances in hydrological applications of DTA with the emphasis on the simulation of the movement of water within landscapes. Pilesjö reports a new algorithm that simulates the flow over a surface defined by a grid DEM. For the study of soil erosion, especially in the Loess Plateau of China, Tian *et al.* report the development of a modern catchment landform evolution model (MCLEM) that describes and simulates the processes of tectonic elevation, weathering, hillslope, and fluvial transportation.

Section 4 "Uncertainty in Terrain Analysis" comprises contributions focusing on the uncertainty and errors in DTA. Zhou and Liu analyse errors in the derived slope and aspect from grid DEMs due to errors in data, algorithm selection, data properties such as precision, grid resolution and orientation, and terrain complexity. A data-independent assessment method is proposed for more objective and accurate evaluation of DTA algorithms. Liu and Bian analyse the impact of spatial autocorrelation of DEM data error, estimate the accuracy of selected slope algorithms accordingly, and then design a Monte Carlo simulation experiment to validate the results. Deng *et al.* study the uncertainty of derived slope field (i.e. farmland with a steep slope) related to the scale of DEM data. On more application oriented studies, Zhu *et al.* examine the combined effect of DEM resolution and neighbourhood size on computed terrain derivatives and its impact on digital soil mapping, while Lees *et al.* analyse the impact of DEM error on the derived indices, which in turn influence predictive vegetation mapping for landuse and land cover classification.

Section 5 "Applications of Terrain Analysis" contains contributions that report the newest developments in DTA applications. Florinsky examines the hypothesis for the existence of hidden global linear (helical) structures, which are tectonically and topographically expressed, using 18 topographic variables derived by DTA based on a 30-arc-minute grid DEM for the entire surface of the Earth. Lindsay and Rothwell present and evaluate a new index of exposure/sheltering to wind, the channelling/deflection index (CDI), which is capable of modelling channelling and deflection of flowlines and shadowing of wind. Tested in 47 different loess landforms represented by a DEM with 5 m resolution, Zhang *et al.* propose a spatial correlation model for nine selected terrain attributes for deriving quantitative terrain parameters and landform recognition. Yang and Xiao show the use of DTA for surface temperature estimation by constructing a terrain reversed model, which estimates surface temperature by simulating insolation on each grid cell of a DEM. Barringer *et al.* report on a project where a national dataset of landform elements derived from a 25 m resolution national DTM has been used to underpin soil-landscape modelling in the complex steepland terrain of New Zealand.

Acknowledgement

The editors would like to express thanks to the organizations and individuals who contributed to the International Symposium on Terrain Analysis and Digital Terrain Modelling and the publication of this volume, especially to the following:

The Key Laboratory of Virtual Geographical Environment, Ministry of Education, Jiangsu Provincial Key Laboratory of Geographical Information Sciences and School of Geographical Science, Nanjing Normal University hosted the Symposium and provided financial and administrative support.

National Natural Science Foundation of China, China Association of Geographical Information System, and Cartography and GIS Special Commission of the Geographical Society of China supported the Symposium.

Professor LU Guonian, director of The Key Laboratory of Virtual Geographical Environment, Ministry of Education, Jiangsu Provincial Key Laboratory of Geographical Information Sciences and School of Geographical Science, Nanjing Normal University provided unreserved support to the Symposium.

Professor LI Zhilin of Hong Kong Polytechnic University provided constructive suggestions and comments on the publication of this volume.

David TAIT of Giffnock Editorial Services of United Kingdom provided services for editing and formatting the volume.

All contributors have shown their enthusiasm and support in the past year for the publication of this volume.

The efforts of anonymous referees who reviewed all contributions of this volume are gratefully acknowledged.

The publication of this book is financially supported by the National Natural Science Foundation of China (Project 40671148 and 40571120).

References

Band, L.E., (1999), Spatial hydrography and landforms, In Longley, P.A., Goodchild, M.F., Maguire, D.J. and Rhind, D.W. (eds.): *Geographical Information Systems*, New York: John Wiley & Sons: 527–542.

Beven, K.J. and Moore, I.D. (eds.), (1991), *Terrain analysis and Distributed Modelling in Hydrology*, Chichester: John Wiley & Sons.

Davis, J.C., (2002), *Statistics and Data Analysis in Geology*, 3rd Edition, New York: John Wiley & Sons.

Hutchinson, M.F. and Gallant, J.C., (1999), Representation of terrain, In Longley, P.A., Goodchild, M.F., Maguire, D.J. and Rhind, D.W. (eds.): *Geographical Information Systems*, New York: John Wiley & Sons: 105–124.

Li, Z., Zhu, Q. and Gold, C., (2005), *Digital Terrain Modelling: Principles and Methodology*, Boca Raton: CRC Press.

Miliaresis, G., (2001), Geomorphometric mapping of Zagros Ranges at regional scale, *Computers and Geosciences*, **27**: 775–786.

Moore, I.D., Grayson, R.B. and Ladson, A.R., (1991), Digital terrain modelling: a review of hydrological, geomorphological and biological applications, *Hydrological Processes*, **5**: 3–30.

Wilson, J.P. and Gallant, J.C. (eds.), (2000), *Terrain Analysis: Principles and Applications*, New York: John Wiley & Sons.

Section 1: Digital Representation for Terrain Analysis

Quantification of Terrain Processes

George Ch. MILIARESIS

Abstract

Terrain processes quantification requires an object terrain segmentation framework allowing the partition of the landscape to either a continuous framework (aspect regions) or a discontinuous framework (landforms). Each object is parametrically represented on the basis of its spatial 3-dimensional arrangement and mapped according to a terrain classification scheme in an attempt to identify regions that include objects with distinct parametric representation. Case studies are presented that include tectonic, fluvial and aeolian, and gravity (landslides) processes quantification in both the Earth and Mars.

Keywords: fluvial, morphotectonic, aeolian, terrain segmentation, terrain pattern recognition, spatial decision making.

1 Processes and Landforms

The Earth's surface is comprised of relief features of different scales. A scale dependent generalized classification assigns continents to the 1st relief order, mountain ranges and basins to the 2nd order, alluvial fans to the 3rd order, while sand dunes are assigned to the 4th order (Pandey 1987).

The relief features are the result of endogenic and exogenic processes that shape the Earth's surface. Exogenic processes include denudation (that is the downwasting of land surfaces due to erosion, gravity forces, weathering, etc.) and deposition (the filling up with sediments) (Summerfield 1996). Endogenic processes are associated with geotectonics and include volcanism, faulting, crustal warping, etc. (Summerfield 2000). The relief features recognized on the Earth's surface, often called landforms, could be the result of different kinds and intensities of processes. Landforms are defined as natural terrain units, which might be developed from the same soil and bedrock or deposited by a similar combination of processes and, under similar conditions of climate, weathering, and erosion, exhibit a distinct and predictable range of visual and physical characteristics (Lillesand and Kiefer 1987).

The study of landforms and the recognition of the various processes acting are of great importance in both geomorphologic and terrain analysis studies for site evaluation and site selection. This fact gave rise to geomorphometry, which involves subdividing a landscape into landforms based on a terrain segmentation methodology and measurement of their size, shape, and relation to each other (Evans 1981). During its initial stages, geomorphometry concentrated mainly on drainage basin analysis from topographic maps, since basins could be defined in a rather continuous way in the majority of geomorphologic landscapes evident in mid-latitudes. The historical steps in the development of geomorphometry involved:

- Orometry, the 19th-Century measurement of mountains was an attempt to interpret landscape evolution and physical process that reflect the interplay of mountain building and erosion in regions of active deformation. Today, the mountain topography (Miliaresis 2001a) is of great significance, not only in tectonic geomorphometry but also in terrain analysis, in navigation of airplanes, and in the InSAR processing chain.
- Physiography corresponds to the regional-scale geomorphologic studies (1st and 2nd order landforms) in the early part of the 20th century (Miliaresis and Argialas 1999). Physiographic analysis was based on the partition of terrain into physiographic units by taking into account the form and spatial distribution of their component features through fieldwork and visual interpretation of topographic maps. Today, physiography is being stimulated by the need to explain enigmatic landscapes, newly explored on the surfaces of other planets through remotely sensed data (Miliaresis and Kokkas 2004).
- Terrain analysis corresponds to large scale geomorphometry and it involves the systematic study of pattern elements relating to the origin, morphologic history and composition of the distinct terrain units, called 3rd and 4th order landforms (Way 1978). Typical pattern elements examined include topographic form, drainage texture and pattern, gully characteristics, soil tone variation and texture, land use, and vegetation cover (Lillesand and Kiefer 1987).

Nowadays, quantitative techniques (Pike 1995, 2000) have been developed and applied in order to automate the interpretation of terrain features from digital elevation models (DEMs) and various geomorphometric parameters were developed in an attempt to characterize the landscape and identify the various processes (Evans *et al.* 2003). This chapter aims to review the physical world terrain partition frameworks at various scales and present how terrain objects' parametric representation, classification, and mapping might be used for tectonic, fluvial, and aeolian processes quantification.

2 Data Analysis Techniques

The quantification of processes requires a partition framework, which transforms the DEM representation of the landscape to elementary objects (Miliaresis and Argialas 2000). Physical processes are scale dependent and define various continuous (Miliaresis and Kokkas 2004) or discontinuous (Miliaresis and Argialas 2002) terrain partition frameworks. A unified terrain partition framework is impossible to achieve; instead various physical processes-objects dependent terrain representation schemes might be established. Thus from the conceptual point of view, processes and scale determine the physical terrain partition framework that should be derived from the DEM data. The derived terrain partition framework defines the objects that are parametrically represented, classified, and mapped towards processes quantification.

2.1 Data

During the initial steps, studies were based on interpretation and measurement performed on topographic maps and imagery. Nowadays, DEMs that are freely available from the WEB represent the Earth's relief at regional to moderate scales (Pike 2002). More specifically:

- The GTOPO (GTOPO30 1996) and the Global land one-kilometre base elevation (GLOBE 2001) DEMs are available, providing a digital representation of the Earth's relief at a 30 arc-seconds sampling interval.
- The Shuttle Radar Topography Mission (SRTM) successfully collected Interferometric Synthetic Aperture Radar data over 80% of the landmass of the Earth between latitudes of 60 degrees North and 56 degrees South in February 2000 (Farr and Kobrick 2000). The Consortium for Spatial Information of the Consultative Group for International Agricultural Research is offering post-processed void-free 3 arc-second SRTM DEM data for the globe (SRTM 2006) that is suitable for 1:250,000 studies.
- A moderate-resolution DEM (500 m spacing) is available for Mars, acquired by the Mars Orbiter Laser Altimeter (MOLA), a 10-Hz pulsed infrared-ranging instrument, which operated in orbit around

the planet from 1997 to 2001 aboard the Mars Global Surveyor (MOLA 2004).

2.2 Terrain segmentation

Terrain segmentation is applied to either DEMs or to derivative products and so slope, aspect, or even curvature could be computed according to the methods proposed by Shary (1995) and Florinsky (1998). The terrain partition framework of the landscape is either continuous (Miliaresis *et al.* 2005) or discontinuous (Miliaresis 2001a).

2.2.1 Continuous segmentation framework

Aspect regions are a paradigm of a continuous terrain segmentation scheme (Miliaresis and Kokkas 2004). In this approach, aspect is computed for every DEM point (Figure 1a). Then aspect is standardized to the eight directions (N, NE, E, SE, S, SW, W, NW) defined in a raster image (Figure 1b). For example, points with aspect in the range 22.5° to 67.5° are considered to slope towards 45° azimuth (NE) and they are labelled with the same integer identifier. Note that zero labels indicate flat terrain (if slope $< 2^{\circ}$, then aspect is undefined). Thus, the points of the resulting aspect image are labelled with 9 integer identifiers corresponding to eight geographic directions defined in a raster image and the aspect undefined label (Miliaresis *et al.* 2005).

The aspect regions are easily interpreted from Figure 1b, since they are formed from points having the same shade of grey. For aspect regions to be explicitly defined, a connected component-labelling algorithm (Pitas 1993) is applied. The algorithm scans the image and identifies the regions formed by adjacent points labelled with the same aspect label. Approximately 30,000 aspect regions were identified. In Figure 1c, each aspect region is assigned to one out of seven classes depending on the mean slope of the DEM points that form the aspect region. In Figure 1d, aspect regions with mean elevation and mean slope in a specific interval are mapped in an attempt to express the following geomorphometric rule: "landslide risk is high if aspect region elevation is low while aspect region slope is high" (Miliaresis *et al.* 2005). So the parametric representation of aspect regions expresses in a quantitative manner the qualitative knowledge that was acquired by domain experts (landslide engineers) (Argialas and Miliaresis 2001).

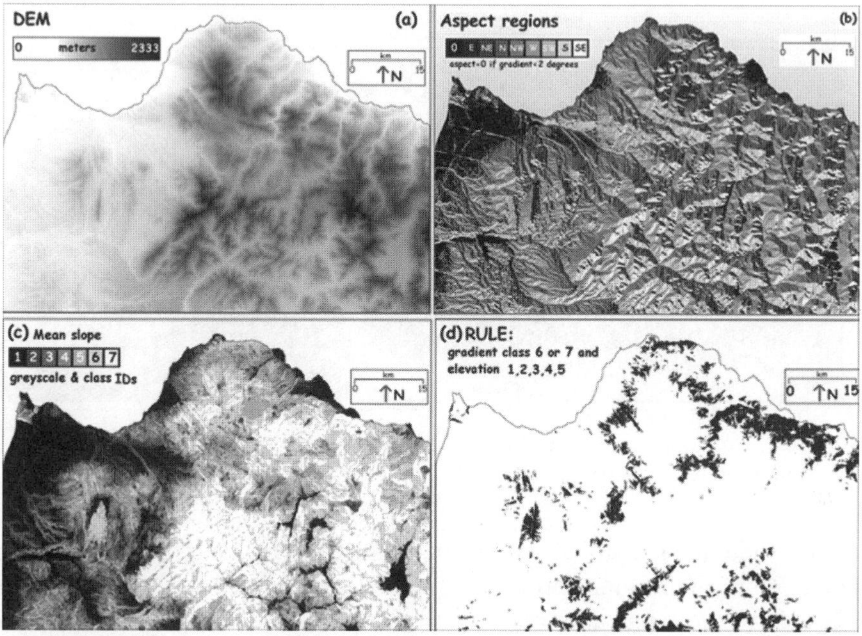

Figure 1. A continuous terrain partition framework (Miliaresis *et al.* 2005).

(a) DEM of the study area in SW Greece; the darker the point, the greater its elevation.
(b) The aspect regions are easily interpreted from the aspect image. Note that aspect was standardized to the 8 geographic directions defined in a raster representation, while an aspect region is formed by adjacent points with the same shade of grey.
(c) Each aspect region is assigned to one of seven classes depending on the mean slope of the DEM points that form the aspect region and mapped by a unique shade of grey.
(d) Aspect regions (black regions) with mean elevation and mean slope in a specific interval are mapped.

2.2.2 Discontinuous segmentation framework

Region growing segmentation is applied in order to isolated specific landforms from the geomorphologic background (Figure 2); thus a discontinuous terrain partition framework is created. This technique uses an initial set of points (seeds) and growing-stopping criteria (Miliaresis and Argialas 1999). The seeds are expanded during successive iterations by checking the neighbouring points. If the region growing criteria are fulfilled, then the neighbouring points are added to the initial set of points. The procedure is repeated until no more points are added.

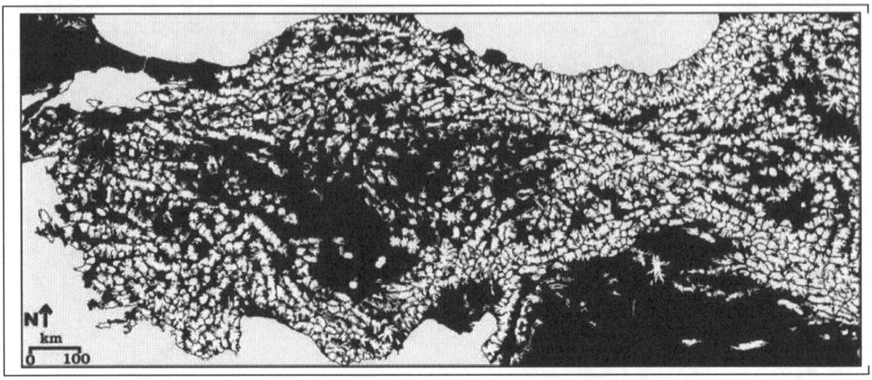

Figure 2. A discontinuous terrain partition framework (Miliaresis 2006) that identifies mountains (white pixels) from the surrounding basins (black pixels) in Asia Minor (SW Asia).

For mountains (Miliaresis 2001a, 2006), seeds correspond to ridge points and the region growing criteria are defined on the basis of slope/elevation range and pixels that belong to the valley network. For alluvial fans (Miliaresis and Argialas 2000), drainage outlet points determine the seeds and the region growing criterion is based on slope (Figure 3). For bajadas (coalescent fans), streams emerging on the basin floor form the set of seeds (Figure 4), while the region growing criterion is based on slope combined with size dependent objects filtering and drainage pixel removal after the first iteration (Miliaresis 2001b).

2.3 Object representation, terrain classification and mapping

The representation of the segmented objects is performed by a set of parameters that are associated either with their planimetric shape (size, elongation, etc.) or their 3-D arrangement (mean elevation, local relief, roughness, mean slope, hypsometric integral, etc.). For example:

- The area of the region occupied by the object is computed as the aggregate of the pixels constituting the object region multiplied by the area extent of each pixel.
- The mean elevation of objects is computed as the average elevation of the pixels that belong to an object's region.
- Roughness corresponds to the standard deviation of elevation and it is a stable measure of the vertical variability of the terrain within an object.

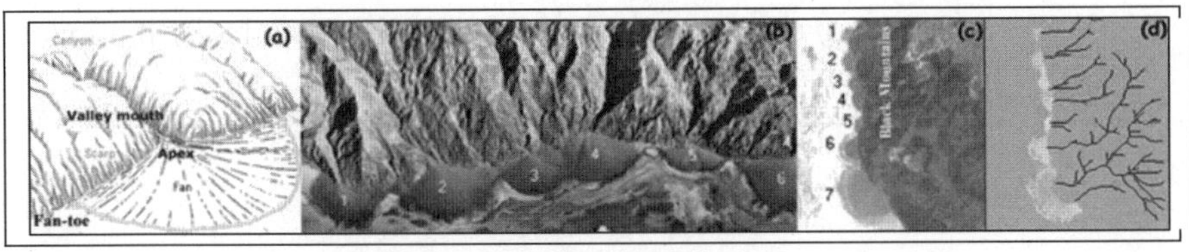

Figure 3. Alluvial fans in Death Valley, California (Miliaresis and Argialas 2000).

(a) Block diagram of an alluvial fan deposited in front of a valley mouth.
(b) 3D view of the study area.
(c) Landsat image of the study area.
(d) Region growing segmentation of alluvial fans on the basis of drainage outlet points.

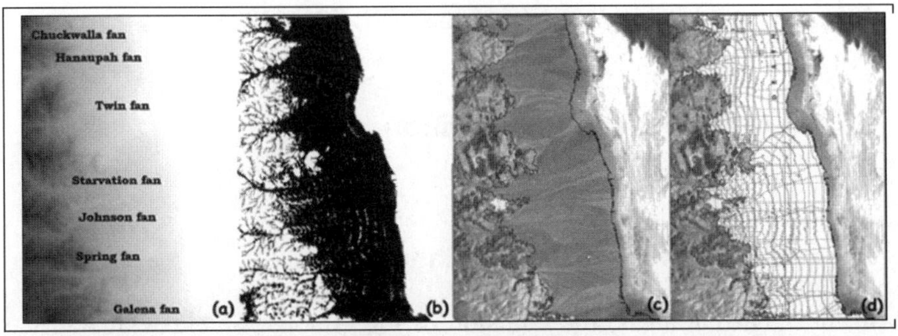

Figure 4. Bajadas segmentation in Death Valley (Miliaresis 2001b).

(a) The DEM of the study area. The elevation values (-86 to 1,904 m) were rescaled to the interval 255 to 0 (the brightest pixels have lowest elevation).
(b) Region growing segmentation (first iteration).
(c) The borderline of the segmented bajadas object superimposed on the TM image (band 5).
(d) Hybrid image. TM band 5 in the background while the map is shown through the segmented bajadas polygon.

These parameters quantify the physical processes since they indicate the elevation and slope variability within the object. The hypsometric integral reflects the stage of landscape development, while the mean slope is associated with the intensity of both erosion and tectonic processes (Miliaresis and Argialas 2002).

Terrain classification is achieved by cluster analysis of the parametric representation of objects. It is based on measurement of the Euclidean distance, which is calculated in a c-dimensional space, where c represents the number of attributes used in the clustering process (Miliaresis and Iliopoulou

2004). Terrain mapping and interpretation of the spatial distribution of clusters will be presented in the following case studies.

2.4 Software for terrain segmentation

Connected components labelling, object filtering, and region growing segmentation can be implemented with Geologic Shell (2001). This software package is freely available for download on the internet from the WEB site of the International Society for Mathematical Geology, and a newer version will be available soon (Miliaresis and Kokkas 2007).

Drainage basins, ridge and valley networks, as well as general geomorphometric parameters could be determined with TAS software (Lindsay 2005) that is available free through the WEB (TAS 2004).

3 Terrain Processes Recognition

3.1 Asia Minor versus Zagros Ranges

In the Zagros Ranges, the collision of the Arabian Shield with Iran has shortened and thickened the crust to produce a spectacular mountainous physiography. The linear topographic highs represent huge folds (NW–SE anticlines), marked by SW facing topographic escarpments, while the geometry of asymmetrical anticlines indicates the existence of basement reverse faults (Berberian 1995). In Asia Minor, horizontal expulsion is taking place and most of the area is extruding westward away from the Arabian-Eurasian collision and towards the small remnant of oceanic crust underlying the Aegean Sea (Miliaresis 2006).

Having decomposed the terrain into the mount and non-mount terrain classes, elevation frequency histograms are computed for each class of Asia Minor and the Zagros Ranges. The underlying idea is that mountains are usually under a different (kind or intensity) physical process than the surrounding basins. These differences should be revealed in the frequency distributions used to describe each class (Miliaresis 2001a).

The elevation frequency histogram of the non-mount terrain class looks very similar to the overall frequency histogram of the study area (Figure 5), both including three major peaks. The resemblance of the two histograms is explained by the fact that the non-mount terrain class occupies 64% of the study area, forcing the histogram of the study area to fit the histogram of the non-mount terrain class. The three major peaks at elevations of 0, 420, and 1,045 m observed in the histogram of the non-mount terrain class indicate the existence of three major peneplains at regional scale. The elevation frequency histogram of the mount terrain class preserves one peak at an elevation of 1,260 m, with the elevation frequency declining gradually away from it.

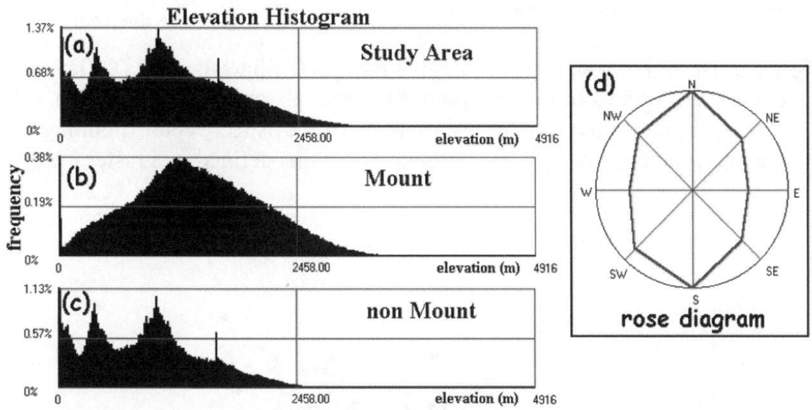

Figure 5. Descriptive statistics of Asia Minor (Miliaresis 2006). Elevation frequency histograms of the decomposed terrain classes of the study area. (a) study area, (b) mount, (c) non-mount, (d) a rose-diagram of the aspect vector (pointing downslope) standardized to 8 geographic directions defined in a raster image.

The aspect vector rose diagram (Figure 5d) indicates that the landscape flows equally in the North and South directions at right angles to the main axes of the mountain ranges (Figure 2).

In the Zagros Ranges, the frequency histograms of elevation (Figure 6) indicate that the extracted mountain objects are developed almost equally on all levels in the elevation domain (Miliaresis 2001a). The greatest frequencies were observed in the range 1,500 m to 2,500 m. The frequency histogram of the non-mount terrain class looks very similar to the overall elevation frequency histogram of the study area because both include the extensive NW coastal plains (Figure 8).

The rose diagram of aspect pointing downslope (Figure 6C) indicates that much of the surface is sloping away from the mountains in a NW to SE direction. This direction is at right angles (Figure 8) to the collision of the Arabian Shield with the Iranian Plateau and verifies the asymmetry of the mountain ranges (huge asymmetrical anticlines).

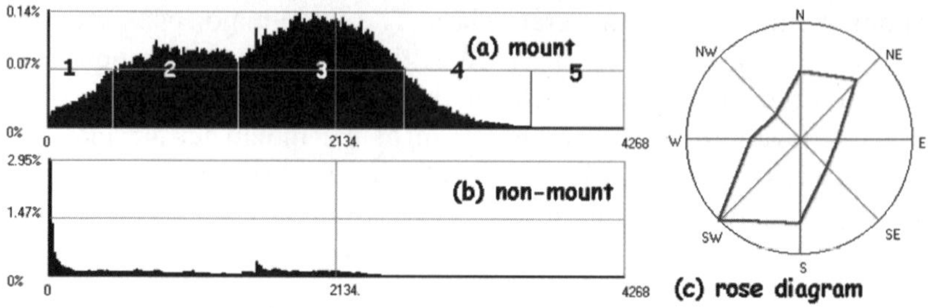

Figure 6. Descriptive statistics of Zagros Ranges (Miliaresis 2001a). Elevation frequency histograms of the decomposed terrain classes of the study area. (a) mount, (b) non-mount, (c) a rose-diagram of the aspect vector (pointing downslope) standardized to 8 geographic directions defined in a raster image.

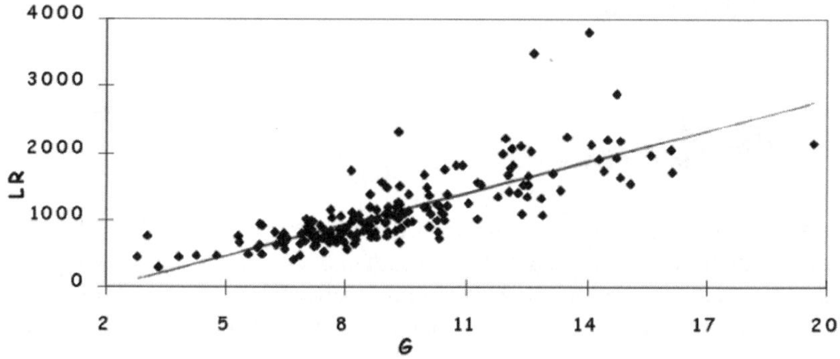

Figure 7. Linear regression of local relief (LR) versus slope (G) for the mountain objects identified in Zagros Ranges (Miliaresis and Iliopoulou 2004).

The correlation between the attributes of the mountain objects is of great significance and might be explored either by computing correlation coefficients, or by assuming the linear regression model (Miliaresis and Iliopoulou 2004). In the case study of the Zagros Ranges, the correlation between Local Relief ($LR= H_{maximun} - H_{minimum}$, within a mountain object) and slope (G) is expressed by the equation (Figure 7):

$$LR = -316.+156.1*G$$

Such models are of great significance since they might prove to be quantitative indicators of landscape development and tools for estimating the intensity of processes.

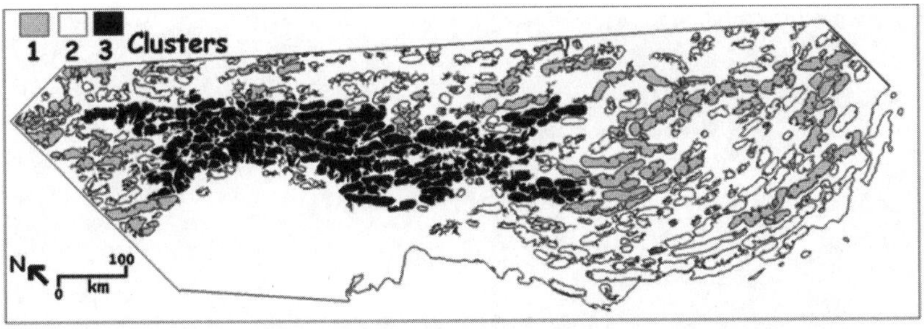

Figure 8. Spatial arrangement of the clusters derived by the centroid method in Zagros Ranges (Miliaresis and Iliopoulou 2004).

3.2 Clustering of Zagros Ranges

The mountain objects segmentation and parametric representation (Miliaresis 2001a) followed by the centroid clustering method revealed clearly the SE–NW stair-step topography observed in the Zagros Ranges, while the steepest and more massive mountains were also observed along this direction (Figure 8).

The zones derived by the mapping of clusters were associated with the existing morphotectonic zones of the study area, while geomorphometric processing proved capable of segmenting morphotectonic zones to subregions with different geomorphometry (Miliaresis and Iliopoulou 2004).

3.3 Tectonic processes identification in Mars

The DEM to Mountain transformation of Valles Marineris in Mars (Miliaresis and Kokkas 2004) revealed numerous tributary valleys originated from the plateau that cross the chasma sides and the mountain features extending to the basin floor (Figure 9).

The observation of segmented chasma downslope borders (fronts) indicates that they are rectilinear. A tentative interpretation is that uplift along front-faults produced the mountain fronts that are relatively straight, since they have not had time to be dissected and embayed by streams. As the range front is eroded, major drainage embays the front and causes it to retreat, depending on the width of the ranges and the length of the steams. This process might be accelerated by landslides that cause the occasional concave in plan hillslope forms (Figure 9).

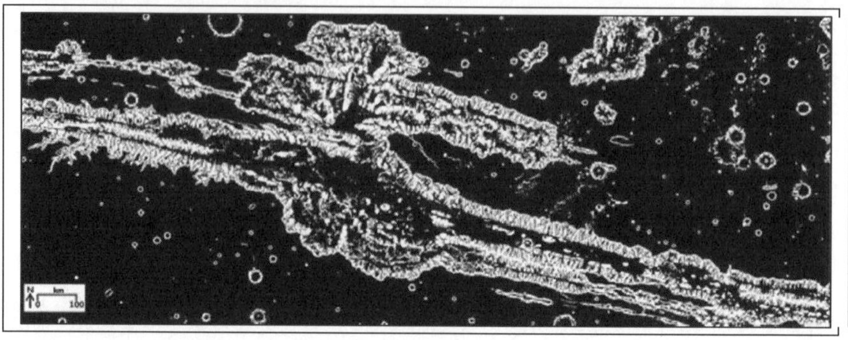

Figure 9. The mountain terrain class in Valles Marineris (Mars). The white pixels represent the pixels labelled by the DEM-to-Mountain transformation (Miliaresis and Kokkas 2004).

Additionally, the segmentation to aspect regions and their representation on the basis of mean elevation and mean gradient revealed the chasma terrain structure, proving that the basin floors of the elementary chasmata are interrupted by regions with higher mean elevation and gradient, due possibly to vertical tectonic movements (Miliaresis and Kokkas 2004).

3.4 Prospects: aeolian landforms segmentation

Desert environments are dominated by dunes that are accumulations of sediment blown by the wind into a mound or ridge. Dunes have gentle upwind slopes on the wind-facing side. The downwind portion of the dune is commonly a steep avalanche slope referred to as a slipface (Bullard 2006). The slipface stands at the angle of repose, which is the maximum angle (30° to 34° for sand) at which loose material is stable.

Dune typical heights and wavelengths (spacing) are in the range of 5 to 30 m and 50 to 300 m, respectively. Linear megadunes, in the Western Desert in between Egypt and Libya (Figure 10) and in the Namib Sand Sea (Namibia), attain even greater dimensions with heights of up to 400 m and wavelengths up to 4 km; the most significant factors determining their morphology are wind regime and sand supply (Summerfield 1996). Megadunes must take hundreds of years to attain an equilibrium form and thus are key landforms in the study of possible severe climatic change that will possibly be expressed by change in the direction and intensity of winds in desert regions.

Towards this end, linear megadune segmentation might be performed from SRTM DEMs (SRTM 2006) acquired in February 2000, while topographic information might be acquired from LANDSAT-SRTM imagery (Levin *et al.* 2004) for the period 1980–2000 and from ASTER imagery for the period 2001–2007. A multi-temporal analysis and change detection of their morphology might provide answers for the global climatic change in desert environments.

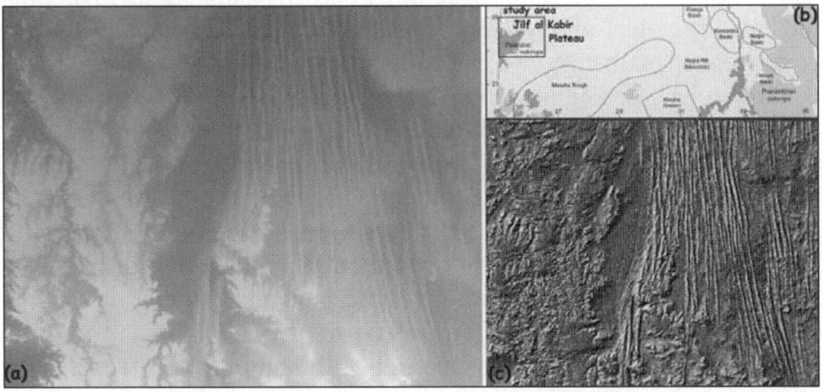

Figure 10. Mega-dunes (aeolian landforms).

(a) The SRTM DEM of the study area (the brighter the pixel, the greater its elevation)
(b) A physiographic map and the location of the study area in SW Egypt.
(c) Shaded relief map of the SRTM DEM of the study area.

An initial experiment on the delineation of megadunes is presented in Figure 11. Slope (Figure 11b) was derived from the DEM (Figure 11a). The initial set of seed points (Figure 11c) was defined by thresholding the upslope runoff image, while region growing criteria were based on both the slope and the valley network (Figure 11d).

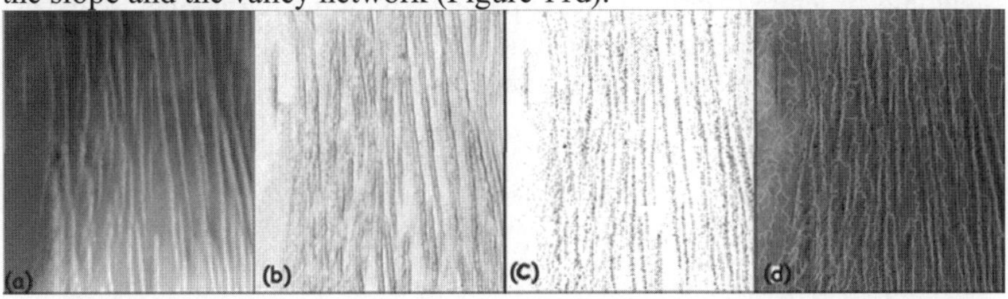

Figure 11. Towards the delineation of sand-dunes from SRTM DEM.

(a) DEM of the study area (the brighter the pixel, the greater its elevation).
(b) Slope image (the darker the pixel, the greater its slope).
(c) Seeds that correspond to points with upslope runoff greater than a threshold.
(d) Valley network extracted by runoff simulation and the seeds superimposed on the Landsat image.

4 Conclusion

Global digital elevation models of earth and other planets have fostered geomorphometric and terrain modelling at broad spatial scales. The broad-scale quantification of topography and the DEM-based analyses transformed geomorphometry into one of the most active and exciting fields in the Earth sciences. Segmentation techniques allowed the partition of the terrain into continuous and discontinuous schemes. The parametric representation of the derived objects combined by object classification schemes allowed the mapping and the quantification of various processes.

More specifically, tectonic processes were quantified on the basis of the discontinuous partition framework based on mountains. The quantification was based on the spatial distribution of the mountain pattern, on the linear regression of mountain attributes, and on the hypsometric and frequency distributions of elevation and aspect. Fluvial landforms (alluvial fans and bajadas) forming zones that are subject to frequent flash flooding were delineated from DEMs.

The aspect regions continuous terrain partition framework allowed the identification of regions with high landslide hazards on the basis of aspect regions parametric representation and knowledge-based rules acquired by domain experts, while in Mars aspect regions modelling revealed the tectonic processes. SRTM DEMs seem to be capable of capturing aeolian processes on the basis of the morphometry of linear megadunes in desert regions.

Geomorphometric analysis provides a quantitative way to compare developed and developing landscapes in areas of both differing and similar geologic structure. Additionally, experience with the Earth's landscape assists the exploration and interpretation of various landscapes in inaccessible areas on other planets from digital elevation data.

References

Argialas, D., and Miliaresis, G., (2001), Human factors in the Interpretation of Physiography by Symbolic and Numerical Representations within an Expert System, In Hoffman, R. and Markman, A. (eds): *Interpreting Remote Sensing Imagery: Human factors,* New York: CRC Press: 59–81.

Berberian, M., (1995), Master "blind" thrust faults hidden under the Zagros folds: active basement tectonics and surface morphotectonics, *Tectonophysics,* **241**: 193–224.

Bullard, J., (2006), Arid geomorphology, *Progress in Physical Geography,* **30:** 542–552.

Evans, I., (1981), General geomorphometry, In *Geomorphological techniques*, Goudie, A. (ed.): British Geomorphological Research Group, London: Allen & Unwin: 31–37.

Evans, I., Dikau, R., Tokunaga, E., Ohmori, H. and Hirano, M. (eds), (2003), *Concepts and modelling in geomorphology, international perspectives*, Tokyo, Japan: TERRAPUB. Available on-line at: http://www.terrapub.co.jp/e-library/ohmori/ (accessed 25 March, 2007).

Farr, T. and Kobrick, M., (2000), Shuttle radar topography mission produces a wealth of data, *Amer. Geophys. Union Eos*, **81**, 583–585.

Florinsky, I., (1998), Derivation of topographic variables from a digital elevation models given by a spheroidal trapezoidal grid. *International Journal of Geographical Information Science*, **12**: 829–852.

Geologic Shell, (2001), *Software for terrain segmentation, version 1*, International Association for Mathematical Geology. Available on-line at: http://207.176.140.93/documents/oldftp/VOL27/v27-10-05.zip (accessed 25 March, 2007).

GLOBE, (2001), One-km spacing digital elevation model. Available on-line at: http://www.ngdc.noaa.gov/mgg/topo/globe.html (accessed 25 March, 2007).

GTOPO30, (1996), One-km spacing digital elevation model. Available on-line at: http://edcdaac.usgs.gov/gtopo30/gtopo30.html (accessed 25 March, 2007).

Levin, N., Ben-Dor, E. and Karnieli, A., (2004), Topographic information of sand dunes as extracted from shading effects using Landsat images, *Remote Sensing of Environment*, **90**: 190–209.

Lindsay, J.B., (2005), The Terrain analysis system: a tool for hydro-geomorphic applications, *Hydrological Processes*, 19: 1123–1130.

Lillesand, T. and Kiefer, W., (1987), *Remote Sensing and image interpretation*, New York, USA: John Wiley and Sons.

Miliaresis, G., (2001a), Geomorphometric mapping of Zagros Ranges at regional scale, *Computers & Geosciences*, **27**: 775–786.

Miliaresis, G., (2001b), Extraction of bajadas from digital elevation models & satellite imagery, *Computers & Geosciences*, **27**: 1157–1167.

Miliaresis, G., (2006), Geomorphometric mapping of Asia Minor from Globe digital elevation model, *Geografiska Annaler Series-A*, **88 A**: 209–221.

Miliaresis, G. and Argialas, D., (1999), Segmentation of physiographic features from the global digital elevation model / GTOPO30, *Computers & Geosciences*, **25**: 715–728.

Miliaresis, G. and Argialas, D., (2000), Extraction & delineation of alluvial fans from DEMs & Landsat TM images, *Photogrammetric Engineering & Remote Sensing*, **66**: 1093–1101.

Miliaresis, G. and Argialas, D., (2002), Quantitative representation of mountain objects extracted from the GTOPO30 DEM, *Int. Journal of Remote Sensing*, **23**: 949–964.

Miliaresis, G. and Iliopoulou, P., (2004), Clustering of Zagros Ranges from the Globe DEM representation, *Int. Journal of Applied Earth Observation & GeoInformation*, **5**: 17–28.

Miliaresis, G. and Kokkas, N. (2004), Segmentation and terrain modeling of extra-terrestrial chasmata, *Journal of Spatial Sciences*, **49**: 89–99.

Miliaresis, G. and Kokkas, N., (2007), Segmentation and object based classification for the extraction of building class from LIDAR DEMs, *Computers & Geosciences*, **33:** 1076–1087.

Miliaresis, G., Sabatakakis, N. and Koukis, G., (2005), Terrain pattern recognition and spatial decision making for regional slope stability studies, *Natural Resources Research*, **14**: 91–100.

MOLA, (2004), *Mars digital elevation model, Planetary data system (PDS), NASA.* Available on-line at: http://pds-geosciences.wustl.edu/missions/mgs/megdr.html (accessed 25 March, 2007).

Pandey, S., (1987), *Principles and applications of photogeology*, New Delhi, India: John Wiley & Sons.

Pike, R., (1995), Geomorphometry-process, practice and prospects, *Zeitshcrift f. Geomorphologie N.F. suppl. Bd.,* **101**: 221–238.

Pike, R., (2000), Geomorphometry - diversity in quantitative surface analysis, *Progress in Physical Geography*, **24:** 1–21.

Pike, R., (2002), *A bibliography of terrain modeling (Geomorphometry), the quantitative representation of topography, Supplement 4.0, Open-file report 02-465*, Menlo Park, California: US Geological Survey.

Pitas, I., (1993), *Digital Image Processing Algorithms*, London, U.K.: Prentice Hall.

Shary, P., (1995), Land surface in gravity points classification by a complete system of curvatures, *Mathematical Geology*, **27**: 373–389.

SRTM, (2006), *Void free 3 arc sec SRTM DEM from the International Centre for Tropical Agriculture.* Available on-line at: http://srtm.csi.cgiar.org/ (accessed 25 March, 2007).

Summerfield, M., (1996), *Global geomorphology*, Essex, England: Longman Group.

Summerfield, M. (ed.), (2000), *Geomorphology and Global Tectonics*, New York, USA: John Wiley & Sons.

TAS, (2004), Terrain analysis system, a software for hydrogeomorphic applications, Available on-line at: http://www.sed.manchester.ac.uk/geography/research/tas/ (accessed 25 March, 2007).

Way, D., (1978), *Terrain analysis*, New York, USA: McGraw-Hill.

Models of Topography

Peter A. SHARY

Abstract

Models of topography, such as smooth and non-smooth, single-sided and double-sided, appear to be of value in choosing algorithms to calculate topographic attributes, and assist in predicting and understanding paradoxes in topography, such as statistically predictable or terrain-specific behaviour of landforms. This results in a new vision of how to use geomorphometry in digital terrain analysis: for example, the dependence of topographic attributes on scale creates results in problems of comparing observations obtained from different scales, while the use of predictable landforms may lead to unrealistic expectations of predictable properties of soil or vegetation patterns. The purpose of this chapter is to study models of topography. To achieve this purpose, direct studies on the dependence of topographic attributes on scale are considered, based on a theoretical background, such as two concepts of scale, Gibbs-like phenomena, an internal smoothing in algorithms, and sub-dividing of topographic attributes into non-intersecting classes. Results include a new approach to the comparison of algorithms, some artefacts from Gibbs-like phenomena, and a method to evaluate a minimal grid mesh from contour DEMs. Based on these results and concepts, methodological issues are discussed of both geomorphometry itself, and of its applications in digital terrain analysis.

Keywords: geomorphometry, scale, landform predictability, topographic attribute and variable, Gibbs phenomenon.

1 Models, Restrictions, and Phenomena

Quantitative terrain description needs topographic attributes that may be ascribed to each terrain location. Such attributes are *variables*, in contrast to measures that refer to some objects, such as area and volume of a depression (that refer to all the depression) or contour line length, which are also topographic attributes, but are not variables (Shary *et al.* 2002). This distinction is important in the context of this chapter, and therefore two terms are used below: topographic variables and topographic attributes.

National traditions have resulted in several meanings of the same term in the literature. Namely, in some Roman-language-based countries (e.g. France, Russia, Slovakia, Spain) the term *relief* is used as a synonym of topography, but this is not the case for Anglo-Saxon countries (e.g. Great Britain, USA), where it has come to denote a difference between maximal and minimal values of elevation. For example, King *et al.* (1999) used the first meaning, in contrast to Evans (1972).

Geomorphometry is a science of quantitative land surface analysis (Pike 1995). The approaches under consideration may be attributed to *general geomorphometry* described by Evans (1972) as follows: '*General geomorphometry as a whole provides a basis for the quantitative comparison even of qualitatively different landscapes, and it can adapt methods of surface analysis used outside geomorphology. Specific geomorphometry is more limited; it involves more arbitrary decisions, and leaves more room for subjectivity in the quantification of its concepts*'.

Any choice of the best model of topography, such as smooth (i.e. differentiable) or non-smooth, single-sided or double-sided ones (see Section 1.1 below), cannot be proven experimentally, because researchers always use only finite sets of points with measured elevations, in contrast to infinite sets of points that are necessary in mathematical proofs. Also, the land surface may be considered as a surface only at scales in which separate fragments that compose it (e.g. gravel, rocks, soil aggregates) are invisible. Consequently, correct mathematical proofs are impossible here and, instead, rules to choose between models include both pragmatic criteria (e.g. dependence of topographic attributes on scale), and the criteria of certain model's ability to predict and explain a wide spectrum of phenomena, such as statistical predictability of landforms.

Specific models of topography related to particular Digital Elevation Models (DEMs), such as elevation grids of Shuttle Radar Topography Mission (SRTM), are not considered here, although they may be useful, for example, in DEM accuracy evaluation. The purpose of this chapter is to study general models of topography, such as smooth and non-smooth, one-sided and orientable (double-sided), together with their influence on principles, methodology, and applications in Digital Terrain Analysis (DTA).

For all calculations and map images, the author's software 'Analytical GIS Eco' (Shary 2005) was used.

1.1 Restrictions of topography

A general method, especially when DEMs are used, is to describe elevations by a function $z = f(x,y)$, where elevation z depends on plan coordinates x, y, and z has only one value at each location (x,y) (Cayley 1859). This restriction is essentially a simplification, which means that researchers usually do not quantitatively describe 'ceilings' (of caves), overhangs, boulders, and so on (Shary 1995), (Figure 1*a*). Another kind of restriction is the characteristic of the land surface as a boundary of two different media (e.g. soil and air), and therefore the land surface is always orientable, or *double-sided* (Shary 1995), in contrast to some *single-sided* mathematical surfaces, like Möbius' strip, for which distinctions between hills and pits cannot be defined (Gauss 1827), (Figure 1*b*).

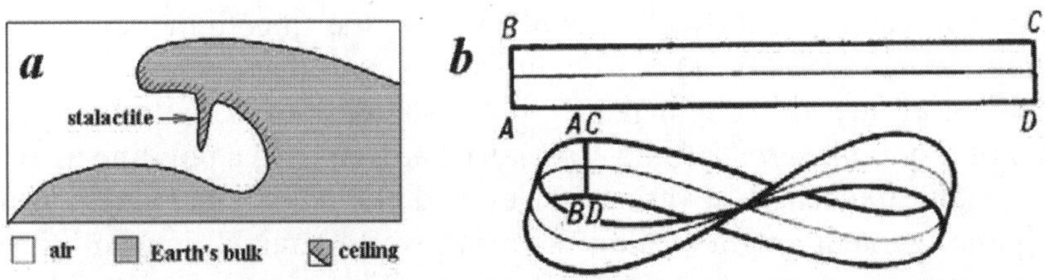

Figure 1. Two kinds of restrictions in models of topography. *a* – a cave with negative slope steepness on its ceiling, *b* – Möbius' strip as an example of a single–sided surface; its model is a sheet of paper *ABCD* connected at its edges *AB* and *CD*.

Although the second restriction is always valid for the land surface (and models of single-sided surfaces are essentially not used in DTA), certain issues from differential geometry of surfaces were developed for single-sided surfaces, and this may result in some misunderstanding. One example is that curvature sign choice is arbitrary for single-sided surfaces (and in differential geometry), while it is commonly chosen to be negative for concave landforms and positive for convex ones. Nevertheless, some authors (e.g. Pachepsky *et al.* 2001) used the opposite sign of curvatures. Another example is as follows. Gauss (1827) introduced two surface form classifications, (i) single-sided and (ii) orientable (double-sided) surfaces; attempts of several authors (e.g. Wood 1996, Schmidt and Hewitt 2004) to describe elliptical forms of Gauss' classification (i) as 'pits' may be misleading (by Gauss' (1827) definition, elliptical forms are both pits and hills), because there is no distinction between hills and pits on a single-sided surface. There is a great distinction between hills and pits in landscapes (Dokuchaev 1886), so that the double-sided model is clearly preferable. Results from differential geometry play an important role in geomorphometry (Koenderink and van Doorn 1994), so that concepts of a single-sided surface should be adapted to those of a double-sided surface before their use.

1.2 Smooth and non-smooth models

First approaches to study topography were based on a smooth model of land surface, in which the surface was considered as differentiable (Gauss 1827, Cayley 1859). This model was developed primarily to study equilibrium capillary surfaces (Finn 1986), where a surface (like a soap film) may be considered as smooth, because particles on it fast change their locations during observation, resulting in a physical smoothing, not observed for the land surface (Maxwell 1870).

Which results are difficult to explain using the smooth model of topography? Richardson (1961) has empirically shown that contour line length tends to infinity as scale increases, but it should have a finite limit in the case of a smooth surface (Sard 1942). Richardson used a polyline to evaluate contour line length at various scales, and plotted polyline length against polyline segment length (scale); the result was a straight line (in logarithmic units), in contrast to a circle. Based on Richardson's results, Mandelbrot (1967) noted that each landform is composed by smaller ones, at any given scale, and that the length of a given contour line should rather be considered as infinite.

Evans (1980) has empirically found a stable positive correlation between independent local variables, plan and profile curvatures, for about 60 terrains (Evans and Cox 1999). Shary *et al.* (2005) have shown that this *Evans' phenomenon* is a consequence of the statistical hypothesis (Shary 1995), according to which the proportion of area occupied by given landform type defined by signs of curvatures is the same for any terrain, and concrete values of these proportions have been predicted in Shary (1995). This was empirically confirmed for a set of DEMs (Shary *et al.* 2005). Such *landform predictability* is not characteristic of a smooth surface. Gauss' (1827) and Troeh's (1964) curvature-sign-based landform types are shown in Figure 2 (numbers indicate predicted probabilities to meet each of them), and results of experimental checks are shown in Figure 3.

Experimental study of oil spills (e.g. Shary *et al.* 2005) have revealed *multiple flow-line branching and confluence* in oil runoff. Indeed, an oil flow encounters small hills and branches on each of them, and is confluent behind them. Since there are many out-of-scale hills at a given fixed scale (Mandelbrot 1967), this process results in multiple flow-line branching and confluence, which is also not characteristic of a smooth surface. Oil traces were satisfactorily described by catchment area from a single grid cell (Shary *et al.* 2005). To prevent interrupts in flows, algorithms of catchment (upslope) area must take depressions into account (Martz and de Jong 1988). Shary *et al.* (2005) concluded that multiple flow-line branching and confluence is physically meaningful in variables like catchment area; they are not simply a tool for diminishing grid bias.

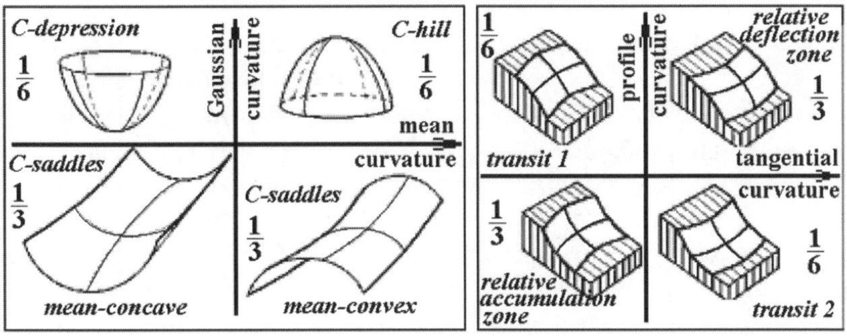

Figure 2. Gauss' (left) and Troeh's (right) landform types. Numbers indicate probabilities to meet specific landforms.

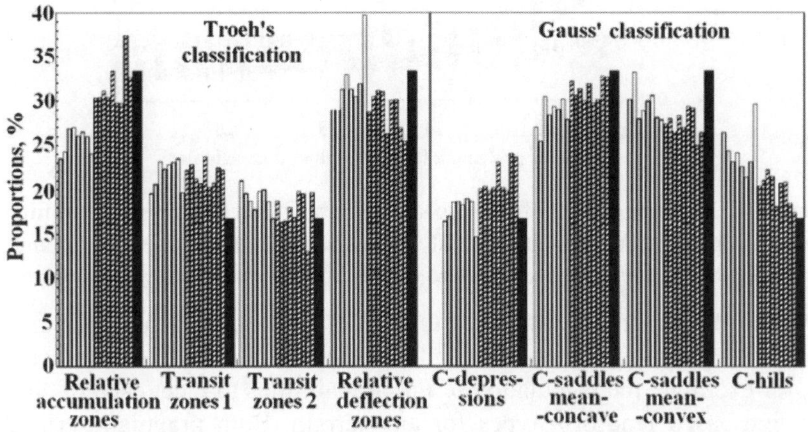

Figure 3. Proportions of terrain area occupied by Thoeh's and Gauss' landform types observed for 17 terrains (Shary and Sharaya 2006). Black columns indicate theoretical values, white – gently sloping, dashed – mountain terrains.

Evans (1975) initiated direct studies of the dependence of topographic attributes on grid mesh (i.e. on scale) using his method of 'matrix thinning' that provided attribute values for grid meshes of *w*, 2*w*, 3*w*, and so on. He has shown that average slope steepness (gradient) increases as scale enlarges. No finite limit was found for the limit case (Figure 4). Shary *et al.*

(2002) have confirmed his results and shown that average slope steepness should be proportional to $1/w$, where w is grid mesh, for a model of quasi-stochastic, noisy surface with a horizontal trend. In logarithmic units, this dependence is described by a straight line (Shary *et al.* 2002). In other words, the dependence of local (i.e. based on derivatives) topographic variables on scale is a 'price to pay' for the land surface behaving rather like a non-smooth surface.

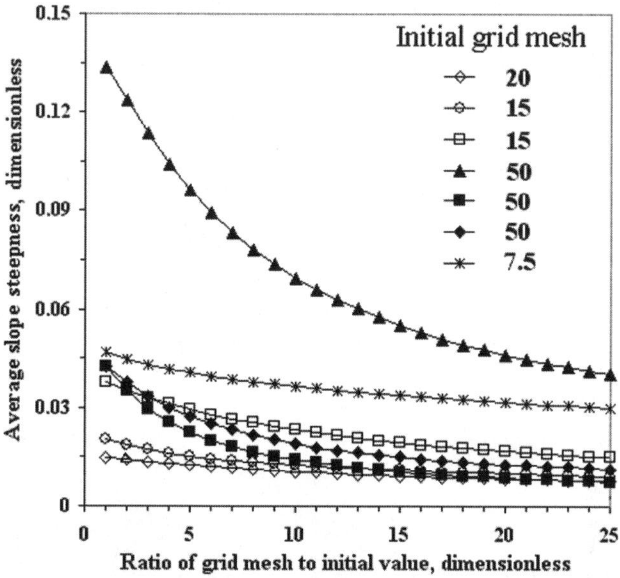

Figure 4. The dependence of average slope steepness on grid mesh w. No finite limit was observed for the limit case of $w \to 0$. See details in Evans (1975) and in Shary *et al.* (2002, 2005).

These phenomena are difficult to predict or explain using the smooth model of topography, in contrast to the non-smooth ones that have permitted, for example, the calculation of the probabilities to meet each of curvature-sign-based landform types for any terrain. Both pragmatic criterion (the dependence of local attributes on scale), and criteria of the model's ability to predict and explain a wide spectrum of phenomena indicate that the non-smooth model of topography is better than the smooth model.

These phenomena may have essential consequences in DTA. Some examples are as follows. The dependence of local topographic attributes on scale prevents results from different scales (authors) to be accurately compared, and this is also an essential source of uncertainty in DTA using these attributes. Curvature-sign-based landform classifications appear statistically predictable, that is, areas occupied by each landform type are predictable, in contrast to landform patterns (Shary *et al.* 2005). When such landform classifications are used in DTA, this may result in corresponding predictability of soil or vegetation that may appear unrealistic. MacMillan and Shary (2007) reviewed current approaches to landform classifications, and classified these classifications themselves into predictable and terrain-specific, objective and subjective, quantitative and qualitative, supervised and unsupervised, and so on.

1.3 Fractal and mathematical models

Another interesting property of topography is its fractal structure. Mandelbrot (1967) was among the first who suggested measures that could be used to describe it. The most important measure is the fractural (i.e. not integer, like 2.5) dimension, also known as *fractal dimension*. The dimension of a smooth line is 1, and that of an area is 2. A non-smooth line may look more or less like a band around a given line, and Mandelbrot (1967) suggested describing and measuring the noise of a line using a single dimension of given line between 1 and 2 (between 2 and 3 for a surface). The fractal theory is also based on the hypothesis of self-similarity (Mandelbrot, 1967), according to which each landform is not only composed by smaller ones, but larger landforms are geometrically similar to smaller ones. Based on these concepts, typical notes on 'fractal models of topography' were like 'Pure self-similarity is not a property of the real land surface' (Goodchild, 1982), or 'Fractal analysis... seems unlikely to capture the essence of complex and diverse landscapes' (Xu *et al.* 1993). Clarke (1988) smoothed DEMs using several first terms of two-dimensional Fourier series and added various proportions of an artificial 'fractal' surface to this smooth one. His results (3D map images) were recognizable by geomorphologists only when these additions were smaller than 10%. He concluded that fractal models of topography are unrealistic. Evans and McClean (1995) noted that such *uni-fractal* models of topography result in pits (closed depressions) being as frequent as summits, in contradiction to the result of Shary *et al.* (2005) in which the ratio of sums of hill to pit volumes is highly terrain-specific, and may vary from 0.6 to 10,000.

The essence of this criticism was that after sub-dividing a contour line (or land surface) into two parts, the fractal dimensions of these parts do not necessarily coincide. Continuing this process, one may conclude that the fractal dimension should vary along a given contour line (or along the land surface), so that a single fractal dimension cannot properly describe a contour line or land surface. Evans and McClean (1995) described land surface as *multi-fractal*, but this concept was rather opposed to the uni-fractal model, and no substantiated multi–fractal models were suggested in the literature.

Attempts to use mathematical models of topography, such as based on Fourier or other orthogonal series, resulted in essential artefacts related to the land surface non-smooth nature. Some general features of non-smooth models of topography, such as resulting from Gibbs-like phenomena, are discussed later in this chapter.

In summary, land surface is a complicated surface that demonstrates statistical predictability of local landforms, and may exhibit lineation or directedness in some regions (such as drumlin fields), with regular spacings of highs and lows. Land surfaces are therefore best summarized in terms of process-related topographic variables, rather than variables of, for example, fractal, spectral, Fourier, or polynomial models, which provide poor fits to any extensive land surface. It is dangerous, therefore, to use such models in digital terrain analysis, because such models ignore essential features of topography.

2 Digital Elevation Models

In computer science, land surface is commonly presented by Digital Elevation Models, which are discrete representations of topography. Indeed, all DEMs fall into two groups: *regular* and *irregular* DEMs (e.g. Li *et al.* 2005), and any type of DEM must store points with known elevations, so that any DEM can be transformed into a *point DEM* composed of a finite set of points (x,y,z), where x and y are plan coordinates, and z is elevation.

Irregular DEMs may use smaller spacing between points for areas with high relief, and a greater one with low relief, so that with the same number of points they may describe geological faults and other sharp elevation changes more accurately than grids. Original data of ground geodetic surveys are irregular point DEMs, contour lines are added only to final documents (such as topographic maps) under the following criterion: the accuracy of a topographic map is 2/3 of the contour interval (this proportion slightly depends on national traditions). *Contour DEMs* contain information on contours and separate points, and elevation values in them are stored one time for a whole contour.

For specific purposes, such as transformation of irregular to regular DEMs (e.g. Akima, 1978), triangulation may be used. This consists of calculating triangles with apices at given points under the Delauney criterion of triangles to be most close to those that have equal angles (Peuquet 1984). Delauney triangulation is complementary to Voronoi (1907) tessellation. In

Figure 5, Delauney triangles are solid, Voronoi tiles are dashed, and circles are irregular DEM points. Every location within a Voronoi tile is closer to the point about which that cell is drawn than it is to any other point.

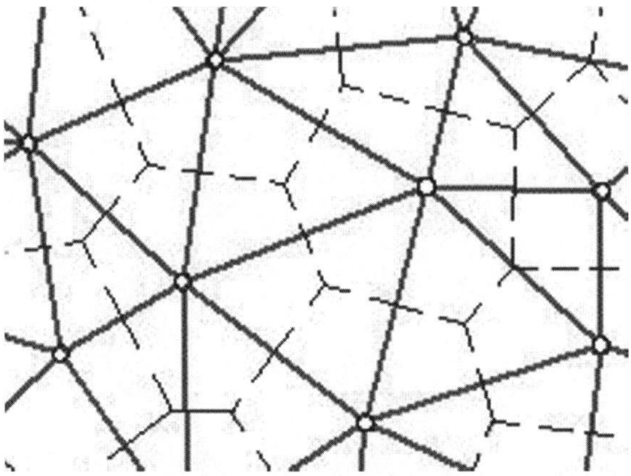

Figure 5. Delauney triangles (solid) as complementary to Voronoi tiles (dashed). Circles are irregular DEM points.

The triangulation process (e.g. Akima 1978) results in Triangulated Irregular Networks (TINs) that are based on point DEMs, to which data on triangle parameters are added. TINs may be drawn in plan, and may be generalized to 3D-structures ('curved TINs') by adding elevation to each triangle apex. TINs need a smaller number of points to describe sharp elevation changes, because they are irregular DEMs to which triangles are added. The only purpose to considering triangulation and tessellation herein is to clarify that TINs and Voronoi's tiles can be transformed into a point DEM.

Regular DEMs are essentially elevation grids (known also as *gridded DEMs*, or elevation matrices) that need no plan coordinates (x,y) of each point to be stored, because they can be calculated 'on-the-fly' due to the regular spacing of grid points. Most applications of topography in DTA are based on gridded DEMs.

The main advantages of gridded DEMs are as follows.

- Grids use a spatially uniform smallest characteristic size, grid mesh, thus making variables from different grid points comparable.
- Grids are more compact, because plan coordinates are not stored in them.
- Calculation of variables using grids is essentially faster (other kinds of DEM may be used also).

The main disadvantages of grids are as follows.

- Re-projecting of a grid is slow, because the initial grid looses its regular structure in a new projection, and should be re-calculated (using an interpolation) for this reason.
- Grids cannot represent features smaller than the grid mesh, which can result in artefacts, such as that shown in Figure 6.

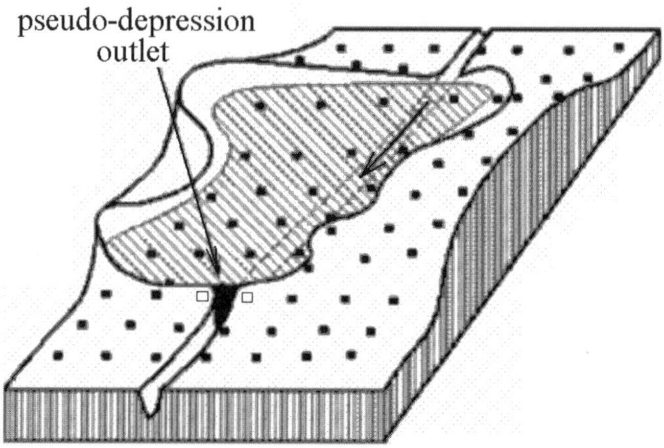

Figure 6. A scheme of a pseudo-depression (dashed), the origin of which is due to that information in given grid was lost because two grid points are at the opposite sides of a river-bed.

Pseudo-depressions are essential for hydrological modelling in DTA. Martz and de Jong (1988) introduced two kinds of catchment (upslope) area: minimal ('local') catchment area, NCA, and maximal ('global') catchment area, MCA. NCA is characterized by flow interrupts in each local elevation minimum; that is, in each depression or pseudo-depression. Pseudo-depressions randomly appear in elevation grids thus making values of NCA arbitrary. In contrast, MCA values are not arbitrary, because all depressions and pseudo-depressions are filled to the elevation value of their lowest outlet, resulting in continuous flows without interrupts (Shary *et al.* 2002). This is important, because many 'morphometric' software packages in fact calculate NCA.

2.1 Gibbs' and Gibbs-like phenomena

As stated above, non-smooth models of topography are of value for correct description of the land surface. In practice, smooth models of topography are used to deduce formulae of local topographic variables in partial derivatives, but after that special algorithms are used to calculate these variables using a DEM for the non-smooth land surface. Such algorithms should at least take into account *Gibbs-like phenomena* and *spatial averaging*. Consider these important concepts.

Gridded DEMs are often generated from contour DEMs by constructing some smooth mathematical surface using something like a Fourier series or splines (at a given scale). Although such smooth interpolators seem good for building land surface models, they can lead to serious artefacts. For example, the value of Fourier series $\sum_{i=1}^{\infty} a_i \cdot \sin(\varphi_i)$ (or any other orthogonal series) at a point of discontinuity of the function it represents is not equal to the value of this function at the same point thus resulting in artificial peaks or overshoots near such points (Figure 7). This is known as Gibbs' phenomenon (e.g. Clarke 1988).

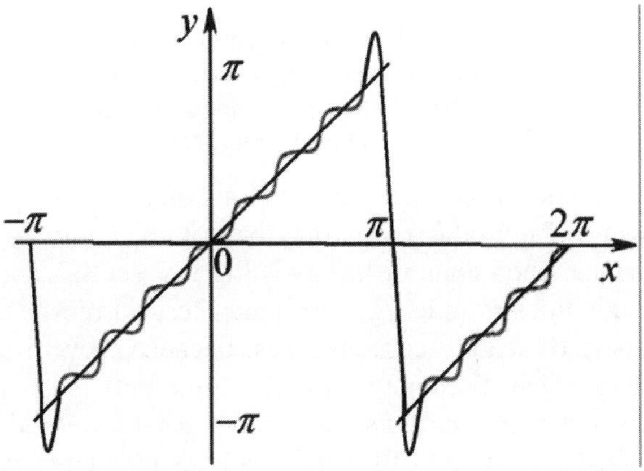

Figure 7. Gibbs' phenomenon: plots for a discontinuous function composed of periodically repeated straight lines ($y = x$ for $-\pi < x \leq \pi$) and its partial Fourier sum $\sum_{i=1}^{n} a_i \cdot \sin \varphi_i$. Even the Fourier series is not equal to the value π in its point of discontinuity $x = \pi$, where the peak is $1.089 \cdot \pi$.

Note that the limit of this function as $n \to \infty$, $x \to 1$ does not exist, because it depends on a path in (n,x) space. To study Gibbs' phenomenon (first discovered empirically by J.W. Gibbs in 1898), one should consider a path that refers to the maximal value of the peak in Figure 7, that is, the limit of the function as $n \to \infty$, $x_n \to 1$, where x_n corresponds to the maximum of the partial Fourier sum. A complete mathematical study of this function (and the general case) is given by Nikol'sky (1975), and this limit is equal to $1.089490 \cdot \pi$. Since Gibbs' phenomenon is essential for partial sums of any orthogonal series, its suppression is important in both algorithms of topographic variables, and in transformations of contour DEMs to gridded DEMs.

Gibbs' phenomenon is observed near elevation discontinuities or sharp elevation changes, and it may result in artificial depressions or peaks near such locations. Besides artefacts from this phenomenon, other artefacts are observed in grids interpolated from contour DEMs that result from discontinuities in slope or from some other reasons. They are (conditionally) referred to as 'Gibbs-like phenomena' (e.g. Gousie and Franklin 2005), (Figure 8).

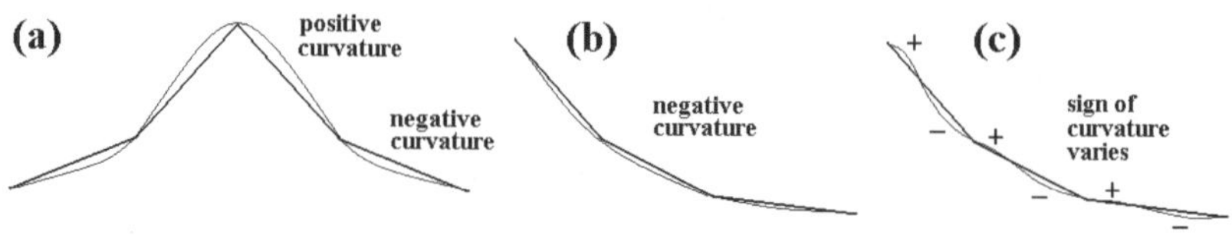

Figure 8. A scheme illustrating Gibbs-like phenomena. Thick lines show the land surface, thin curves illustrate an interpolated surface. (a) Positive and negative curvature of straight slopes, (b) negative curvature of straight slopes, (c) variations in curvature sign of straight slopes.

Although a cubic spline will demonstrate negative curvature for the polyline shown in Figure 8(b), this may be not so (Figure 8(c)) for an interpolated surface when contour line twisting is essential, and the distance between contour lines is much greater than the grid mesh. Such artefacts are characteristic of wiggly contour lines, because they result in interpolated smooth surface flattening (smaller gradient) near contours and steeper slopes between contours. With triangulation-based interpolators (e.g. Akima 1978), all apices of several triangles may have equal elevation values, introducing artificial flat areas at high contour curvature radius because they appear on the same contour line. As a result, profile curvature oscillates with zero values in the middle of each contour and halfway between adjacent contours (see Figure 9).

This effect is essentially weaker after replacing wiggly contours with smoothed ones, as in Figure 10. Gousie and Franklin (2005) noted that manual adding of contour lines diminishes artefacts in similar situations. Favalli and Pareshi (2004) suggested a solution based on a modified Delauney triangulation algorithm 'DEST' to avoid Gibbs-like phenomena. These authors tested their algorithm using both DEMs and model surfaces with discontinuous slopes, and compared the results from DEST and those from TOPOGRID interpolation. They concluded that DEST provides better interpolation because it diminishes artificial flat areas where contour lines demonstrate high curvature.

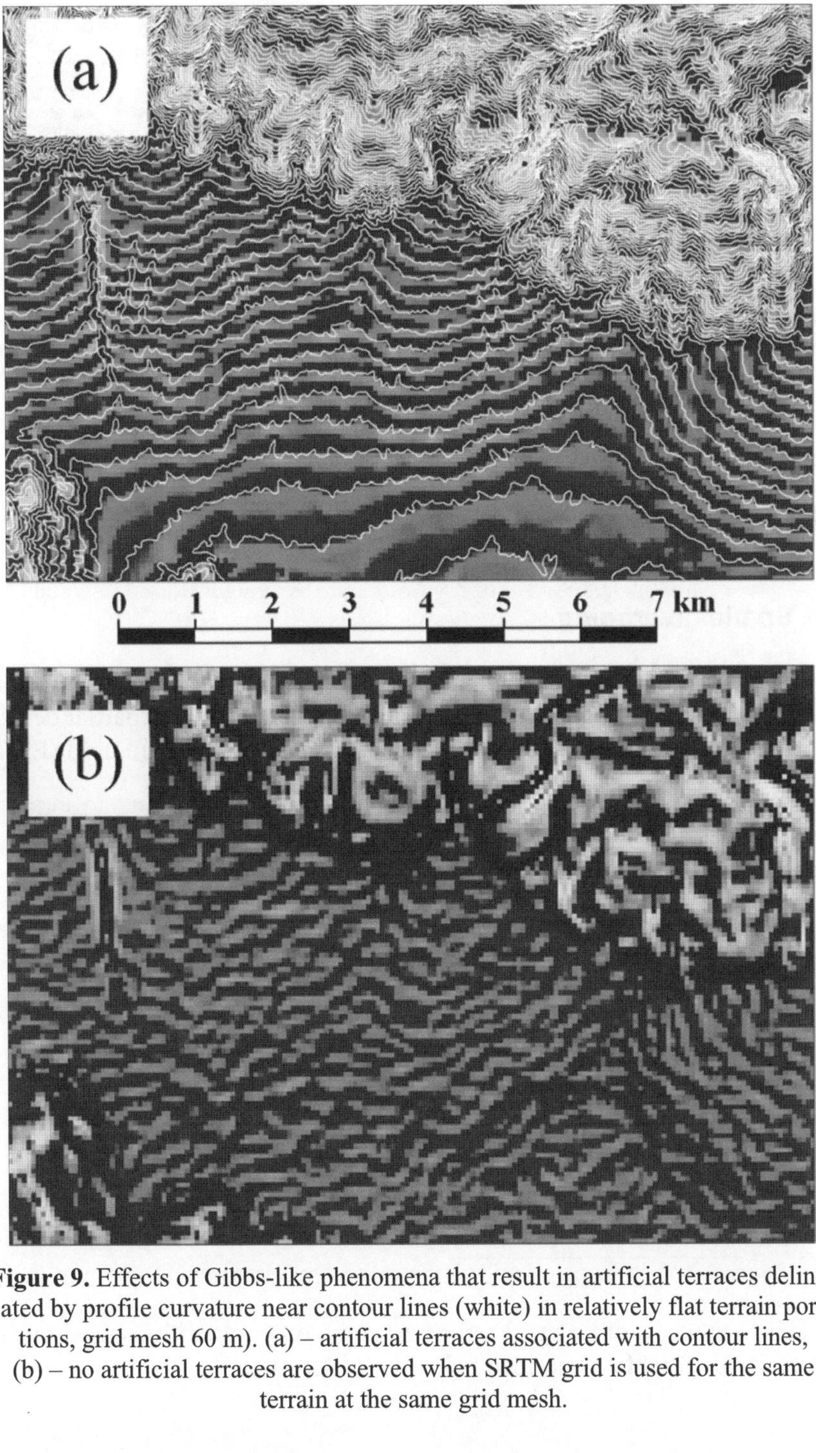

Figure 9. Effects of Gibbs-like phenomena that result in artificial terraces delineated by profile curvature near contour lines (white) in relatively flat terrain portions, grid mesh 60 m). (a) – artificial terraces associated with contour lines, (b) – no artificial terraces are observed when SRTM grid is used for the same terrain at the same grid mesh.

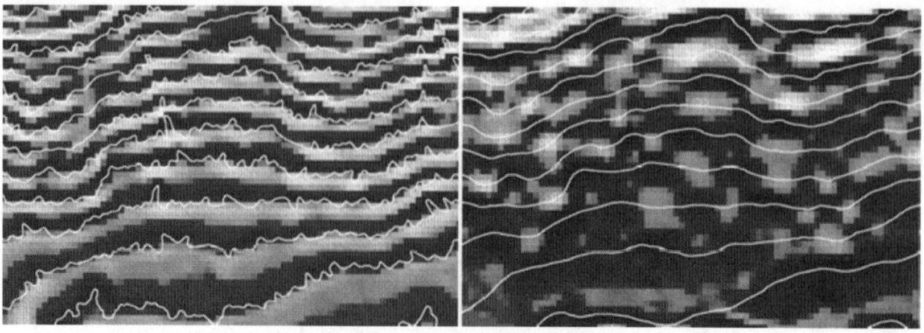

Figure 10. Replacing of wiggly contour lines (left) with smoothed ones diminishes regular artefacts from Gibbs-like phenomena (right). Light colours refer to positive profile curvature, dark ones – to negative values.

Popular algorithms to calculate local topographic variables (described in the next section) normally do not use a single surface, but rather 'sheets' of smooth surface pieces that are not connected at their boundaries (Dikau 1988).

2.2 Spatial averaging

A gridded DEM is a set of regularly spaced (in plan) points with elevations ascribed to each (Figure 11). Grid points are discrete, so that partial derivatives of $z = f(x,y)$ are not defined; instead *finite differences* are used (Evans 1972).

A simple example of a finite difference at the central (5-th) point is

$$p_5 = \frac{z_6 - z_4}{2w};$$

for a smooth surface, the value of p_5 should tend to partial derivative $p_5 = \partial z/\partial x$ at the point 5 as w tends to zero, but this limit may be absent in non-smooth models of topography. To apply spatial averaging, one may use the average of finite differences evaluated at points 2, 5, and 8:

$$p = \frac{1}{3}(p_2 + p_5 + p_8) = \frac{1}{3}\left(\frac{z_3 - z_1}{2w} + \frac{z_6 - z_4}{2w} + \frac{z_9 - z_7}{2w}\right)$$

$$= \frac{z_3 + z_6 + z_9 - z_1 - z_4 - z_7}{6w}.$$

This is the case of Evans–Young and some other algorithms used to calculate local variables described below in this section.

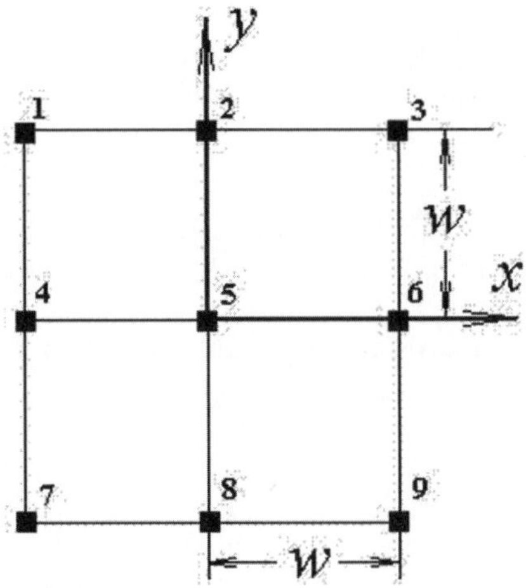

Figure 11. A sub-grid 3 × 3 of a regular grid of grid mesh *w*, axes, locations and numbers of points. Elevation values at points are z_1 to z_9.

Local variables are described by their formulae for a smooth model of the land surface, to which an algorithm is applied that takes the non-smooth nature of topography into account. Such algorithms are based mostly on the fitting of some polynomial by the least squares method to nine grid points of a sub-grid 3 × 3, and its partial derivatives are used to calculate local variables at the central point of this sub-grid. Then the sub-grid moves to the next grid point, and this is repeated many times to compute values of local variables for all grid points. Schmidt *et al.* (2003) and Sousa *et al.* (2003) noted that the most popular algorithms are Evans–Young (Evans 1972), Shary (1995), Zevenbergen–Thorne (Zevenbergen and Thorne 1987), and the modified Evans–Young (Shary *et al.* 2002) methods.

A first such algorithm, known as Evans–Young (Evans 1972; Young 1978, see also Pennock *et al.* 1987), consists of the following. The second order polynomial

$$z = \frac{rx^2}{2} + sxy + \frac{ty^2}{2} + px + qy + z_0$$

where p, q, r, s, t, z_0 are coefficients to be defined, is fitted by the least squares method to the sub-grid 3 × 3, with grid mesh *w* and elevations in nodes $z_1,\ldots,z_9$ (Figure 11).

This gives the following formulae for coefficients p, q, r, s, t, z_0 of the polynomial:

$$p = \frac{z_3 + z_6 + z_9 - z_1 - z_4 - z_7}{6w},$$

$$q = \frac{z_1 + z_2 + z_3 - z_7 - z_8 - z_9}{6w},$$

$$r = \frac{z_1 + z_3 + z_4 + z_6 + z_7 + z_9 - 2(z_2 + z_5 + z_8)}{3w^2}, \quad (1)$$

$$s = \frac{-z_1 + z_3 + z_7 - z_9}{4w^2},$$

$$t = \frac{z_1 + z_2 + z_3 + z_7 + z_8 + z_9 - 2(z_4 + z_5 + z_6)}{3w^2},$$

$$z_0 = \frac{5z_5 + 2(z_2 + z_4 + z_6 + z_8) - (z_1 + z_3 + z_7 + z_9)}{9}.$$

See Young (1978), and Pennock *et al.* (1978) for the derivation of these formulae. In accordance with the polynomial formula, coefficients p, q, r, s, t approximate the following partial derivatives:

$$p = \frac{\partial z}{\partial x}, \quad q = \frac{\partial z}{\partial y}, \quad r = \frac{\partial^2 z}{\partial x^2}, \quad s = \frac{\partial^2 z}{\partial x \partial y}, \quad t = \frac{\partial^2 z}{\partial y^2}. \quad (2)$$

The Evans–Young polynomial does not necessarily pass through all the nine original elevations; its elevation at the central point is given by z_0.

In the algorithm of Shary (1995), the following polynomial is used:

$$z = \frac{rx^2}{2} + sxy + \frac{ty^2}{2} + px + qy + z_5,$$

where p, q, r, s, t are coefficients to be defined that approximate the partial derivatives (2); its elevation at the central point is z_5. Fitting by the least squares method to the sub-grid 3 × 3, one obtains:

$$p = \frac{z_3 + z_6 + z_9 - z_1 - z_4 - z_7}{6w},$$

$$q = \frac{z_1 + z_2 + z_3 - z_7 - z_8 - z_9}{6w},$$

$$r = \frac{z_1 + z_3 + z_7 + z_9 - 2(z_2 + z_8) + 3(z_4 + z_6) - 6z_5}{5w^2}, \quad (3)$$

$$s = \frac{-z_1 + z_3 + z_7 - z_9}{4w^2},$$

$$t = \frac{z_1 + z_3 + z_7 + z_9 - 2(z_4 + z_6) + 3(z_2 + z_8) - 6z_5}{5w^2}.$$

This differs from the Evans–Young algorithm in that this polynomial passes through the central point but is close to Evans–Young one (Schmidt *et al.* 2003), with differences only in r and t.

In the Zevenbergen–Thorne algorithm (Zevenbergen and Thorne 1987) the following polynomial is used:

$$z = Ax^2y^2 + Bx^2y + Cxy^2 + \frac{rx^2}{2} + sxy + \frac{ty^2}{2} + px + qy + D,$$

where $A, B, C, D, p, q, r, s, t$ are coefficients. Here there are nine coefficients and nine elevations, so that coefficients of this Lagrange polynomial are:

$$p = \frac{z_6 - z_4}{2w},$$

$$q = \frac{z_2 - z_8}{2w},$$

$$r = \frac{z_4 + z_6 - 2z_5}{w^2},$$

$$s = \frac{-z_1 + z_3 + z_7 - z_9}{4w^2}, \quad (4)$$

$$t = \frac{z_2 + z_8 - 2z_5}{w^2},$$

$$A = \frac{(z_1 + z_3 + z_7 + z_9) - 2(z_2 + z_4 + z_6 + z_8) + 4z_5}{4w^4},$$

$$B = \frac{(z_1 + z_3 - z_7 - z_9) - 2(z_2 - z_8)}{4w^3},$$

$$C = \frac{(-z_1 + z_3 - z_7 + z_9) - 2(z_6 - z_4)}{4w^3},$$

$$D = z_5,$$

where p, q, r, s, t approximate the same partial derivatives (2).

Schmidt *et al.* (2003) experimentally compared these three algorithms. These authors used plots with curvature values on both axes that were calculated using different algorithms. They noted also that the modified Evans–Young algorithm (Shary *et al.* 2002) diminishes grid bias in maps of curvatures. Their conclusion was that the Evans–Young and Shary algorithms provide close results for curvatures (i.e. smallest deviations from straight lines at their plots), in contrast to Zevenbergen–Thorne. Florinsky (1998) compared the same algorithms theoretically. He assumed that deviations in topographic variables can be described by the first term of a polynomial series, and used Root Mean Square Error (RMSE) criterion to compare the algorithms. But he did not evaluate the remaining terms of the series, which may not be small. On this assumption, he concluded that the accuracy in second derivatives determining these algorithms is diminishing in the following order: Zevenbergen–Thorne, Evans–Young, Shary.

The Evans–Young and Shary algorithms use additional internal smoothing of the land surface in contrast to the Zevenbergen–Thorne algorithm. The latter is based on the simplest form of finite difference; the former use averaged finite difference. For example, the simplest finite difference

$$p = \frac{z_6 - z_4}{2w}$$

found in the Zevenbergen–Thorne algorithm (4) is replaced with the average of the three finite differences along the axis x,

$$p = \frac{1}{3}\left(\frac{z_3 - z_1}{2w} + \frac{z_6 - z_4}{2w} + \frac{z_9 - z_7}{2w}\right)$$

in both the Evans–Young (1) and Shary (3) algorithms.

A grid bias may appear essential in the case of curvatures (Shary *et al.* 2002). In this regard, Shary *et al.* (2002) have suggested a modified Evans–Young algorithm that consists of the following. A smoothing filter is used that obeys the following criteria:

- an inclined plane piece of surface is transformed into the same plane,
- the smoothing filter is isotropic,
- filter weights decrease linearly with the distance from the centre of the sub–grid.

This results in the following formula for the filter (Shary *et al.* 2002):

$$z_5^* = \frac{k(z_2 + z_4 + z_6 + z_8)}{9} + \left(1 - \frac{4k}{9}\right) z_5$$

$$\text{if } k \in \left[0, 1 - \frac{1}{\sqrt{2}}\right] = [0, 0.293],$$

$$z_5^* = \frac{\left[1 - (1-k)\sqrt{2}\right] \cdot \left(z_1 + z_3 + z_7 + z_9\right)}{9} + \frac{k \cdot \left(z_2 + z_4 + z_6 + z_8\right)}{9} + \left(1 - \frac{4\left[1 + k - (1-k)\sqrt{2}\right]}{9}\right) z_5$$

$$\text{if } k \in \left[1 - \frac{1}{\sqrt{2}}, 1\right] = [0.293, 1],$$

where k is the smoothing parameter. Its value of 1/5 provides good results, so that

$$z_5^* = \frac{z_2 + z_4 + 41 z_5 + z_6 + z_8}{45}, \tag{5}$$

where elevation z_5 at the central point is replaced with the new value z_5*. After all the nine elevations of the sub-grid 3 × 3 are replaced with their smoothed values, the original Evans–Young algorithm is applied to calculate the derivatives *p*, *q*, *r*, *s*, *t*. The modified Evans–Young algorithm is based on 5 × 5, rather than 3 × 3, sub-grids.

So, the averaging in 3 × 3 sub-grids in these algorithms increases in the following order:

- Zevenbergen–Thorne,
- Evans–Young and Shary,
- modified Evans–Young.

Tomer and Anderson (1995) noted that the land surface is considered to be composed of deterministic (smooth) and noisy (non-smooth) components. This is a conceptual remark, not a mathematical one, because it is difficult to sub-divide topography into these components as they have similar internal properties.

The sensitivity of local topographic variables to the noisy DEM component (and dependence of variables on grid mesh) increases in the following order (Shary *et al.* 2002):

1. functions of first derivatives (slope steepness, aspect, and insolation),
2. linear functions of second derivatives (simple curvatures),
3. square functions of second derivatives (total curvatures).

Functions of higher order derivatives may encounter major problems due to their excessive sensitivity to DEM noises.

2.3 Two concepts of scale

Traditional topographic maps are corresponding images printed on a sheet of paper, to which information on scale is added; for example, 1:100,000 indicates the ratio of length on this paper to the corresponding length in the terrain (in plan). This is a *geographical concept* of scale. When an elevation grid is measured, the geographical scale appears arbitrary (it depends on screen pixel size and map image magnification), but now the grid mesh describes the *detail concept* of scale. Any change of given topographic map does not change the contour interval, so that measures like grid mesh or contour interval remain constant as the geographical scale changes.

Understanding of interrelationships between these two concepts of scale is of value because it permits the deduction of the minimal grid mesh from a given contour DEM.

2.3.1 Minimal grid mesh

Grid mesh is a basic concept of a gridded DEM or a regular point DEM. A gridded DEM with any grid mesh may be obtained from a contour or irregular point DEM using an interpolation process (e.g. Akima 1978). In practice, there is no way to restore information on the shapes of ant hills using 1:250,000 scale topographic maps, because they are out-of-scale. This is because geodetic surveys use an agreement specifying some standardized contour intervals for use for each geographical scale (e.g. about 20 metres for 1:250,000 scale). Such practices result in some interdependence between geographical and detail scale concepts. Considering these practices, it is possible to deduce a formula of minimal grid mesh for transformations of contour DEMs into gridded DEMs.

There exists some minimal grid mesh that corresponds to a given elevation data resolution. Indeed, sufficiently small landforms, of size w_0 in plan, are not described at the given elevation data detail, so that values of grid mesh w smaller than w_0 do not provide new information. One may use w values greater than w_0, with some loss of information on smaller landforms.

When contour lines are digitized, an operator records points on each contour line, the distance between which is greater in straight portions of contour line and smaller in their curved portions. A grid is calculated only from these and separate points, with the total number of points being N. The size of the study area is always known (X along the x–axis, and Y along the y–axis), so that each point occupies an average area equal to XY/N, with the average distance between points equal to $L = (XY/N)^{1/2}$. Since L is greater for spatially dense contour lines and smaller in the opposite case, it may be used for automated minimal grid mesh w_0 determination using the formula

$$w_0 = \sqrt{\frac{X \cdot Y}{\lambda \cdot N}}, \tag{6}$$

where X, Y, w_0 are in the same units of length, and λ is a dimensionless empirical coefficient. In the author's experience, $\lambda = 2.5$ is a reasonable choice for most cases. As explained above, any kind of DEM can be transformed into an irregular point DEM, so that Equation (6) is applicable to any irregular DEM, excluding cases describes below.

Some examples of elevation data result in artificial terrain features that are due to regular or discontinuous features of the given terrain (Figure 9). This problem is not a subject of algorithms for computing local topographic variables, but rather of the algorithms for the transformation of irregular point DEMs to grids. There is no general rule for all such special cases. For example, when distance between points is relatively small in each of the parallel lines, but is much greater between these lines, one may add points of an intermediate elevation between adjacent lines to diminish artefacts in the resulting grid. In fact, it is not clear which landforms are in areas with no elevation data, so that attempts to restore out-of-scale landforms do not produce information on them. In such cases, one may replace w_0 calculated from Equation (6) with some greater value.

3 Regional Topographic Attributes

To calculate derivatives or finite differences at a given point, one needs only a restricted vicinity of this point, so that the formula Variable = $f(p,q,r,s,t)$ may represent only local (derivative-based) topographic variables. To describe the relative position of a given point in the landscape, one should use integrals or sums, that is, to test extended surroundings of this point defined by corresponding terrain features (Speight 1974). For example, catchment area at a given point needs to sum inputs from all upslope portions of a basin that may lie up to thousands of kilometres from this point. Correspondingly, formulae are used only for local variables; to compute regional variables, one needs to use algorithms instead of formulae (Shary *et al.* 2002). In general, algorithms to describe regional variables need program codes (such as those of Martz and de Jong (1988) and Freeman (1991) for catchment area). For this reason, algorithms of regional variables are not described here.

The tendency of local variables to depend on grid mesh may appear not to be characteristic of some regional variables resulting in so-called scale-free variables (in contrast to scale-specific ones) that tend towards limit values as the grid mesh tends to zero (Shary *et al.* 2002, 2005). For example, the calculated depth of the Lake Baikal at a given point tends to the value one can measure from a boat (this is the limit value) as grid mesh tends to zero; note that corresponding regional variable is the depression depth, and that direct measurements from a boat are impossible for dry depressions (or for not completely filled ones, such as the Caspian depression) characteristic, for example, of large deserts (Shary *et al.* 2005).

3.1 Kinds of topographic attributes and landforms

Some topographic variables describe water flow (e.g. catchment area) and therefore need the gravitational field to be used; others describe thermal regimes of slopes and need solar irradiation to be taken into account; and yet others describe the land surface itself, thus ignoring all geophysical vector fields. Only the latter are studied in differential geometry of surfaces (Koenderink and van Doorn 1994). In other words, these variables fall into two non–intersecting groups, one of which describes the system '*land surface + vector field*', and another describing the land surface itself. On the other hand, all variables fall into two other non-intersecting groups (Speight 1974), one describing the 'local geometry' of land surface (*local* variables), and others describing the relative position in the landscape (*regional* variables). Regional variables need integrals or sums for their determination. This leads to a classification of all (both discovered and not) topographic attributes into several non-intersecting classes (Shary 1995; Shary *et al.* 2002), shown in Table 1.

Shary *et al.* (2002) have also introduced two global (planetary) classes of topographic attributes and landforms that are not considered here because no variables are currently introduced in these classes (Shary 2006a). The list of local variables is mostly complete (Shary 1995), while regional topographic variables are not yet fully described (Shary *et al.* 2002, Shary 2006a).

In contrast to local landform classifications, regional ones (see MacMillan and Shary 2007) include integral-based landforms, such as non-intersecting closed depressions, hills (mountains), and saddles, and the corresponding natural hierarchy of these landforms, such as first and higher continental (or island's) level and landform scale–free characteristics (see Section 3.2 for the meaning of the 'scale–free' term), such as boundaries, area, volumes, some others (Shary *et al.* 2005).

Table 1. The four non-intersecting classes of topographic attributes and landforms.

Topographic attributes	Local	Regional
Field-specific	Local field-specific (class A)	Regional field-specific (class B)
Field-invariant	Local field-invariant (class C)	Regional field-invariant (class D)

3.2 Scale-specific and scale-free topographic attributes

Results in Figures 3 and 4 demonstrate that local topographic attributes and landforms may be scale-specific, and that curvature-sign-based landform classifications appear to be statistically predictable. This may not be the case for regional attributes and landforms, because they are based on integrals that may exist even for non-smooth surfaces. The scale-free properties of some regional landforms were shown experimentally in Shary *et al.* (2002, 2005).

According to the definitions in these papers, scale-free topographic variables are interpreted not as variables that do not depend on scale (e.g. on grid mesh), but rather as variables that have limit values as scale becomes finer (e.g. grid mesh diminishes). Such topographic attributes do exist; examples are maximal catchment area in thalwegs (where it is an integral value), closed depression volume, and the maximal value of maximal catchment area that is reached at grid boundaries (Shary *et al.* 2005).

In contrast to statistically predictable local landforms, regional landforms appear terrain-specific; that is, their characteristics cannot be predicted in advance (MacMillan and Shary 2007), so that, for example, the ratio of sums of hills' / mountains' volumes to those of depressions are changed from 0.6 to 10,000 depending on the geological terrain features (Shary *et al.* 2005). Currently, we have only a few regional variables in contrast to local ones (Shary 2006a), so that most variables of general geomorphometry are local and demonstrate an essential dependence on scale. Even 'secondary' regional variables (i.e. composed by basic variables (Wilson and Gallant 2000), which are essentially variables of specific geomorphometry) demonstrate scale-specific behaviour (Shary *et al.* 2002).

4 Discussion

The non-smooth model of topography means that the land surface cannot be considered as smooth, not just at separate breaks or points, but rather at almost any point. There are many consequences from this model for local derivative-based topographic variables and landforms that demonstrate scale-specific behaviour with no 'reference' values at fine resolutions. In contrast, regional integral-based variables and landforms may have limit values as grid mesh tends to zero. On the other hand, the set of local variables is almost completed (Shary 1995, Shary *et al.* 2002), in contrast to the regional variables that are not yet fully described (Shary 2006a).

Correspondingly, there is a set of unsolved tasks in geomorphometry, mostly in the regional classes (Shary 2006b), such as the absence of substantiated regional slope profile descriptions that are very important in such DTA applications as predictive soil and vegetation mapping based on multiple regression models (e.g. Scull *et al.* 2003), or the problem of quantitative description of open depressions (Shary 2006b).

The statistical predictability of local landforms results in a stable belief that any process or phenomenon in landscape is best described at its own scale or grid mesh defined by the characteristic size of that process or phenomenon (e.g. Klemeš 1983, Phillips 1988). However, this is not so with regional integral-based variables and landforms that are best described at fine resolutions. For example, volume and area of a closed depression (e.g. of a lake) tend to their limit values as grid mesh diminishes (Shary *et al.* 2005). Regional integro-differential variables (i.e. based on both integrals and derivatives), such as maximal catchment area, form an intermediate case. Here, the non-smooth nature of the land surface results in water divides not being lines, but rather bands that may be determined at some probability level, and they usually have greater widths in gently sloping terrain than in mountains.

Several authors consider multiple flow-lines branching and confluence as resulting from discrete land surface representation in gridded DEMs (e.g. Tarboton 1997), but this phenomenon is a generalization obtained from direct observation in landscapes, where grids are absent. Shary *et al.* (2005) concluded that multiple flow branching and confluence is a physical phenomenon that is independent of DEMs, so that corrections to catchment area suggested by several authors (e.g. Freeman 1991) describe essential physical properties of terrain rather than simply efforts to diminish the grid bias.

Also, land surfaces demonstrate natural hierarchies in landform arrangement in regional classes, so that hills or mountains of the first continental (or island's) level of hierarchy essentially differ from those of the second and higher levels. In DTA, this results in an ambiguous meanings of the same concept, and even in some discussions around the questions 'What is a mountain?' or 'Do mountains exist?' (e.g. Smith and Mark 2003, Mark and Smith 2004). In regional class B, they may be described as closed depressions of an inverted surface (for which elevations Z are replaced with –Z). This definition applied to a continent results in all the continent being a single 'hill' of the zero hierarchical level surrounded by a closed contour line. The first continental level is defined by the largest closed contour lines inside the continent (normally hills and mountains of this level have geographical names). Continuing this process, one may calculate nested hills/mountains of 2-nd, 3-rd,… hierarchical levels. It is important that these landforms are integral-based, and therefore have limit (reference) values at large scales (Shary et al. 2005). Boundaries of closed depressions, hills/mountains, and saddles (i.e. the remaining area) cannot have intersections (*ibid.*), thus providing a stable regional hierarchical landform classification.

Geomorphometry of regional variables is still at its beginning, but it demonstrates limit (reference) values of its variables and terrain-specific landforms, even for the non-smooth models of topography. In contrast to predictable areas, curvature-sign-based landform patterns appear not to be predictable because they are considered as regional entities in human perception, and might therefore be generalized to regional entities in a similar way as maximal catchment area generalize the concept of flow-line convergence/divergence by tangential or plan curvature.

Acknowledgements

The author is grateful to Prof. George Miliaresis (University of Patras, Greece) and to Dr. Harris Pappas (Technical University of Serres, Greece) for their data used in Figure 9, to Ian S. Evans (University of Durham, England), Robert A. MacMillan (LandMapper Environmental Solutions Inc., Canada), Richard J. Pike (U.S. Geological Survey, USA), and other colleagues for their useful comments and references from our discussion of the topic.

References

Akima, H. (1978), A method of bivariate interpolation and smooth surface fitting for irregularly distributed data points, *ACM Transactions on Mathematical Software*, **4**: 148–159.

Cayley, A. (1859), On contour and slope lines, *Philosophical Magazine*, Series 4, (**18**)120: 264–268.

Clarke, K.C., (1988), Scale-based simulation of topographic relief, *The American Cartographer*, **15**: 173–181.

Dikau, R., (1988), Case studies in the development of derived geomorphic maps, *Geologisches Jahrbuch*, **A104**: 329–338.

Dokuchaev, V.V., (1886), Razbor glavneishikh pochvennykh ponyatiy, In *Dokuchaev, Izbrannye Sochineniya*, Vol.3, 1949: 161–239 (Moscow: Sel'hozgiz, in Russian, original work published 1886).

Evans, I.S., (1972), General geomorphometry, derivatives of altitude, and descriptive statistics, In Chorley, R.J. (ed.): *Spatial Analysis in Geomorphology*, London: Methuen & Co., Ltd.: 17–90.

Evans, I.S., (1975), The effect of resolution on gradients calculated from an altitude matrix, *Report 3 on Grant DA-ERO-591-73-G0040*, '*Statistical characterization of altitude matrices by computer*', Department of Geography, University of Durham, England.

Evans, I.S., (1980), An integrated system of terrain analysis and slope mapping, *Zeitschrift für Geomorphologie N.F.*, Suppl.–Band **36**: 274–295.

Evans, I.S. and McClean, C.J., (1995), The land surface is not unifractal: variograms, cirque scale and allometry, *Zeitschrift für Geomorphologie N.F., Suppl.-Bd.*, **101**: 127–147.

Evans, I.S. and Cox, N.J., (1999), Relations between land surface properties: altitude, slope and curvature, In Hergarten, S. and Neugebauer H.J. (eds.): *Process Modelling and Landform Evolution*, Berlin: Springer: 13–45.

Favalli, M. and Pareshi, T.T., (2004), Digital elevation model construction from structured topographic data: The DEST algorithm, *Journal of Geophysical Research*, **109**, F04004.

Finn, R., (1986), *Equilibrium Capillary Surfaces*, New York: Springer-Verlag.

Florinsky, I.V., (1998), Accuracy of local topographic variables derived from digital elevation models, *International Journal of Geographical Informational Science*, **12**: 47–61.

Freeman, T.G., (1991), Calculating catchment area with divergent flow based on a regular grid, *Computers and Geosciences*, **17**: 413–422.

Gauss, K.F., (1827), Disquisitiones generales circa area superficies curves, *Göttingen gelehrte Anziegen*, No.177: 1761–1768 (In Latin).

Goodchild, M., (1982), The fractional Brownian process as a terrain simulation model, *Modelling and Simulation*, **13**: 1133–1137.

Gousie, M.B. and Franklin, W.R., (2005), Augmenting grid-based contours to improve thin-plate DEM generation, *Photogrammetric Engineering and Remote Sensing*, **(71)**1: 69–79.

King, D., Bourennane, H., Isampert, M. and Macaire, J.J., (1999), Relationship of the presence of a non-calcareous clay-loam horizon to DEM attributes in a gently sloping area, *Geoderma*, **89**: 95–111.

Klemeš, V., (1983), Conceptualization and scale in hydrology, *Journal of Hydrology*, **65**: 1–23.

Koenderink, J.J. and van Doorn, A.J., (1994), Two–plus–one–dimensional differential geometry, *Pattern Recognition Letters*, **15**: 439–443.

Li, Z., Zhu, Q., and Gold, C., (2005), *Digital Terrain Modeling: Principles and Methodology*, Florida, Boca Raton: CRC Press.

MacMillan, R.A. and Shary, P.A., (2007). Extraction and classification of landforms and landform elements, In Hengl, T. and Reuter, H.I. (eds.): *Geomorphometry: Concepts, Software, Applications*, Ch. 9, Joint Research Centre: European Commission: (in press).

Mandelbrot, B., (1967), How long is the coast of Britain? Statistical self-similarity and fractional dimension, *Science*, **156**(3775): 636–638.

Mark, D.M. and Smith, B., (2004), A science of topography – from qualitative ontology to digital representations, In Bishop, M.P., and Shroder Jr., J.F. (eds.): *Geographic Information Science and Mountain Geomorphology*, Chichester UK, Praxis Scientific Publishing: Springer-Verlag: 75–110. Available online at: http://ontology.buffalo.edu/smith//articles/topography.pdf (accessed 29 October 2007).

Martz, L.W. and de Jong, E., (1988), CATCH: a Fortran program for measuring catchment area from digital elevation models, *Computers and Geosciences*, **14**: 627–640.

Maxwell, J.C., (1870), On hills and dales, *Philosophical Magazine*, Series 4, **(40)**269: 421–427.

Nikol'sky, S.M., (1975), *Kours Matematicheskogo Analiza*, 2: 199–203, Moscow: Nauka (in Russian).

Pachepsky, Ya.A., Timlin, D.J. and Rawls, W.J., (2001), Soil water retention as related to topographic variables, *Soil Science Society of America Journal*, **65**: 1787–1795.

Pennock, D.J., Zebarth, B.J. and de Jong, E., (1987), Landform classification and soil distribution in hummocky terrain, Saskatchewan, Canada, *Geoderma*, **40**: 297–315.

Peuquet, D.J., (1984), A conceptual framework and comparison of spatial data models, *Cartographica*, **21**: 66–113.

Phillips, J.D., (1988), The role of spatial scale in geomorphic systems, *Geographical Analysis*, **20**: 308–317.

Pike, R., (1995), Geomorphometry – progress, practice, and prospect, *Zeitschrift für Geomorphometry N.F. Suppl.–Band*, **101**: 221–238.

Richardson, L., (1961), The problem of contiguity: An appendix to statistics of deadly quarrels, In *Yearbook of the Society for General Systems Research*, Vol.6, Michigan: Ann Arbor, 140–187.

Sard, A., (1942), The measure of the critical values of differentiable maps, *Bulletin of the American Mathematical Society*, **48**: 883–890.

Schmidt, J., Evans, I.S. and Brinkmann, J., (2003), Comparison of polynomial models for land surface curvature calculation, *International Journal of Geographical Information Science*, **17**: 797–814.

Schmidt, J. and Hewitt, A., (2004), Fuzzy land elements classification from DTMs based on geometry and terrain position, *Geoderma*, **121**: 243–256.

Shary, P.A., (1995), Land surface in gravity points classification by a complete system of curvatures, *Mathematical Geology*, **27**: 373–390.

Shary, P.A., Sharaya, L.S. and Mitusov, A.V., (2002), Fundamental quantitative methods of land surface analysis, *Geoderma*, **107**: 1–32.

Shary, P.A., (2005), *Geomorphometry and 'Analytical GIS Eco' software description and demonstration version*, Available on-line at: http://www.giseco.info/ (accessed 29 October 2007).

Shary, P.A., Sharaya, L.S. and Mitusov, A.V., (2005), The problem of scale-specific and scale-free approaches in geomorphometry, *Geografia Fisica e Dinamica Quatrenaria*, **28**: 81–101.

Shary, P.A., (2006a), Variables of geomorphometry: The current state-of-art, In *International Symposium on Terrain Analysis and Digital Terrain Mapping* (TADTM 2006), November 23–25 in Nanjing, China, Proceedings of TADTM 2006, full paper: CD–ROM.

Shary, P.A., (2006b), Unsolved tasks of geomorphometry. In *International Symposium on Terrain Analysis and Digital Terrain Mapping* (TADTM 2006),

November 23–25 in Nanjing, China, Proceedings of TADTM 2006, full paper: CD-ROM.

Shary, P.A. and Sharaya, L.S., (2006), Predictable and terrain–specific landform classifications. In *International Symposium on Terrain Analysis and Digital Terrain Mapping* (TADTM 2006), November 23–25 in Nanjing, China, Proceedings of TADTM 2006, full paper: CD–ROM.

Smith, B. and Mark, D.M., (2003), Do mountains exist? Towards an ontology of landforms, *Environment & Planning B (Planning and Design)*, **30**: 411–427. Available on-line at: http://ontology.buffalo.edu/smith/articles/Mountains.pdf (accessed 29 October 2007).

Sousa, M.C., Foster, K., Wyvill, B. and Samavati, F., (2003), Precise ink drawing of 3d models, *EuroGraphics*, **22**: 47–61.

Speight, J.G., (1974). A parametric approach to landform regions. In *Progress in Geomorphology*. Institute of British Geographers special publ. No.7, Oxford: Alden & Mowbray Ltd. at the Alden Press: 213–230.

Tarboton, D.G., (1997), A new method for the determination of flow directions and upslope areas in grid Digital Elevation Models, *Water Resources Research*, **33**: 309–319.

Tomer, M.D. and Anderson, J.L., (1995), Variation of soil water storage across a sand plain hillslope, *Soil Science Society of America Journal*, **59**: 1091–1100.

Troeh, F.R., (1964), Landform parameters correlated to soil drainage, *Soil Science Society of America Proceedings*, **28**: 808-812.

Voronoi, G., (1907), Nouvelles applications des paramètres continus à la théorie des formes quadratiques. *Journal für die Reine und Angewandte Mathematik*, **133**: 97–178 (in French).

Wood, J., (1996), *The geomorphological characterization of Digital Elevation Models*, PhD thesis, University of Leicester, Available on-line at: http://www.soi.city.ac.uk/~jwo/phd/ (accessed 27 October 2007).

Wilson, J. and Gallant, J. (eds.), (2000), *Terrain Analysis: Principles and Applications*, New York: John Wiley & Sons, Ltd.

Xu, T.B., Moore, I.D. and Gallant, J.C., (1993), Fractals, fractal dimensions and landscapes – a review, *Geomorphology*, **8**: 245–262.

Young, M., (1978), Terrain analysis: program documentation. Report 5 on Grant DA-ERO-591-73-G0040, 'Statistical characterization of altitude matrices by computer', Department of Geography, University of Durham, England.

Zevenbergen, L.W. and Thorne, C.R., (1987), Quantitative analysis of land surface topography, *Earth Surface Processes and Landforms*, **12**: 47–56.

Multi-Scale Digital Terrain Modelling and Analysis

LI Zhilin

Abstract

Scale is one of the most important but unsolved issues in various scientific disciplines. This chapter tackles systematically the issues related to scale in terrain modelling and analysis. It starts with some discussions on the notions of *scale* and *multi-scale*. Then three approaches for the multi-scale representation of terrain surface are presented, i.e. critical-points-based, smoothing-based, and scale-driven. Principles are explained and algorithms presented. Two issues of multi-scale terrain analysis are discussed, i.e. the modelling of multi-scale effects, and the optimization of scale. Examples are given to illustrate the concepts and principles discussed. Finally, some concluding remarks are made.

Keywords: scale, multi-scale, multi-resolution, variable resolution, terrain representation, terrain analysis, scale effect, optimization.

1 Introduction

The depiction of terrain surface with accuracy has always been a challenge for scientists and engineers in geography, geomorphology, civil engineering, geology, and surveying. The terrain surface can be represented in different ways. A painting (or drawing) (e.g. Figure 1) is perhaps the oldest way of terrain representation. Such a representation looks intuitive but offers very low metric quality (accuracy).

A development from the simple drawing is the shading technique. Two types of shading are available, i.e. oblique (hill) and vertical (slope). In oblique shading, those surface facets that are facing the light source are depicted with lighter tones, while those facing away from the light source are drawn with darker tones. In slope shading, the darker tone is used to depict those facets with steeper slopes. Hachuring is another technique to depict the terrain slope with line symbols.

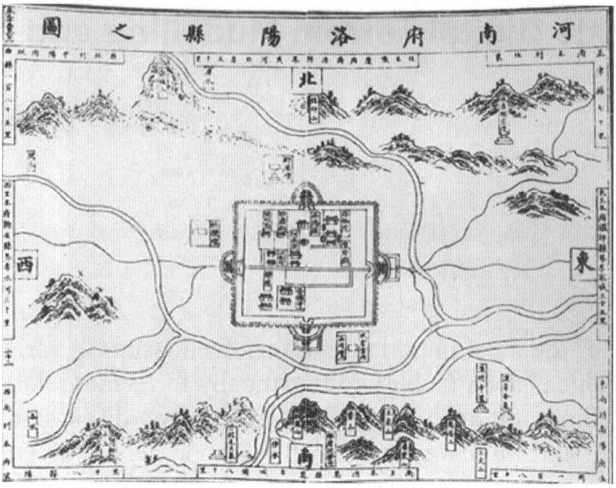

Figure 1. A map of Luoyang County (China) (produced in 1403-1407) with the terrain surface represented by simple drawing.

A topographic map with contour (level set, isopleth, isoline, isogram, or isarithm) lines is perhaps the most popular way of representation since its invention in 1701, as such a map offers high metric quality (Wikipedia, http://en.wikipedia.org/wiki/Contour_line#History). However, such maps do not look very intuitive to many users. To produce a better visual effect, inclined contours, shadowed contours, and illuminated contours have been proposed. Hypermetric tints (using colour layers) have been widely used.

A physical model of the terrain surface is perhaps the most intuitive representation. Such a model could be made in clay or plastic. However, it is time-consuming to build such a model. A digital terrain model (DTM), since its introduction in 1958 (Miller and Laflamme 1958), has found wide application, as it offers great flexibility in many ways. It has been utilized for terrain analysis, such as the computation of slope and aspect, area and volume, roughness parameters and hydrological parameters, the derivation of viewsheds and the analysis of inter-visibility between points on terrain surfaces. A volume with a collection of articles on DTM-based terrain analysis has been edited by Wilson and Gallant (2000).

A DTM, like a topographic map, is associated with scale, though many researchers have argued that there is no scale with digital data sets. Then there is a scale issue in the representation and analysis of terrain surfaces.

Scale is an old issue in geo-sciences. In cartography, maps are produced at certain scales, e.g. 1:10,000 and 1:100,000. At different scales, the level of detail represented on a map is different. The derivation of smaller-scale maps from larger-scale maps through some operations is a multi-scale issue and is called "map generalization". In geography, there is a similar issue. Normally, geographical data are sampled in small enumeration units. And in some applications, such data need to be aggregated to a larger enumeration unit. However, the statistical results will be different when the aggregation is carried out in different ways. This issue is called the "modifiable areal unit problem" (MAUP) (Openshaw 1994), although some researchers have reservation on this issue (Tobler 1989). Indeed, there is a similar issue in all geo-sciences, such as geomorphology (e.g. de Boer 1992), oceanography (e.g. Stommel 1963), soil science (e.g. Hillel and Elrick 1990), biology (e.g. Haury *et al.* 1977), biophysics (e.g. Friedl 1997), social sciences (e.g. Dovers 1994), hydrology (e.g. Blöschl and Sivapalan 1995), environmental sciences (e.g. Bian 1997), etc. Indeed, the issue of scale has attracted great attention from the geoscience communities. As a result, a number of books on scale issues have recently been published, including *Scaling Up in Hydrology Using Remote Sensing* edited by the Institute of Hydrology and Stewart (1996), *Scale in Remote Sensing and GIS* edited by Quattrochi and Goodchild (1997), *Scale Dependence and Scale Invariance in Hydrology* edited by Sposito (1998), *Modelling Scale in Geographical Information Science* edited by Tate and Atkinson (2001) and *Algorithmic Foundation of Multi-Scale Spatial Representation* authored by Li (2007).

In spite of international efforts for many years, scale is a still major unsolved issue in geographical information related sciences. This chapter will address two scale issues, i.e. multi-scale digital terrain representation and multi-scale digital terrain analysis.

2 Basic Concepts of Scale and Multi-Scale

In order to facilitate the discussion on multi-scale terrain modelling and analysis, some basic concepts need to be clarified.

2.1 The concept of scale

Scale is a term not well defined. "Of all words that have some degree of specialized scientific meaning, 'scale' is one of the most ambiguous and overloaded" (Goodchild and Quattrochi 1997).

There is a spatial scale and a temporal scale. In both domains, the scale may range from micro to macro (see Figure 2) and may be in different levels of measurement (i.e. nominal, ordinal, interval and ratio). In this chapter, only spatial scale is discussed.

Terrain specialists are not interested in reality at either the micro or micro scales. Instead, they are interested in the middle part of the scale spectrum, i.e. the geo-scale.

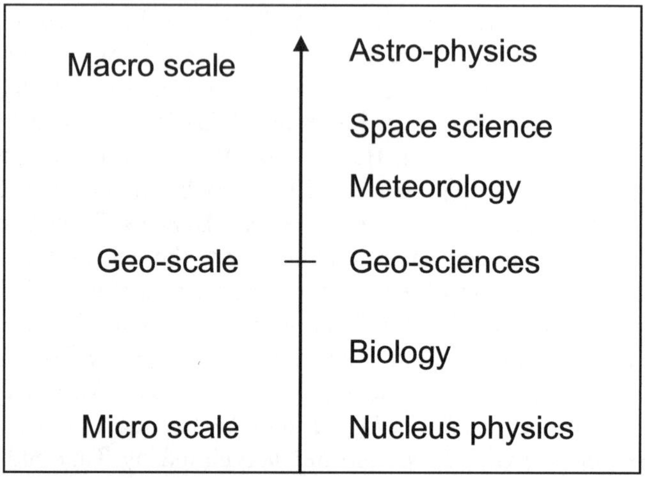

Figure 2. Spectrum of spatial scale, from micro to macro.

In different contexts, the term scale may mean different things. From a dictionary, one could find nearly 20 different meanings. In digital terrain analysis, scale may mean:

- degree of abstraction,
- degree of detail,
- ratio of distance (cartographic ratio), and
- magnitude of the study area.

Many researchers insist that "digital data sets do not have any scale but have only resolution". This is true in a sense that digital data sets can be plotted (and manipulated) at any scale one wishes. However, scale is also implied by the accuracy level of the digital data sets. For example, the height accuracy of the 1:50,000 DEM of China is expected to range from 4 m (flat areas), 7 m (hilly areas), 11 m (mountainous areas) to 19 m (highly mountainous areas). On the other hand, accuracy is affected by resolution. For example, the resolution of the 1:50,000 DEM of China is 25 m. If the resolution is too coarse, then the accuracy cannot be guaranteed. Therefore, the scale of a set of DEMs should be defined by a set of parameters, possibly:

- magnitude of study area (or cartographic ratio),
- accuracy, and
- resolution.

2.2 The concept of multi-scale scale

Multi-scale is also a term that is not well defined. By this term one may mean different things such as multi-resolution, multi-ratio (cartographic ratio), variable ratio and variable resolution. In addition, level of details (LoD) is another term used for multi-scale. It is therefore essential to clarify these terms.

For a single presentation, it is possible to use the following four combinations (Figure 3):

- Uniform cartographic ratio with uniform resolution,
- Uniform cartographic ratio but with multi-resolution,
- Variable cartographic ratio but with uniform resolution, and
- Variable cartographic ratio with variable resolution.

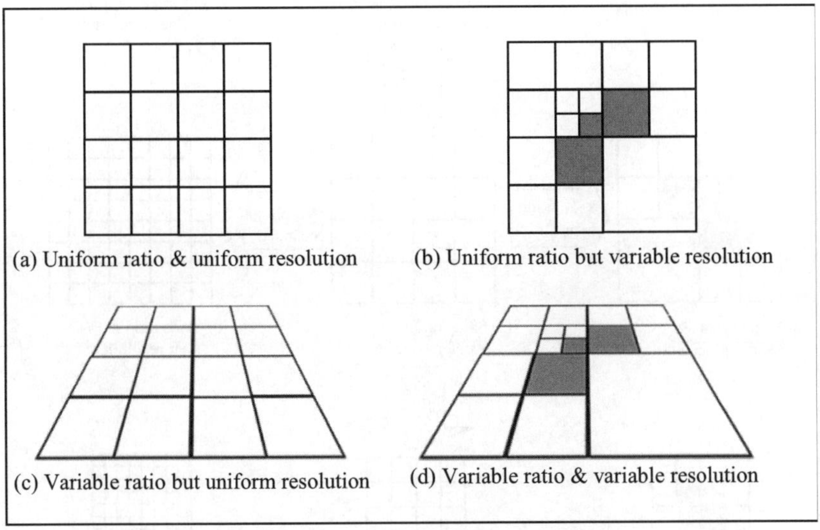

Figure 3. Combinations of cartographic ratio and resolution

For multiple presentations, cartographic ratio and resolution are still the two parameters to be considered. As in the case of single representation, resolution could be varied or uniformly changed (i.e. finer or coarser) and cartographic ratio could also be varied or uniformly changed (i.e. enlarged or reduced). As a result, there are nine possible types of changes for any given representation, as shown in Figure 4:

1. Same cartographic ratio and same resolution (i.e. as the original),
2. Same cartographic ratio but with different (but uniform) resolution,
3. Same cartographic ratio but with variable resolution,

4. Different (but uniform) cartographic ratio but with same resolution,
5. Different (but uniform) cartographic ratio with different (but uniform) resolution,
6. Different (but uniform) cartographic ratio with variable resolution,
7. Variable cartographic ratio with same resolution,
8. Variable cartographic ratio with differcnt (but uniform) resolution, and
9. Variable cartographic ratio with variable resolution.

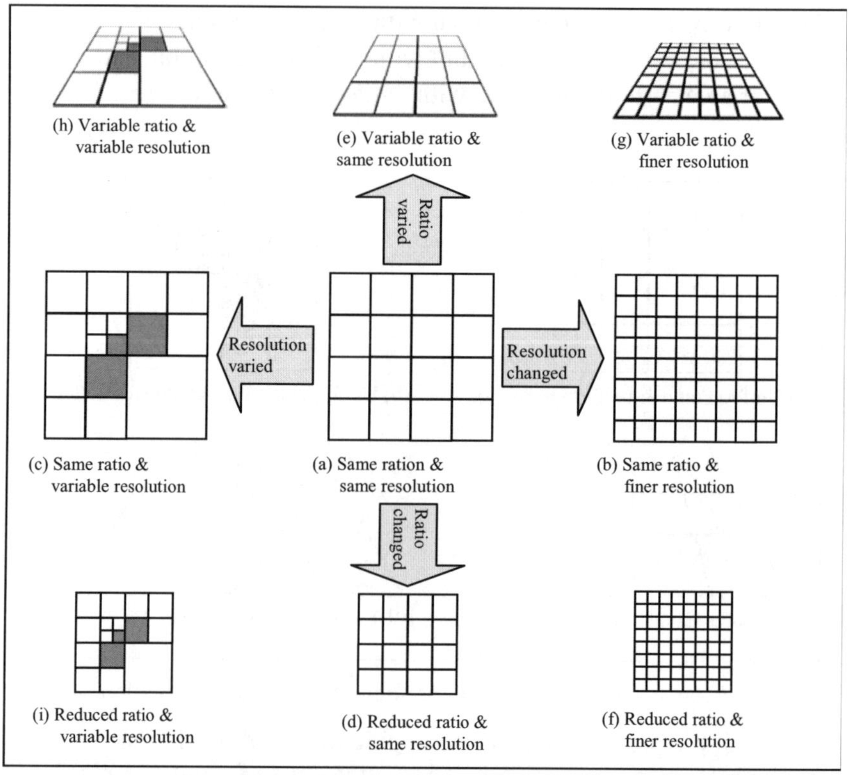

Figure 4. Nine types of possible scale changes for a representation.

A representation with variable cartographic ratio is normally generated for visualization with a perspective view and not for numerical analysis. Variable resolution is usually employed to represent different levels of detail, if there is a need. However, the most commonly used multi-scale representation for terrain analysis is either "different (but uniform) cartographic ratio with different (but uniform) resolution", or "different (but uniform) cartographic ratio with variable resolution".

3 Multi-Scale Terrain Representation (Modelling)

Based on different kinds of philosophy, three types of approach can be identified to the generation of multi-scale representations for a given DTM, i.e. critical-points-based, smoothing-based, and scale-driven.

3.1 Critical-points-based multi-scale representation

The critical-points-based approach is built upon the assumption that some points on a terrain surface are more important than others and the fidelity of the terrain surface is largely preserved even if only those important points are kept. The resultant data set may not be in a regular grid anymore. Another reason behind this approach is that some height values (such as the peaks of hills) are not allowed to change in some types of terrain analysis. A critical-points-based approach will produce a data set with variable resolution.

To detect critical points, some kind of significance measures should be used. Various parameters such as length between two points, perpendicular distance to a line, angle between two lines, local maxima and minima in height, curvature, and so on have been used.

Those algorithms developed for the detection of critical points along a line (e.g. Douglas and Peucker 1973, Li 1988, Visvalingham and Whyatt 1993) can be used to detect critical points along terrain profiles. In this case, the profiles in both X and Y directions (and along diagonals) should be examined.

The popular Douglas-Peucker algorithm makes use of perpendicular distance as a criterion. It means that if the perpendicular distance from a point under consideration to a line joining two already selected points is smaller than the criterion, then this point can be neglected. The working principle is illustrated in Figure 5. It starts from two end points and always compares the maximum perpendicular distance (among all possible perpendicular distances) with the given criterion. In the end, a set of points at different levels (e.g. four levels in Figure 5) of significance will be selected.

The Douglas-Peucker algorithm is computationally expensive. Li (1988) made efforts to improve the speed. He suggested that critical points be detected simply by selecting local maxima and minima in height. After the first round selection, those already selected points will be used to establish new local coordinate systems. Through a coordinate rotation, new local maxima and minima at the next level can then be selected. Figure 6 shows the principle. If the values of local maxima or minima are smaller than a criterion, then the point is not selected. Alternatively, one could specify the number of rounds for selection.

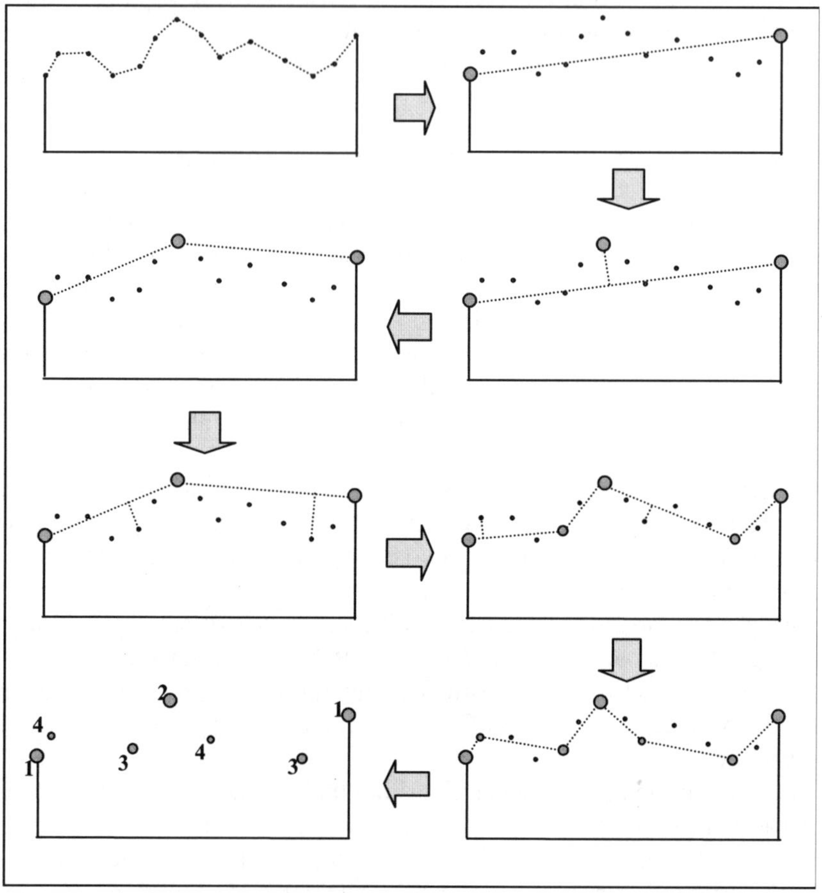

Figure 5. Douglas-Peucker algorithm based on perpendicular distance.

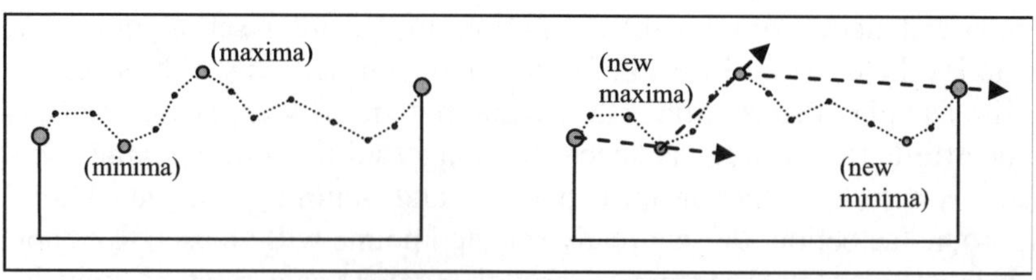

Figure 6. Li algorithm based on local minima and maxima.

It has also been found that large distortion may be caused by using the points selected by the Douglas-Peucker algorithm. Visvalingham and

Whyatt (1993) suggested making use of area as a criterion to minimize shape distortion. The principles of the Visvalingham-Whyatt algorithm and many other algorithms have been explained in a recent book by Li (2007).

Chen and Guevara (1987) designed a so-called VIP (very important points) procedure for the selection of critical points in a regular grid form. They made use of the sum of the second differential values at a point in all four directions to represent the degree of significance for the point. Suppose the height (H) of a point along a profile is a function of its position (x), as shown in Figure 7. The horizontal distance between X_{i-1}, X_i and X_{i+1} is equal because of the regular grid. Let the mathematical function of this profile be

$$H = f(x) \tag{1}$$

then its second differential value at point X_i is

$$\frac{d^2H}{dX^2} = f''(X_i) = 2\left(f(X_{i-1}) - \frac{f(X_{i-1}) + f(X_{i+1})}{2}\right) \tag{2}$$

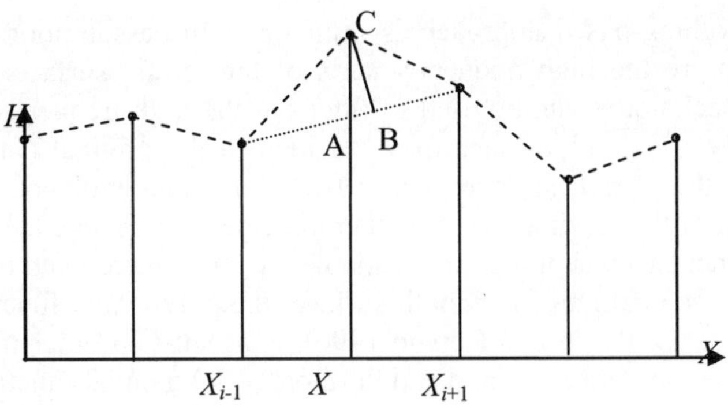

Figure 7. A terrain profile and its second differential value.

The distance AC in Figure 7 is the second differential value at point X_i, although they made use of CB. In their procedure, the number of points to be selected is specified first and then those points with greatest significance are selected. However, Li (1990) and Li *et al.* (1998) have argued that the selection of points should be related to the required accuracy of the resulting DTM, instead of a pre-defined number of points. They investigated the relationship between the degree of significance ($Sig_{Threshold}$)

used as threshold and the resultant DTM accuracy loss (σ_{loss}) after VIP selection and found:

$$\sigma_{loss} = \frac{Sig_{Threshold}}{3} \tag{3}$$

Suppose that the required accuracy of the final DTM (after the VIP selection) in terms of variance is σ^2_{after} and the accuracy of initial DTM (before the VIP selection) is σ^2_{before}, then

$$\sigma^2_{after} = \sigma^2_{before} + \sigma^2_{loss} \tag{4}$$

By combining Equations (3) and (4), the relationship between the threshold for VIP selection and the final DTM accuracy is as follows:

$$Sig_{Threshold} = 3\sigma_{loss} = 3\sqrt{\sigma^2_{after} - \sigma^2_{before}} \tag{5}$$

3.2 Smoothing-based multi-scale representation

The smoothing-based approach is built upon the assumption that small variations are the high-frequency parts of the terrain surface. Therefore, filtering techniques can be used to filter out the high-frequency parts. Alternatively, a smooth surface may be fitted to the original data points to represent the terrain surface. These two techniques will not change the resolution of the original DTM. If there is a need to change the resolution, then another two techniques are available, i.e. aggregation and resampling.

For the generation of a smooth surface, the polynomial function (Table 1) is the basis (Petrie and Kennie 1990). Normally, in order to avoid unpredictable oscillation, second and third order polynomial functions are the best choices. A least squares solution is obtained when the number of reference points used is larger than the number of coefficients in the polynomial function. Wavelets and other methods have also been in use.

A least squares solution defines the surface that makes the sum of all the square errors (e^2) (residuals) at the reference points the smallest among all possible surfaces. Mathematically,

$$\sum_{i=1}^{n} e_i^2 = \min \tag{6}$$

Table 1. Polynomial function used for surface reconstruction.

Individual Terms	*Order*	*Descriptive Terms*	*No. of Terms*
$Z = a_0$	Zero	Planar	1
$+ a_1X + a_2Y$	First	Linear	2
$+ a_3XY + a_4X^2 + a_5Y^2$	Second	Quadratic	3
$+ a_6X^3 + a_7Y^3 + a_8X^2Y + a_9XY^2$	Third	Cubic	4
$+ a_{10}X^4 + a_{11}Y^4 + a_{12}X^3Y$ $+ a_{13}X^2Y^2 + a_{14}XY^3$	Fourth	Quartic	5
$+ a_{15}X^5 + a_{16}Y^5 + \ldots\ldots$	Fifth	Quintic	6

Using the second-order polynomial for illustration:

2nd order polynomial:

$$z = f(x, y) = a_0 + a_1x + a_2y + a_3xy + a_4x^2 + a_5y^2 \tag{7}$$

where $a_0, a_1, a_2, \cdots, a_5$ are the six coefficients. They need to be determined by making use of reference points. If there are n (>6) reference points, then there are n equations as follows:

$$\begin{bmatrix} z_1 \\ z_2 \\ \vdots \\ z_n \end{bmatrix} = \begin{bmatrix} 1 & x_1 & y_1 & x_1y_1 & x_1^2 & y_1^2 \\ 1 & x_2 & y_2 & x_2y_2 & x_2^2 & y_2^2 \\ \vdots & \vdots & \vdots & \vdots & \vdots & \vdots \\ 1 & x_n & y_n & x_ny_n & x_n^2 & y_n^2 \end{bmatrix} \begin{bmatrix} a_0 \\ a_1 \\ \vdots \\ a_5 \end{bmatrix} \tag{8}$$

The error functions can be written as follows:

$$\begin{bmatrix} v_1 \\ v_2 \\ \vdots \\ v_n \end{bmatrix} = \begin{bmatrix} 1 & x_1 & y_1 & x_1y_1 & x_1^2 & y_1^2 \\ 1 & x_2 & y_2 & x_2y_2 & x_2^2 & y_2^2 \\ \vdots & \vdots & \vdots & \vdots & \vdots & \vdots \\ 1 & x_n & y_n & x_ny_n & x_n^2 & y_n^2 \end{bmatrix} \begin{bmatrix} a_0 \\ a_1 \\ \vdots \\ a_5 \end{bmatrix} - \begin{bmatrix} z_1 \\ z_2 \\ \vdots \\ z_n \end{bmatrix} \tag{9}$$

and simplified as

$$\underset{n\times 1}{V} = \underset{n\times 6}{X}\,\underset{6\times 1}{A} - \underset{n\times 1}{Z} \tag{10}$$

According to the least squares solution,

$$\underset{6\times1}{A} = \left(\underset{6\times n}{X^T}\ \underset{n\times6}{X}\right)^{-1}\left(\underset{6\times n}{X^T}\ \underset{n\times6}{X}\ \underset{n\times1}{Z}\right) \tag{11}$$

After the coefficients are computed, the height z_p of any point P at location (x, y) can be obtained by substituting x, y into Equation (7).

The commonly used filtering techniques are low-pass filters. Simple averaging is the simplest and perhaps most efficient filter, and is thus most commonly used. Figure 7 illustrates the principle of a moving-averaging low-pass filter. In the filtering processing, a template (e.g. a 3 x 3 window) is moved over the DTM cells one by one in the row and column directions, the average of the heights within the template is used to represent the height value at this position in the output DTM. For example, if the points in a 3 x 3 window centred at P are selected for averaging, their heights are 9, 9, 9, 8, 7, 9, 4, 3 and 6, then the average is 7.

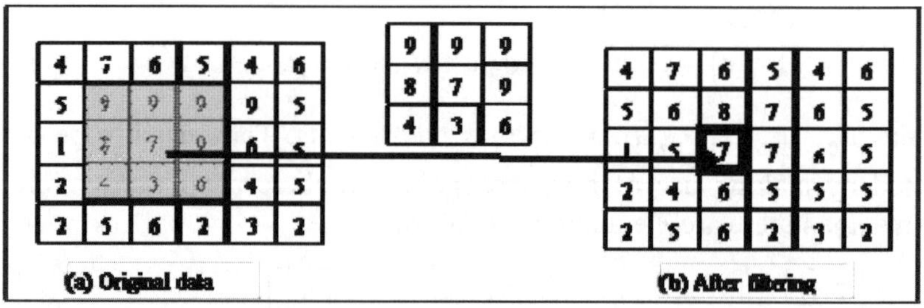

Figure 7. Moving averaging as low-pass filter for smoothing a DTM surface.

A median filter is also effective, but is much less efficient. The working principle of a median filter is similar to that of a low-pass filter except that the median filter takes the median of the values in the window (instead of the average). Other more complicated filters, such as the Kalman filter, have also been used.

To make a change in resolution, aggregation and resampling techniques can be employed. Aggregation makes multiple cells (e.g. a 2 x 2 window) into a new cell. Figure 8 shows the working principle. In this case, a 3 x 3 window is aggregated into a new cell. Simple averaging is most widely used to estimate the value for the new cell, which in fact may take the mode or median within the window. It is also possible to select only the pixels at the nth row and column from the original DTM grid. Polynomial functions may even be used for estimation if the window is very large.

Sometimes, the required new resolution is not an integral times the original resolution (e.g. 2 m to 3 m); therefore, one cannot use aggregation. In that case, resampling is the choice. Figure 9 illustrates the sampling of a 3 x 3 DTM into a 2 x 2 DTM. In this case, both nearest neighbour and bilinear interpolations have been employed. Indeed, aggregation might be regarded as a special case of resampling.

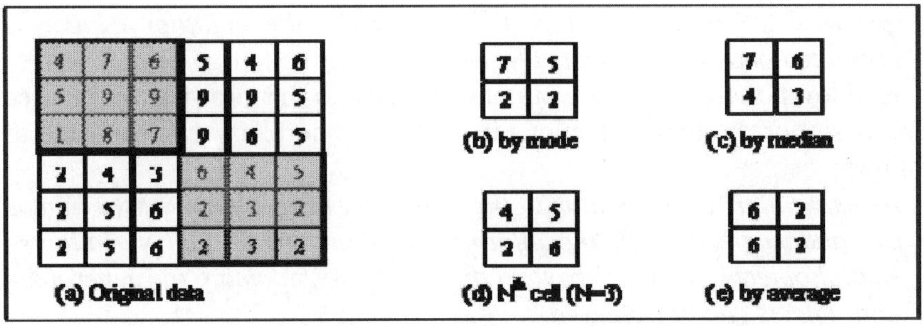

Figure 8. Aggregation for smoothing a DTM surface.

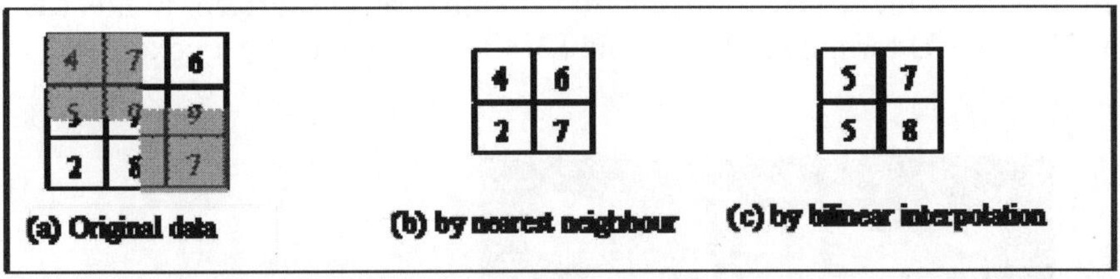

Figure 9. Resampling for smoothing a DTM surface.

3.3 Scale-driven multi-scale representation

In the case of smoothing, there is a need to determine the window size. In other words, the window size is not directly related to the scales. To solve this problem, Li and Openshaw (1993) formulated a *Natural Principle* for objective generalization, which objectively links the window size to the source (input) data scale and result (output) data scale.

Li and Openshaw (1993) made use of the terrain surface viewed from different height levels as an example to illustrate the *Natural Principle*. If one views the terrain surface from the moon, all terrain variations disappear. If one views the terrain surface from a satellite, then the terrain surface becomes very smooth. When one views the terrain surface from an aircraft, small details do not appear but the main characteristics of the terrain variations are very clear. This is also due to the limitation in the resolution of the human eye. When the viewpoint is higher, the ground area corresponding to the human eyes' resolution becomes larger, thus the ground surface appears to be more abstract. These examples underline a universal principle, a *natural principle,* according to Li and Openshaw (1993). It can be stated as follows:

For a given scale of interest, all details about the spatial variations of geographical objects (features) beyond a certain limitation cannot be presented and can thus be neglected.

It follows, therefore, that a simple corollary to this process can be used as a basis for the transformations in scale. The corollary can be stated as follows:

By using a criterion similar to the limitation in resolution of the human eye and neglecting all the information about the spatial variation of spatial objects (features) beyond this limitation, zooming (or generalization) effects can be achieved.

Li and Openshaw (1992) also term such a limitation as the smallest visible object (SVO), or smallest visible size (SVS) in other literature. Figure 10 illustrates the natural principle to generalize a 3-D surface. In this case, a piece of 3-D terrain surface within SVS is represented by a voxel.

Figure 10. The natural principle: Spatial variations within a smallest visible size (SVS) to be neglected.

Figure 11 illustrates the working example of applying this natural principle to a terrain surface. Figure 11a shows the view of a terrain surface at two different heights, resulting in representations at two different scales. Figure 11b shows the result viewed at level L_A and Figure 11c shows the result viewed at level L_B. In these two figures, the generalization effects are very clear.

To apply this natural principle, the critical element to be considered is the value of this "certain limitation" or the value of SVS, beyond which all spatial variations (no matter how complicated) can be neglected. Li and Openshaw (1992, 1993) suggested the following formula:

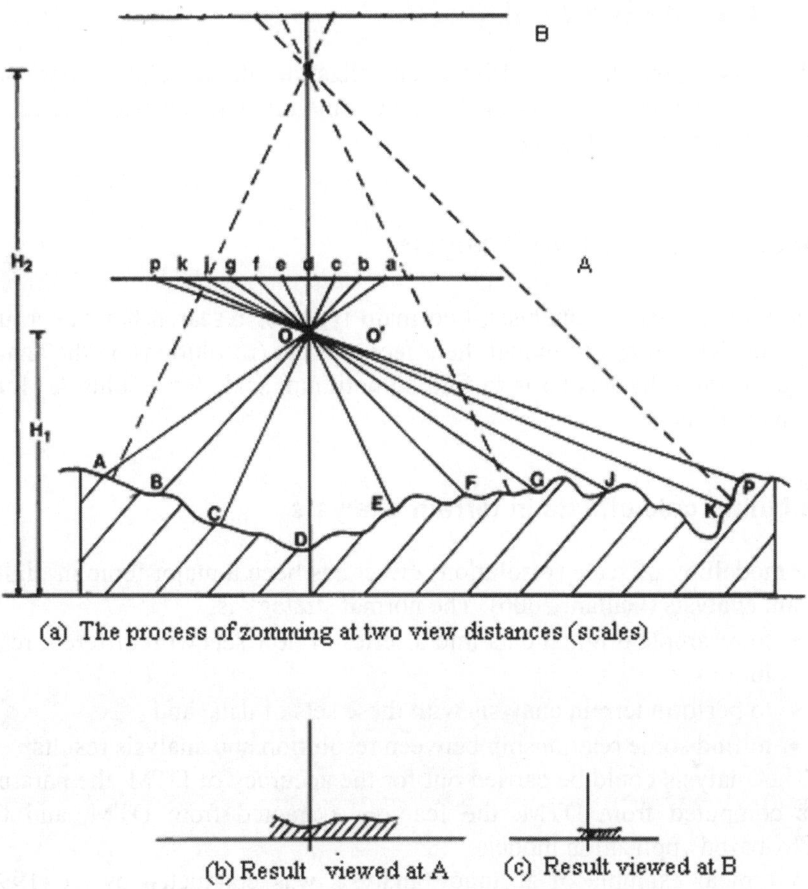

Figure 11. The *natural principle* applied to a terrain surface (Li and Openshaw 1993).

$$K = k \times S_T \times \left(1 - \frac{S_S}{S_T}\right) \tag{12}$$

where S_T and S_S are the scale factors of the target and source data, respectively; k is the SVS value in terms of map distance on target scale. Through intensive experimental testing, Li and Openshaw (1992) found that a value between 0.5 mm and 0.7 mm will enable them to produce generalization results similar to that by manual generalization. Therefore, it is recommend that

$$k = \{0.5mm, 0.7mm\} \tag{13}$$

In carrying out the scale-driven generalization, the K value is first computed and then this value is used to determine the window size. The rest is similar to low-pass filtering.

4 Multi-Scale Terrain Analysis

In multi-scale terrain analysis, two main types of research have been undertaken. The first is to model the effect of scale (resolution) on the analysis results and the second is to find an optimum scale (or resolution) for a given analysis.

4.1 Multi-scale effects in terrain analysis

The modelling of scale (resolution) effect has been a major topic in digital terrain analysis (Gallant 2006). The normal strategy is

- to resample original data into a series of new sets with different resolutions,
- to perform terrain analysis with these sets of data, and
- to find some relationship between resolution and analysis results.

The analysis could be carried out for the accuracy of DTM, the parameters computed from DTM, the features extracted from DTM, and the DTM-based application models.

A typical example of accuracy analysis was conducted by Li (1990, 1992). Figure 12 shows his experimental results, which illustrate the relationship between the grid interval *(d)* (resulting from regular-grid sampling) and the accuracy of the final DTM (σ) in terms of standard deviation, for three different test areas. It is clear that the relationship is linear for composite data (i.e. grid plus feature data, including peaks, pits, break lines, ridge lines, and ravines) as shown in the right column but is second order for regular grids only as shown in the left column. Mathematically, the relations could be written as follows (Li *et al.* 2005):

For composite data, $$\sigma_{DTM-c} = k_1 \times \sigma_{Data} + k_2 \times d \tag{14}$$

For grid data only, $$\sigma_{DTM-g} = k_1 \times \sigma_{Data} + k_2 \times d + k_3 \times d^2 \tag{15}$$

where σ_{Data} is the accuracy of measured data points and k_i is a coefficient.

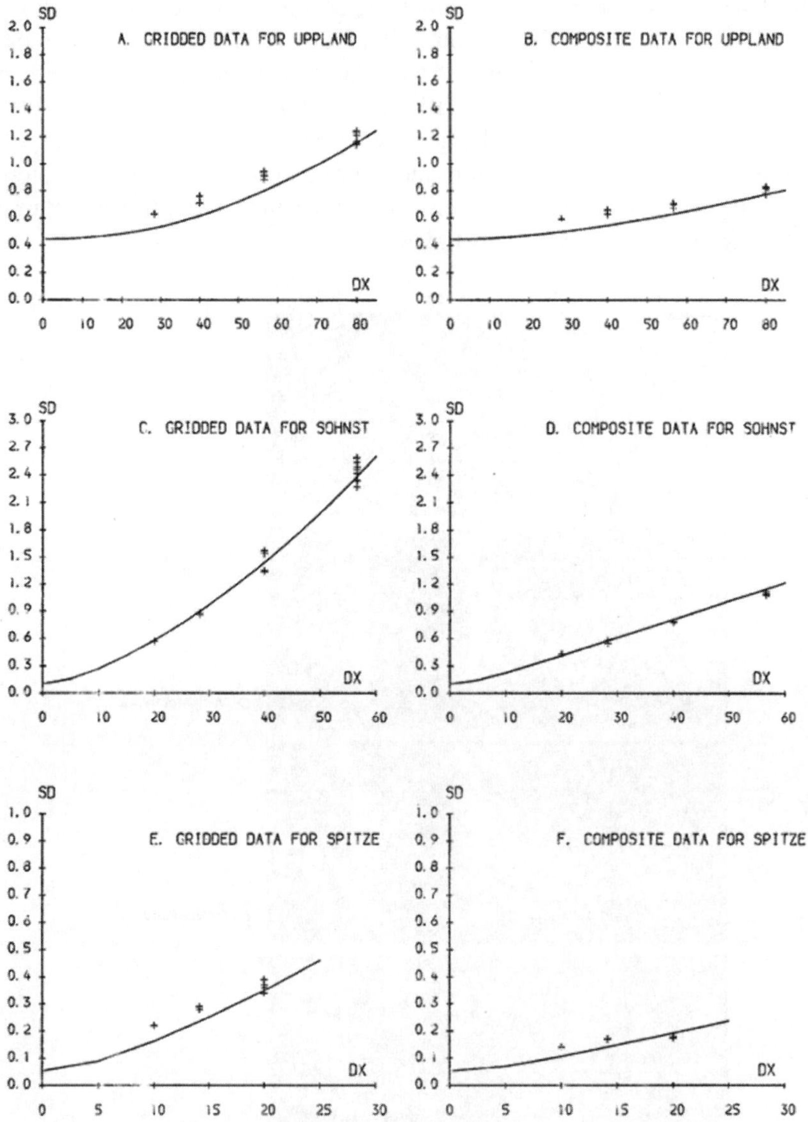

Figure 12. Variation of DTM accuracy with sampling interval (Li 1990, 1992).

k_2 is related to the slope angle of the terrain. The estimation of these parameters was addressed in detail by Li (1990, 1992). Similar mathematical models have also been derived through theoretical analysis by Li (1990, 1993) (also see Li et al. 2005).

In a similar way, the effect of DEM resolution on the various terrain parameters and DTM-based applications have been investigated, e.g. on slope (Chen *et al.* 2006), on a number of terrain parameters and attributes (Gao 1998, Albani *et al.* 2004, Zhu 2006, Deng *et al.* 2007), on watershed model (Wolock and Price 1994), on digital soil mapping (Zhu 2006), on the accuracy of a distributed rockfall model (Dorren and Heuvelink 2004), and on terrain classification (Cheng *et al.* 2004). Figure 13 shows the relationship between DTM resolution and four terrain parameters (i.e. slope, wetness index, profile curvature, and plan curvature), which was obtained by Deng *et al* (2007). It was concluded that:

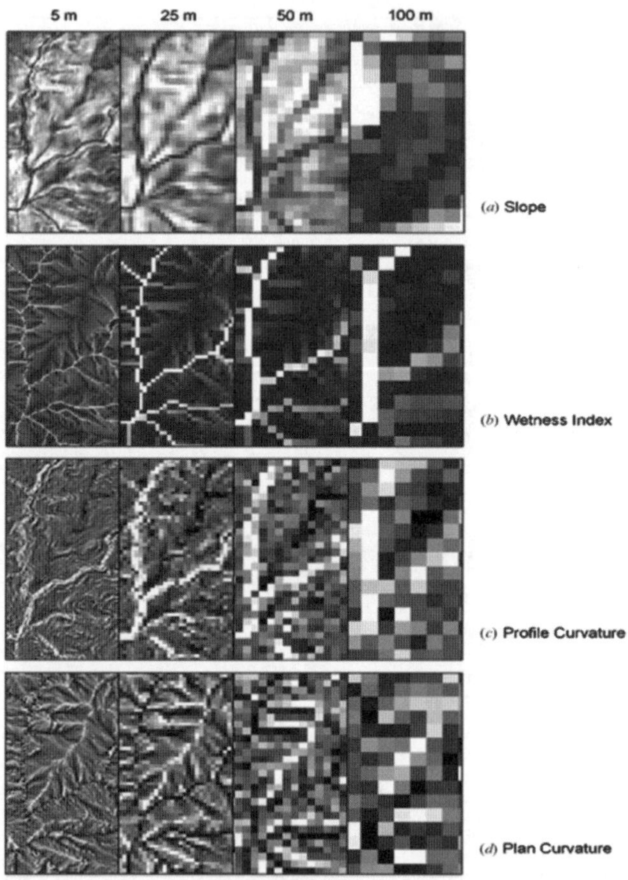

Figure 13. Relationship between DTM resolution and four terrain parameters (Deng *et al.* 2007).

- Terrain attributes respond to resolution change in characteristically different ways, especially when the resolution is coarsened in the range 5–50 m. Plan and profile curvatures are the most sensitive among these parameters, whereas slope is the least sensitive; and

- Consistently smaller correlations were observed across all terrain attributes when the pairs of DEM resolution compared were increasingly distant from each other.

These conclusions were obtained from a study in the eastern Santa Monica Mountains, California, in an area of 962 km^2. The original grid interval (resolution) was 5 m, thus resulting in 38,484,560 cells.

Figure 14 shows the relationship between DEM resolution and terrain classification results, which were reported by Cheng *et al.* (2004). They tested the fuzzy classification of morphological objects, i.e. foreshore (lower than -6.0 m), beach (-6.0 ~ -1.1 m) and fore-dune (-1.1~ +2 m), using Ameland Island of the Netherlands as a test area. The original DEM was in a 60 m x 60 m grid. They aggregated data using a series of window sizes (i.e. 3 x 3, 5 x 5, …, 25 x 25). They found that it was difficult to accurately predict the effect of scale on fuzzy spatial objects, although the change with the scale exhibits a linear relationship in some statistical indicators.

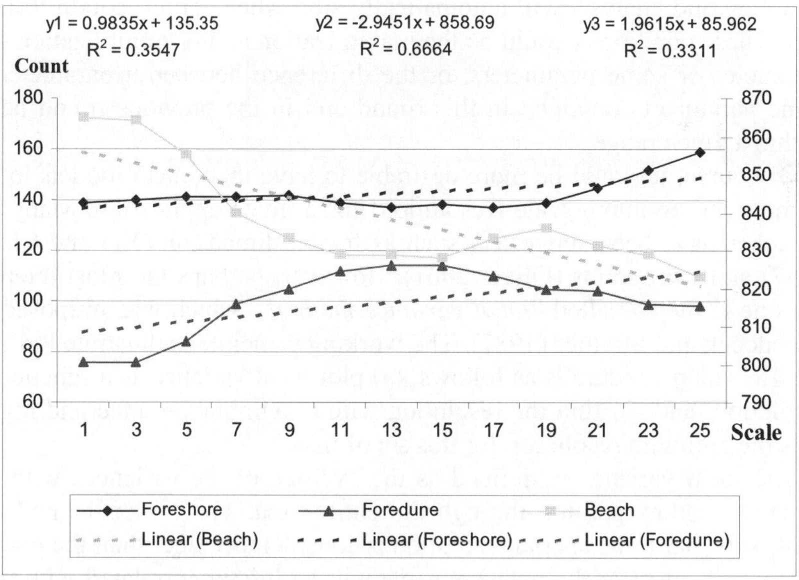

Figure 14. Effect of resolution on the classification of fuzzy objects (in terms of area) (Cheng *et al.* 2004).

4.2 Optimization of scale in terrain analysis

As one can understand, it is not enough to model the scale effects. Indeed, it would be more desirable to find an optimum scale (resolution) for a particular analysis or application, if it exists. This could be carried out through experimental testing or through theoretical analysis.

In experimental analysis, one could first make a systematic classification of terrain types, then conduct terrain analysis with DTM data resampled into various resolutions, and finally compare all these results with a benchmark. The best result will correspond to the optimum resolution. For example, Florinsky and Kuryakova (2000) produced the variation of correlation coefficients between terrain attributes (slope, aspect, curvature, etc.) and targeted biophysical properties (e.g. soil moisture) and considered the resolution range where such a variation was relatively stable as the optimum resolution. In this way, some empirical models may be obtained. Currently, however, no such empirical models have been established, although a number of empirical studies have been carried out (see previous section).

This process could also be made self-adaptive. This means that the resampling and analysis will automatically stop when certain conditions are met. These conditions could be the maximization and/or minimization of a parameter or some parameters, or the difference between a parameter or some parameters obtained in this round and in the previous round being within a given range.

Of course, it would be more desirable to have theoretical models to determine the optimum scale (resolution) for a given application. Many approaches have been suggested, such as fractal dimension (Xia and Clarke 1997) and variograms (Oliver 2001). However, perhaps the most interesting one is the so-called "*local variance method*", which was proposed by Woodcock and Strahler (1987). The working principle is illustrated in Figure 15. The procedure is as follows: (a) plot local variance as a function of resolution; and (2) find the resolution with maximum variance and regard it as the optimum resolution for this set of data.

The local variance is defined as the average of the variances within a moving window passing through the entire area. The reason behind this method is that if the spatial resolution is considerably finer than the objects in the area, most of the measurements will be highly correlated with their neighbours and thus the local variance will be low. On the other hand, if the objects studied approximate the size of the resolution, the values tend to be different from each other and therefore the local variance increases. (Cao and Lam 1997).

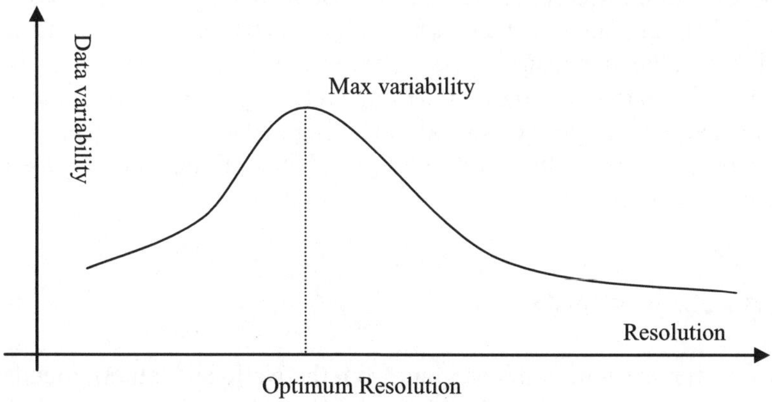

Figure 15. Data variability for the determination of optimum resolution.

This method was originally used for remote sensing image data and it should be equally applicable to DTM data. However, this method can only indicate the scale range of the terrain properties but not the optimum resolution for the analysis of a given terrain parameter. In this respect, the "*Natural Principle*" for objective generalization by Li and Openshaw (1993) might be customized for this purpose in different applications.

5 Concluding remarks

"Scale is a confusing concept, often misunderstood, and meaning different things depending on the context and disciplinary perspective" (Quattrochi and Goodchild 1997, p395). In this chapter, a number of issues on multi-scale modelling and analysis have been tackled. First, a discussion on the concepts of scale and multi-scale was conducted. Second, three approaches for the multi-scale representation (modelling) of terrain surfaces are presented, i.e. critical-points-based, smoothing-based and scale-driven. Then, two issues on multi-scale analysis are examined. The first issue is the modelling of scale effect and the other is the optimization of scale. Hopefully, the confusing concepts *scale* and *multi-scale* have been clarified somehow.

"Scale is undoubtedly one of the most fundamental aspects of any research" (Quattrochi and Goodchild 1997). Cola (1997) argues that scale is such a pervasive issue in the analysis of spatial data as to require it to be at the level of a basic dimension, along with space, time and theme. This view is similar to the view previously expressed by this author (Li 1994, Li

1996), but from different angles. Indeed, Quattrochi and Goodchild (1997) in their Epilogue also re-iterated a similar view: "Scale is a fundamental and inescapable dimension of geographic data". Therefore, scale is not only an issue in digital terrain modelling and analysis but also a fundamental dimension in all types of spatial modelling and analysis. Indeed, scale is everywhere in science like God in religion. The optimization of scale is an issue of some urgency.

Acknowledgements

This research is supported by a grant from the National Natural Foundation of China (40329001) and a grant from the Hong Kong Polytechnic University (G-U045).

References

Albani, M., Klinkenberg, B., Ansison, D.W. and Kimmins, J.P.K., (2004), The choice of window size in approximating topographic surface from digital elevation models, *International Journal of Geographical Information Science*, 18(6): 577–593.

Bian, L., (1997), Multi-scale nature of spatial data in scaling up environmental models, In Quattrochi, D. and Goodchild, M. (eds.): *Scale in Remote Sensing and GIS*, CRC Press: 13–26.

Blöschl, G. and Sivapalan, M., (1995), Scale issues in hydrological modelling: A review. *Hydrological Processes*, 9: 251–290.

Cao, C. and Lam, N.S.-N., (1997), Understanding the scale and resolution effects in Remote Sensing and GIS. In Quattrochi, D. and Goodchild, M. (eds.): *Scale in Remote Sensing and GIS*, CRC Press: 57–72.

Chen, N., Tang, G.A. and Wang, Y.M., (2006), The influence of resolutions of DEM on the accuracy of the slope information, *International Symposium on Terrain Analysis and Digital Terrain Modelling*, 23–25 November 2006, Nanjing, China, (CD-ROM).

Chen, Z. and Guevara, J., (1987), Systematic selection of very important points (VIP) from digital terrain model for constructing triangular irregular networks, *Auto Carto 8*: 50–56.

Cheng, T., Fisher, P. and Li, Z.L., (2004), Double vagueness: Uncertainty in multi-scale fuzzy assignment of duneness, *Geo-spatial Information Science*, 7(1): 58–66.

Cola, L., (1997), Multiresolution covariation among Landsat and AVHRR vegetation indices. In Quattrochi, D. and Goodchild, M. (eds.): *Scale in Remote Sensing and GIS*, CRC Press: 73–91.

de Boer, D.H., (1992), Hierarchies and spatial scale in process geomorphology: a review, *Geomorphology*, **4**: 303–318.

Deng, Y., Wilson, J.P. and Bauer, B.O., (2007), DEM resolution dependencies of terrain attributes across a landscape, *International Journal of Geographical Information Science*, **21**(1–2): 187–213.

Dorren, L.K. and Heuvelink, G.B.M., (2004), Effect of support size on the accuracy of distributed rockfall model, *International Journal of Geographical Information Science*, **18**(6): 595–609.

Douglas, D.H. and Peucker, T.K., (1973), Algorithms for the reduction of the number of points required to represent a digitized line or its caricature *The Canadian Cartographer*, **10**(2): 112–122.

Dovers, S.R., (1994), A framework for scaling and framing policy issue in sustainability. *Ecological Economy*, **12**: 93–106.

Florinsky, I.V. and Kuryakova, G.A., (2000), Determination of grid size for digital terrain modeling in landscape investigations - exemplified by soil moisture distribution at a micro-scale, *International Journal of Geographical Information Science*, **14**: 795–814.

Friedl, M.A., (1997), Examining the effects of sensor resolution and sub-pixel heterogeneity of spectral vegetation indices: Implications for biophysical modelling, In Quattrochi, D. and Goodchild, M. (eds.): *Scale in Remote Sensing and GIS*, CRC Press: 113–140.

Gallant, J., (2006), Multiscale methods in terrain analysis. Keynotes Speech, *International Symposium on Terrain Analysis and Digital Terrain Modelling*, 23–25 November 2006, Nanjing, China. (CD–ROM).

Gao, J., (1998), Impact of sampling intervals on the reliability of topographic variables mapped from grid DEMs at a micro-scale, *International Journal of Geographical Information Science*, **12**: 875–890.

Goodchild, M. and Quattrochi, D., (1997), Introduction: scale, multiscaling, remote sensing and GIS, In Quattrochi, D. and Goodchild, M. (eds.): *Scale in Remote Sensing and GIS*, CRC Press: 1–12.

Haury, L., McGowan, J. and Wiebe, O., (1977), Patterns and processes in time-space scales of plankton distribution, In Steele, J. (ed.): *Spatial Pattern in Planton Communities*, NY and London: Plenum Press: 277–327.

Hillel, D. and Elrick, D., (1990), *Scaling in Soil Physics: Principles and Applications*, Vol.25, Soil Science Society of America.

Institute of Hydrology and Stewart, J. (ed.), (1996), *Scaling up in hydrology using remote sensing*, John Wiley & Sons.

Li, Z.L., (1988), An algorithm for compressing digital contour data, *The Cartographic Journal*, **25**(2): 143–146.

Li, Z.L., (1990), *Sampling Strategy and Accuracy Assessment for Digital Terrain Modelling*, Ph.D. Thesis, The University of Glasgow.

Li, Z.L., (1992), Variation of the the accuracy of digital terrain models with sampling interval, *Photogrammetric Record*, **14**(79): 113–128.

Li, Zhilin, (1993), Mathematical models of the accuracy of digital terrain model surfaces linearly constructed from gridded data, *Photogrammetric Record*, **14**(82): 661-674.

Li, Z.L., (1994), Reality in Time–Scale System and Cartographic Representation, *The Cartographic Journal*, **31**(1): 50 51.

Li, Z.L., (1996), Transformation of spatial representation in scale dimension: A new paradigm for digital generalisation of spatial data, *International Archives for Photogrammetry and Remote Sensing,* **XXXI**(B3): 453–458.

Li, Z.L., (2007), *Algorithmic Foundation of Multi–scale Spatial Representation,* CRC Press (Taylor & Francis Group).

Li, Z.L. and Openshaw, S., (1992), Algorithms for automated line generalisation based on a natural principle of objective generalisation, *International Journal of Geographic Information Systems,* **6**(5): 373–389.

Li, Z.L. and Openshaw, S., (1993), A Natural Principle for Objective Generalisation of Digital Map Data, *Cartography and Geographic Information System,* **20**(1): 19–29.

Li, Z.L., Lam, K. and Li, C.M., (1998), Effect of compression on the accuracy of DTM, *Geographic Information Science,* **4**(1–2): 37–43.

Li, Z.L., Zhu, Q. and Gold, C., (2005), *Digital Terrain Modelling: Principles and Methodology*, CRC Press (Taylor & Francis Group).

Miller, C. and Laflamme, R., (1958), The digital terrain model - theory and applications, *Photogrammetric Engineering*, **24**: 433–442.

Oliver, M.A., (2001), Determining the spatial scale of variation in environmental properties using the variogram, In Tate, N.J. and Atkinson, P.M. (eds.): *Modelling Scale in Geographical Information Science*, Chichester: Wiley: 193–220.

Openshaw, S., (1994), *The Modifiable Areal Unit Issue CATMOG # 38,* Norwick, England: Geo Books.

Petrie, G. and Kennie, T. (eds.), (1990), *Terrain Modelling in Surveying and Civil Engineering*, Caithness, UK: Whittles Publishing.

Quattrochi, D.A. and Goodchild, M.F. (eds.), (1997), *Scale in Remote Sensing and GIS*, CRC Press.

Sposito, G. (ed.), (1998), Scale Dependence and Scale Invariance in Hydrology, Cambridge University Press.

Stommel, H., (1963), Varieties of oceanographic experience, *Science*, **139**: 572–576.

Tate, N. and Atkinson, P. (eds.), (2001), Modelling Scale in Geographical Information Science, Chichester: Wiley.

Tobler, W., (1989), Frame independent spatial analysis, In Goodchild, M. and Gopal, S. (eds.): *Accuracy of Spatial Databases,* London: Taylor and Francis: 115–122.

Visvalingham, M. and Whyatt, J., (1993), Line generalization by repeated elimination of points, *The Cartographic Journal*, **30**(1): 46–51.

Wilson, J.P. and Gallant, J. (eds.), 2000. *Terrain Analysis: Principle and applications,* Singapore: John Wiley & Sons, Inc.

Wolock, D.M. and Price, C.V., (1994), Effects of digital elevation model and map scale and data resolution on a topography-based watershed model, *Water Resources Research*, **30**: 3041–3052.

Woodcock, C.E. and Strahler, A.H., (1987), The factor of scale in remote sensing, *Remote Sensing of Environment*, 21: 311–332.

Xia, Z.G. and Clarke, K., (1997), Approaches to scaling of geo-spatial data, In Quattrochi, D.A. and Goodchild, M.F. (eds.), *Scale in Remote Sensing and GIS* CRC Press: 309–360.

Zhu, A.X., (2006), Neighborhood size of terrain derivatives and its impact on digital soil mapping, *International Symposium on Terrain Analysis and Digital Terrain Modelling*, 23–25 November 2006), Nanjing, China. (CD-ROM).

A Seamless and Adaptive LOD Model of the Global Terrain Based on the QTM

ZHAO Xuesheng, BAI Jianjun and CHEN Jun

Abstract

In this paper, a new seamless and adaptive LOD (Level of Details) modelling method of global terrain based on the QTM (Quaternary Triangular Mesh) is described. Our approach starts with QTM tessellation based on the WGS84 ellipsoidal surface. Thus, the global terrain can be represented by an array of elevation values of QTM vertices. Next, an adaptive idea of a *Binary Triangle Tree* is inserted in this QTM model to form an adaptive, continuous, and uniform LOD triangular mesh in the DEM grids and their vertices completely coincide with the vertices of QTM. The experiment and analysis are carried out with the global terrain data, GTOPO30. The results illustrate that: (1) the global DEM model based on QTM is seamless, hierarchical, and regular over the whole Earth, and the dataset occupies only about half of its initial size; (2) an adaptive LOD (Level of Details) model of the DEM data blocks is constructed and the problem of discontinuity in different subdivision levels is overcome; and (3) the number of the triangles in this adaptive LOD model decreases greatly as their error tolerances increase, but there is no obvious change in the results.

Keywords: global terrain, QTM, adaptive partition, LOD.

1 Introduction

The representation of the physical surface of the Earth in digital systems is a subject of considerable current attention. As the area of coverage of such systems increases, it becomes necessary to provide methods to model very large, continuous surface conglomerates in a manner that does not violate surface integrity. The regular or irregular Digital Elevation Model (DEM) grids based on the idea of map projections are effective traditional methods for modelling the terrain on the local or small-scale spherical surfaces as a flat surface, and many corresponding algorithms have been presented over the last decade (Lindstrom *et al.* 1996, Lindstrom and Pascucci 2002,

Duchaineau *et al.* 1997, Kolar 2004). However, a spherical surface is not topologically equivalent to planar Euclidean space in geometry. Although the traditional DEM grid may support an individual project or small area terrain visualization modelling, they do have some significant drawbacks for modelling large areas or the whole Earth surface, such as geometric distortions, data discontinuity, space inefficiency, and difficulties with data sharing between projects caused by a lack of good global representation schemes (Lukatela 2000).

The Quaternary Triangular Mesh (QTM), first proposed by Dutton (1989), is a tessellation of the Earth's surface with non-overlapping (or broken) triangular cells. In QTM partition, an octahedron is used as a basis. It can be readily aligned with the conventional geographic grid consisting of longitude and latitude. When this is done, QTM vertices occupy cardinal points and their edges assume cardinal directions, following the equator, the prime meridian, and the 90th, 180th and 270th meridians, making it simple to determine in which facet a point on the planet is located. It can simulate the surface of the Earth when it is recursively partitioned to a certain degree. In addition, the QTM grid structure is seamless, hierarchical, and numerically stable everywhere on the spherical surface. Its hierarchical grid structure can be used to efficiently manage multi-resolution global data, and it allows spatial phenomena to be studied at different levels of detail in a consistent fashion across extensive regions of the sphere (Lee and Samet 2000, Chen *et al.* 2003). It is possible to resolve problems, such as geometric distortions, spatial data overlapping (or breaking), and space inefficiency in managing the global multi-scale spatial data by using QTM instead of the raster mode on a planar surface (Dutton 1999). However, it cannot preserve continuity between different resolutions if the elevations of these vertices are taken into account and, as a result, a uniform adaptive LOD (Level of Details) model of the global terrain, which is very important in multi-resolution operations and realistic visualization, cannot be constructed using a QTM directly.

To overcome these deficiencies, a new seamless and adaptive LOD modelling method of the global terrain based on the idea of the *Binary Triangle Tree* will be introduced in this chapter. Following this introduction is a section critically reviewing current methods of modelling and visualization of the global DEM. Then, a hierarchical and seamless modelling method of the global terrain, based on QTM, is demonstrated. This is followed by the presentation of an adaptive simplification rule of hierarchical triangles in *Diamond* data blocks, after which the GTOPO30 data of the whole Earth and STRM data in some regions are used to test the methods and algorithms presented in this chapter. Finally, conclusions and recommendations for future work will be presented.

2 A Critical Examination of Related Works

A number of approaches to subdividing the surface of the Earth in a continuous and hierarchical way for global terrain visualization have been suggested. The approaches can be classified as regular latitude and longitude grids, adjusted latitude and longitude grids, and adaptive subdivision grids.

Regular latitude & longitude grids. The familiar latitude/longitude grid is the most common basis for global DEMs in use today, such as GTOPO30 and ETOPO5 data (supplied by the US Geological Survey), and the JGP95E5' data (compiled by The Defence Mapping Agency and NASA/Goddard Space Flight Centre). Elevation values may be associated with points spaced at equal-angle intervals of latitude and longitude (for example, the 30" × 30" spacing of the GTOPO30 global elevation data sets and the 5' × 5' in the ETOPO5). In February 2000, the Shuttle Radar Topography Mission (SRTM) successfully collected Interferometric Synthetic Aperture Radar (IFSAR) data over 80% of the landmass of the Earth between 60ºN and 56ºS latitudes and SRTM-1 values are spaced at 3". The 2.5-dimensional Earth terrain is expressed using JGP95E 5' data (Sun *et al.* 2000). In Gerstner (2003), the regular grid DEM data, subdivided by latitude and longitude, can be compressed and used for interactive visualization of the whole globe. The global DEMs based on the latitude/longitude graticule have numerous practical advantages and have been used to develop sound survey sampling designs on the Earth's surface. These are based on a 2-dimensional Cartesian coordinate system and such systems have long been a foundation of scientific inquiry on spatial domains. Such square-based grids can also be easily mapped into common display devices. The latitude/longitude system itself has been used extensively since well before the computer era and is therefore the basis for a wide array of existing data sets, processing algorithms, and software.

But the latitude/longitude grids become increasingly distorted in area and/or shape as one moves north or south from the equator (Sahr *et al.* 2003). The grid cells of the north and south poles are in fact triangles, not "squares" as appear elsewhere. These polar singularities have forced global visualization modelling to make use of special grids for the polar region. Another defect is that DEMs based on the latitude/longitude grid do not have equal area cells, which is important for many applications and much redundancy of data occurs near the poles. For example, the elevation value of one pole point is repeated 43,200 times in the GTOPO30 data file.

Adjusted latitude and longitude grids. In order to tessellate the spherical surface into approximately equal area grids with latitude and longitude, another set of attempts based on adjusted latitude and longitude grids have been made in recent years (NIMA 2003, Bjørke *et al.* 2004). For example, DTED data (supplied by US National Imagery and Mapping Agency) keeps the latitude space constant (3"), and the longitude space is changed from equator to pole, such that the longitude space is 3" in latitudes 0°–50°, 6" in latitudes 50°–70°, 9" in latitudes 70°–75°, 12" in latitudes 75°–80°, and 18" in latitudes 80°–90° (NIMA 2003). Bjørkc *et al.* (2004) presented a similar grid tessellation method in which the grid areas are almost equal. Compared to regular latitude and longitude grids, the adjusted latitude and longitude grids are almost equal in area and the amount of redundant data occurring near the poles is decreased greatly. However, these schemes achieved more regular grid areas at the cost of more irregular grid shapes and more complex cell adjacencies (Sahr *et al.* 2003).

Adaptive subdivision grids. Lukatela (1987, 2000) constructed seamless global triangular networks in the Hipparchus System similar to the planar TIN commonly used to facilitate terrain modelling and volumetric analysis. This is based on global coordinates and a planetary surface tessellation using spherical Voronoi polygons. The Hipparchus System makes no attempt to produce a "regular" grid. Instead, the grid is designed so that any particular implementation of the grid can match the density of the data that inhabits it, and accepts this "irregular" nature while attempting to make computations based on it as fast and as precise as possible. Kolar (2004) presented a solution of subdivision grids in order to deal with possibly huge amounts of terrain data with multiple LOD. The geographic index (geoindex) of subdivision grids is based on Voronoi grids on the sphere, providing multiple levels of the division scheme. Each level of geoindex is given by a set of centroids distributed semi-regularly around the unit sphere. The division scheme is composed of cells. The cells are defined by a set of points with radial distances to a particular centroid lower than to any other centroid. However, in these irregular grid systems, the hierarchy of space is created by grouping and organizing spatial objects according to some pre-defined relations. In this case, changes are referred to spatial objects themselves and the hierarchy of the spatial object is maintained using explicitly defined relations among spatial objects instead of recursive decomposition of space. When a spatial process results in changes to spatial objects at one particular level, these changes cannot be propagated to its adjacent levels (Pang and Shi 1998). Hence, these irregular grid systems are difficult to manage with the large volume of global DEM data and cannot manipulate multi-resolution data visualization efficiently. It would be desirable to have grids consisting of highly regular regions with evenly distributed elevation points.

In this chapter, the global DEM data are assigned according to the vertex positions of QTM triangles and elevation points are arrayed at regular and almost equal space intervals everywhere on the globe. For the seamless and efficient construction of a global DEM, the idea of the *Binary Triangle Trees* is inserted into the QTM structure. A seamless and adaptive LOD modelling method of global terrain visualization and the corresponding algorithms will be given in detail in the following sections.

3 A Seamless QTM Tessellation on an Ellipsoidal Surface

Our approach starts with an extension method from the traditional QTM, i.e. the global tessellation is based on the WGS84 ellipsoidal surface instead of a spherical surface. Meanwhile, the distortion of areas and lines of triangles at different levels between the ellipsoidal QTM and spherical QTM are compared. The elevation values of the ellipsoidal QTM vertices are obtained through the method of *double linear interpolation* (Li & Zhu 2000) from the DEM data expressed by longitude and latitude grids.

3.1 QTM partition on an ellipsoidal surface

The QTM structure used in Fekete (1990), Dutton (1989, 1999), and Goodchild and Yang (1992) are all based on inscribed octahedra. The reason is that it can be readily aligned with the conventional geographical grid of longitude and latitude. Here, the ellipsoidal surface of WGS84 is tessellated by 8 triangles, which is the same as the initial partition in QTM (see Figure 1). Its vertices occupy cardinal points and its edges assume cardinal directions, following the equator, the prime meridian, and the 90th, 180th and 270th meridians, making it simple to determine which facet a point on the planet occupies. In addition, each facet is a right spherical triangle and one subdivision line of each face is parallel to the equator. When a triangle is subdivided, the latitude/longitude pairs of any two of its three vertices are averaged to yield edge midpoint locations. Clearly after each level of subdivision, the triangles become smaller, and at the 21st level of subdivision, their size is approximately 1 m, going down to 1 cm at the 28th level. Figure 1 illustrates subdivisions at levels 1, 2 and 3.

The vertices, which make up *QTM,* are called *QTM vertices* whose position can be identified by latitude/longitude coordinates. In term of the regularity of the subdivision, the QTM structure has several attractive properties, for example, the topology of the *QTM* makes neighbour search easy, and the tessellation is regular, i.e. the coordinates of the points can be implicitly figured out according to their storage location and the levels of the subdivision. Thus, the global terrain can be represented by the array of elevation values of the vertices.

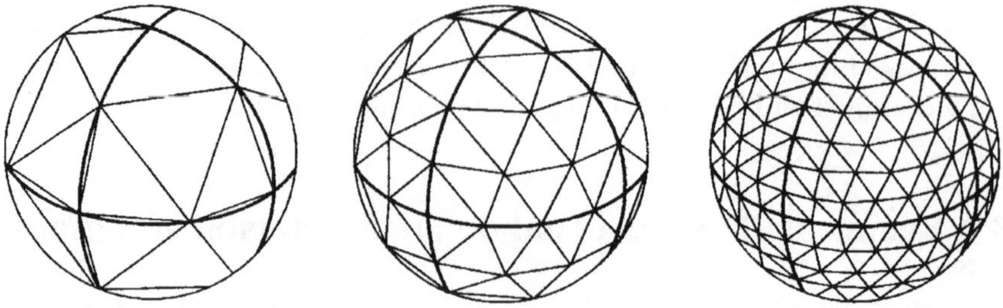

(a) Level 1 (b) Level 2 (c) Level 3

Figure 1. QTM partition of spherical facet based on octahedron (Dutton 1989).

3.2 Approximate calculations of area and length in the QTM triangles

The precise calculations of area and length of triangles on the ellipsoidal surface are more complex. In this section, in order to analyse the distribution of distortion in the triangular grid, the approximate calculation of area and length is both simple and sufficient. The details are as follows.

$P_1(u_1, \lambda_1)$, $P_2(u_2, \lambda_2)$ are two different points on the ellipsoidal surface, the distance S between these two points can be calculated by (Bian *et al.* 2005)

$$\begin{aligned} S &= b\int_{\sigma_1}^{\sigma_2} \sqrt{1+e^{-2}\sin^2\sigma}\, d\sigma \\ &= b\left[\alpha\sigma + \beta\sin 2\sigma + \gamma\sin 4\sigma + \delta\sin 6\sigma\right]\Big|_{\sigma_1}^{\sigma_2} \end{aligned} \tag{1}$$

where α, β, γ and δ are respectively calculated by

$$\begin{cases} \alpha = 1 + \dfrac{1}{4}e^{-2} - \dfrac{3}{64}e^{-4} + \dfrac{5}{256}e^{-6} \\ \beta = -\dfrac{1}{8}e^{-2} + \dfrac{1}{32}e^{-4} - \dfrac{15}{1024}e^{-6} \\ \gamma = -\dfrac{1}{256}e^{-4} + \dfrac{3}{1024}e^{-6} \\ \delta = -\dfrac{1}{3072}e^{-6} \end{cases}$$

where σ_1 and σ_2 can be calculated by

$$\begin{cases} \sin\sigma_1 = \sin u_1 / \cos A_0 \\ \sin\sigma_2 = \sin u_2 / \cos A_0 \end{cases}$$

In order to calculate the triangular area on an ellipsoidal surface, the triangle is partitioned into many long and narrow strips according to latitude lines (see Figure 2).

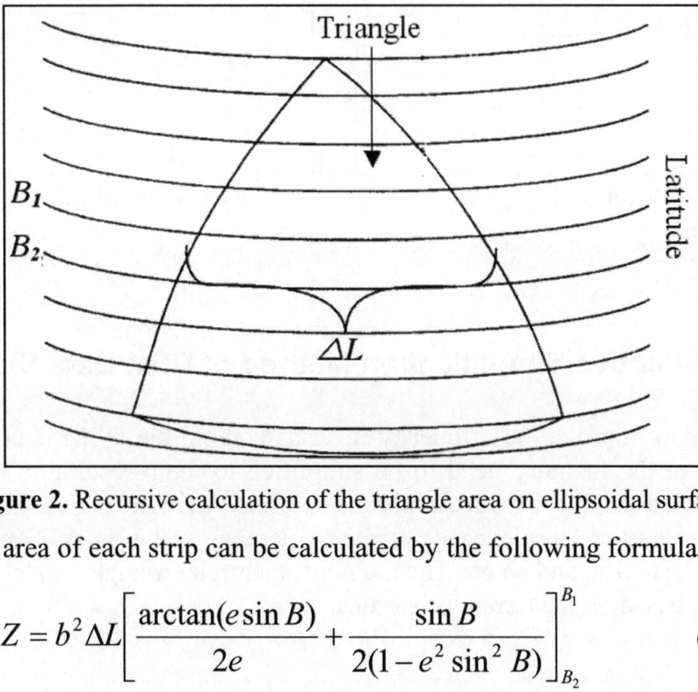

Figure 2. Recursive calculation of the triangle area on ellipsoidal surface.

The area of each strip can be calculated by the following formula:

$$Z = b^2 \Delta L \left[\frac{\arctan(e \sin B)}{2e} + \frac{\sin B}{2(1 - e^2 \sin^2 B)} \right]_{B_2}^{B_1} \quad (2)$$

where b is the polar radius, e is the flattening of the ellipsoid, and ΔL is the longitude interval. Here the precision of S and Z can be controlled by n (the number of strips).

3.3 Distortion of triangles in the recursive subdivision

The area and length of the QTM triangles can be approximately calculated by Equations (1) and (2). From the results in Table 1, it can be seen that the ratios of S_{max}/S_{min} and L_{max}/L_{min} become larger with increasing partition level, but the increment becomes less and less, at last converging to 1.73 and 1.89, respectively. This means that the triangles are almost the same in size and shape, which is very useful in the hierarchical index and extended operations on global data. In addition, the distortion of the ellipsoidal triangles is almost equal to that of the spherical triangles at every level (see Figure 3).

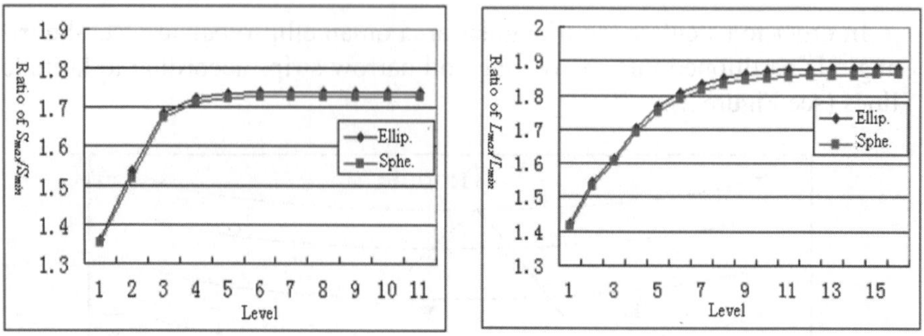

(a) Ratio of S_{max}/S_{min} (b) Ratio of L_{max}/L_{min}

Figure 3. Distortion comparison between ellipsoidal triangle and spherical triagle.

4 An Adaptive Simplification Method of DEM Data Block

In order to improve the efficiency in visualization, the DEM data blocks located in the memory need to be simplified to multi-resolution triangle grids based on the terrain properties, i.e. any triangle region in QTM, whose elevation error exceeds a given tolerance value, is split into four smaller triangles, and so on. Then, a multi-hierarchy triangle model is constructed based on the terrain properties.

Table 1. Ratio changes of S_{max}/S_{min} and L_{max}/L_{min} in different levels.

Level	*Number of triangles*	*Longest edge (m)*	*Shortest edge (m)*	L_{max}/L_{min}	*Largest area* (m^2)	*Smallest area* (m^2)	S_{max}/S_{min}	*Scale*
1	4	7.09622e+06	4.98494e+06	1.42353	1.87148e+13	1.38242e+13	1.35377	
2	16	3.84501e+06	2.48917e+06	1.54473	4.88296e+12	3.21733e+12	1.51770	
3	64	2.00924e+06	1.24412e+06	1.61499	1.32029e+12	7.89190e+11	1.67297	
4	256	1.06044e+06	622 000	1.70489	3.36458e+11	1.96352e+11	1.71354	
5	1 024	549 456	310 993	1.76678	8.45161e+10	4.90290e+10	1.72380	
6	4 096	281 100	155 495	1.80777	2.11542e+10	1.22540e+10	1.72637	
7	16 384	142 618	77 747	1.83437	5.29012e+09	3.06316 e+09	1.72701	
8	65 536	71 964	38 873	1.85124	1.32263e+09	7.65776 e+09	1.72717	1:100M
9	262 144	36 194	19 436	1.86216	3.30663e+08	1.91443 e+08	1.72721	1:50M
10	1 048 576	18 164	9 718	1.86909	8.26661e+07	4.78607 e+07	1.72722	1:20M
11	4 194 304	9 103	4 859	1.87347	2.06665e+07	1.19652 e+07	1.72726	1:10M
12	16 777 216	4 558	2 429	1.87623				1:5M
13	67 108 864	2 281	1 214	1.87798				1:2M
14	2.6844e+08	1 141	607	1.87908				1:1M
15	1.0737e+09	570	303	1.87977				1:500K
16	4.2950e+09	285	151	1.88021				1:250K

4.1 *Diamond* data organization based on QTM

White (2000) notes that pairs of adjacent triangle faces may be combined to form a *Diamond*, and this *Diamond* may be recursively sub-divided in a fashion analogous to the square quadtree subdivision. According to this idea, the whole Earth surface can be represented as a quadtree with the root having four child nodes (four base-*Diamonds* as seen in Figure 4), and the interior node has four child nodes (four child-*Diamonds*). Each *Diamond* is assigned a quadcode, and the leaf nodes can be labelled according to the *Z* space-filling curve. The global DEM data is organized on levels of *Diamond* subdivision as the basic storage unit, which correspond to different resolutions of traditional mapping. In the file system, these are the binary terrain data files in which the height values of the *QTM vertices* within the *Diamond* region are stored in Binary Large Objects (BLOB).

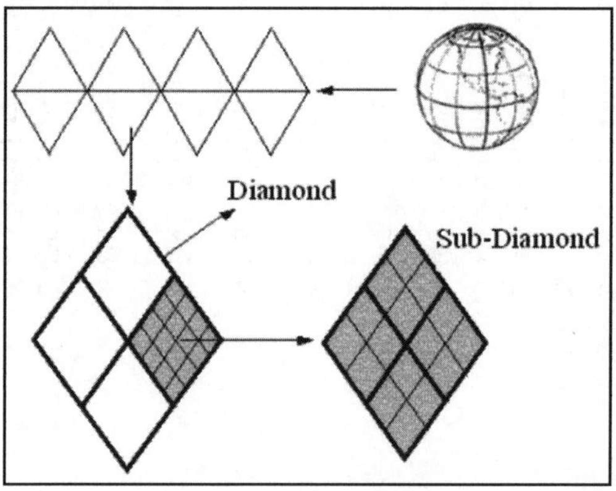

Figure 4. The DEM data organizing based on the levels of *Diamonds*.

In QTM, each triangle can be divided into four smaller triangles by breaking each edge into 2 pieces and connecting the midpoints with lines. Like the quadtree subdivision of the square, each base *Diamond* in the octahedron can be divided into four smaller *Diamonds* (see Figure 5). These two kinds of subdivision are essentially the same tessellation and they share the same vertices in the spherical surface, so the QTM can be regarded as *Diamond* meshes.

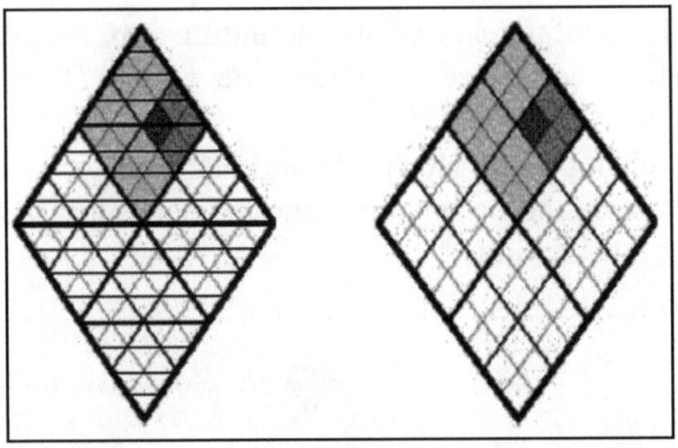

Figure 5. QTM and the *Diamond* tessellation at the third level.

4.2 The dynamic paging method of *diamond* blocks

As the dataset is too large to be stored in the main memory at one time, within the hierarchical operations of terrain data only a few Diamonds in a given level of data are paged in according to the proportion between the screen space distance and its corresponding object space distance. When the project area of given objects on the screen is larger, the object space distance per pixel is smaller; thus higher resolution data is needed, or *vice versa*. The Diamond-cell interval required is selected by calculating the object space distance per pixel according to a projection relationship in the visualization, thus the level of data to page in can be determined.

For all Diamonds located in the same level, only the Diamonds inside the view-frame are added, while the Diamonds located outside the view are not required for efficient rendering. Diamonds are chosen as the active areas based on considerations of PC memory availability and the desired range of view. This controls the amount of terrain data in the main memory at any given time and available for operations. When the operation algorithm is initialized, the viewpoint is centred on the four Diamonds. A bounding box is established around the screen centre as shown in Figure 6. When the user reaches the bounding box in any direction, memory space is freed in the direction opposite of travel, Diamonds terrain is paged in the direction of travel, and the bounding box moves. The dynamic algorithm of paging Diamonds data in the database is as follows:

- The required QTM cell resolution is calculated according to the projection relationship, and then the level of *Diamond* is determined in the visualization.

- For the level of *Diamond*, the bounding box and the scope of the *Diamonds* is compared, to determine which *Diamonds* should bc added.
- With the changing of viewpoint and angle of view, some *Diamonds* far from the bounding box are removed and other *Diamonds* near the bounding box are added.

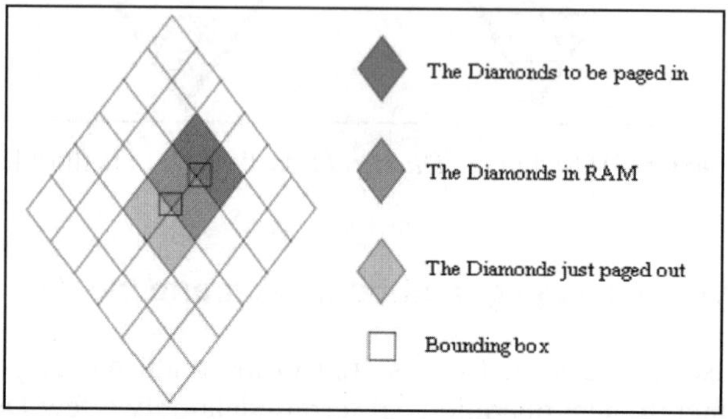

Figure 6. The dynamic adding or removing of *Diamonds.*

4.3 Data discontinuity in different subdivision levels

This mode cannot preserve the continuity of the terrain surface along the edges between the different subdivision levels (see Figure 7). Traditionally, the continuity of the terrain can be kept by tessellating the triangles to a higher level (see Figure 8).

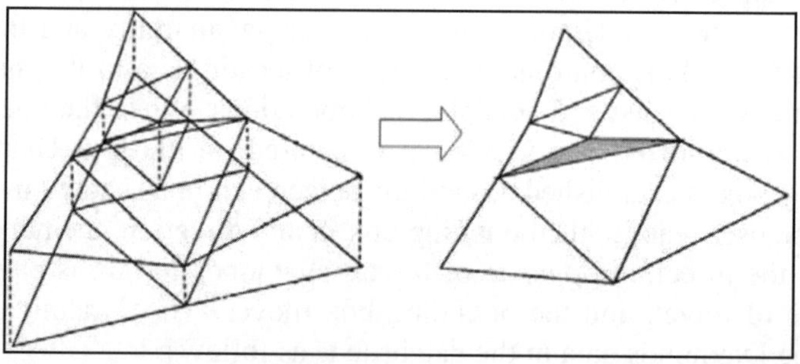

Figure 7. Hierarchical QTM and the crack caused by different subdivisions.

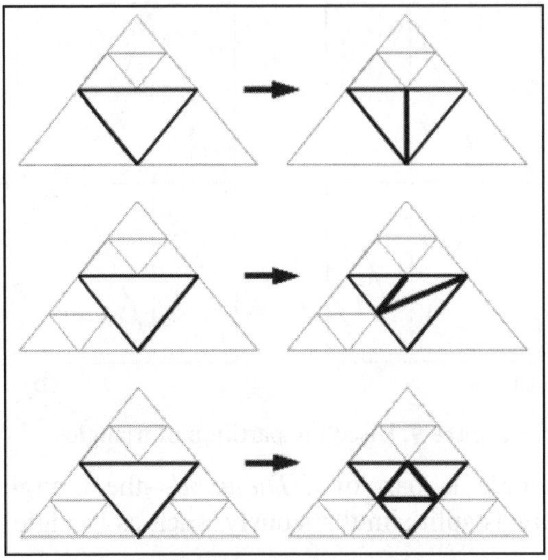

Figure 8. Eliminating cracks by splitting the triangle in a higher level.

However, the traditional method has some deficiencies. Firstly, the subdivisions need to be controlled manually and the operation can be done only when the difference between the adjacent levels are no larger than one. Secondly, the regularity of tessellation grids cannot be preserved, in which case some grids are QTM and others are irregular. In order to overcome this problem, an adaptive idea of a *Binary Triangle Tree* (Lindstrom and Pascucci 2002) is introduced in the next section.

4.4 An adapting multi-hierarchy grids based on *Binary Triangle Tree*

There are two triangles in one diamond. For each initial triangle, the pole and the midpoint of its subtense are connected to generate two smaller triangles (the latitude-edge is bisected). Next, in every subdivision, the subtense of the newly generated vertex is bisected. In one triangle, the edge that needs to be subdivided is called the *base edge*, the two triangles that share the same *base edge* are made up of a pair of triangles (shown as triangle *a* and *b* in Figure 9a). The simplification principle of the adapting grid is as follows (see Figure 9):

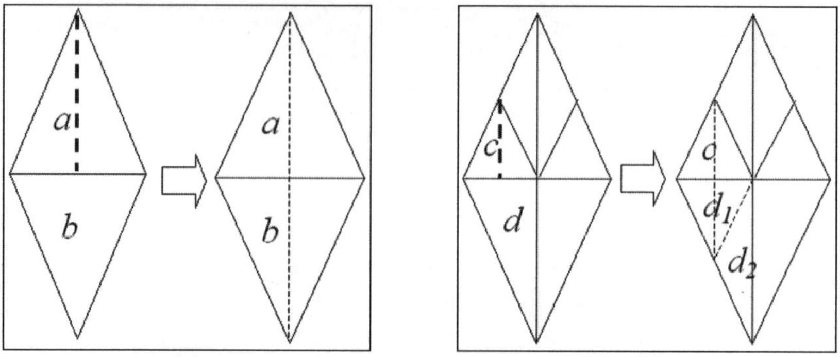

(a) (b)

Figure 9. Bisecting partition of triangles.

Step 1. The triangle is part of a *Diamond*---the triangle and its *Base Neighbour* is split simultaneously, such as triangle *a* and *b* in Figure 9a.

Step 2. The triangle is not part of a *Diamond*--- Firstly, its *Base Neighbour* is split and then do step 1. For example, in Figure 9b, if triangle *c* is to be split, its *base neighbour d* must be split into d_1, d_2 and then the triangle *c* and d_2 are split simultaneously.

Step 3. The process is to be done recursively until its error exceeds the tolerance value.

By recursive subdivision based on the *Binary Triangle Tree*, an adapting, continuous, and hierarchical triangular mesh is formed in the data blocks and their vertices coincide with the vertices of QTM exactly. The problem of discontinuity between different subdivision levels is overcome and the number of triangles of paged diamond in the visualization is greatly reduced.

5 Experiments and Analysis

The experimental results of the hierarchical algorithm for global terrain visualization are analysed in this section. The computations were performed on a processor 2.2 GHz Pentium IV PC, with 512MB of RAM and 64M GeoForceIV graphics.

Each triangle vertex can be determined by its longitude/latitude coordinates. The elevation values of the ellipsoidal QTM vertices are obtained through the method of *double linear interpolation* from the DEM data expressed by GTOPO30 (see Figure 10). *P* is a grid point of QTM, and *A*, *B*, *C*, *D* are interpolation points which are nearest point *P*. Let the function of *double linear interpolation* be

$$Z = a_0 + a_1 X + a_2 Y + a_3 XY \tag{3}$$

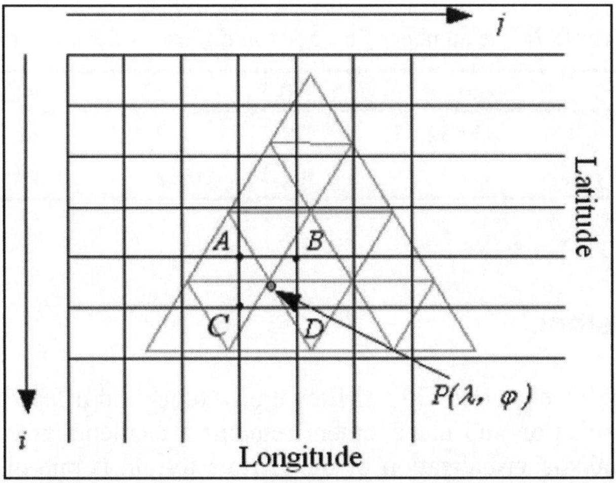

Figure 10. Elevation data conversion from GTOPO30 to QTM grids.

In order to be organized into a Diamond based file, the GTOPO30 dataset (Courtesy of US Geological Survey) had to be translated into a QTM data structure before the visualization operation. It is interpolated into 4 Diamond, 16 Diamond, 64 Diamond and 256 Diamond datasets; each includes 2,049 × 2,049 grid vertices. The four Diamond dataset levels make up a multi-resolution pyramid, each marked by a Diamond code. Each elevation value in the Diamond file represents the height value in metres above sea level and is stored using 2 bytes as a signed integer. Ocean areas are marked by no data values. Compared to the GTOPO30 Data, the volume of the QTM based DEM data is reduced to about half of its initial volume. This is illustrated by Equation (4) (Q is the volume of the QTM-based DEM data, G is the volume of the GTOPO30 DEM data, while $n = 2m + 1$, m is the partition levels):

$$Q/G = \frac{4n^2 - 4(2n-1)}{(2n-1)\times(n-1)\times 4} = \frac{(n-1)^2}{2n^2 - 3n + 1} \approx \frac{1}{2} \tag{4}$$

The number of triangles on the Earth is $N = 33{,}445{,}532$ before simplification and this number decreases greatly by using different error tolerances τ_i (i = 1, 2, 3) (see Table 2). The corresponding visualized results are shown as Figure 11 (ocean area are not displayed). From the results, it can be seen that the number of triangles is decreased greatly (left figures of Figure 11a and 11b) with their error tolerances increasing, but the visualized results show no obvious changes (right figures of Figures 11a and 11b). That is to say, in order to improve visualized efficiency, a simplified model can be applied by selecting a suitable error tolerance τ.

Table 2. The number of triangles at different tolerances τ..

τ_i	τ_0 =0	τ_1 =0.5	τ_2 =1	τ_3 =2
N	33,554,432	99,529	70,912	33,786
N_i / N_0 (%)		0.30	0.21	0.10

6 Conclusions

A global multi-resolution Digital Elevation Model and a feasible solution for its visualization and management remains a challenging vision. Real-time and realistic visualization of the Earth's terrain is one of the fundamental problems in Geography and Spatial Information Science. In this chapter, a seamless and adaptive LOD model of the global terrain based on QTM is presented.

In this model, the DEM grids are tessellated on an ellipsoid directly without any cartographic projection and its regularity makes it convenient for data organization and compressed storage. Not only the deficiencies of data overlapping (or breaking) caused by traditional DEM are overcome, but the distortion in area and/or shape and much of the redundant data found in the latitude/longitude grids are avoided as well. Compared with the GTOPO30 Data expressed in longitude/latitude coordinates, the volume of the QTM-based DEM data is about half of its original volume.

Another contribution of this chapter is the presentation of an adaptive simplification model of hierarchical triangles in DEM data blocks based on the *Binary Triangle Tree*. This adaptive triangular mesh is regular and uniform. The problem of discontinuity between different subdivision levels is overcome, greatly reducing the number of triangles required for visualization, and ensuring that triangle vertices coincide with the vertices of QTM exactly. Our analysis indicates that the QTM approach with LOD model provides an efficient, smooth and acceptable representation of global terrain.

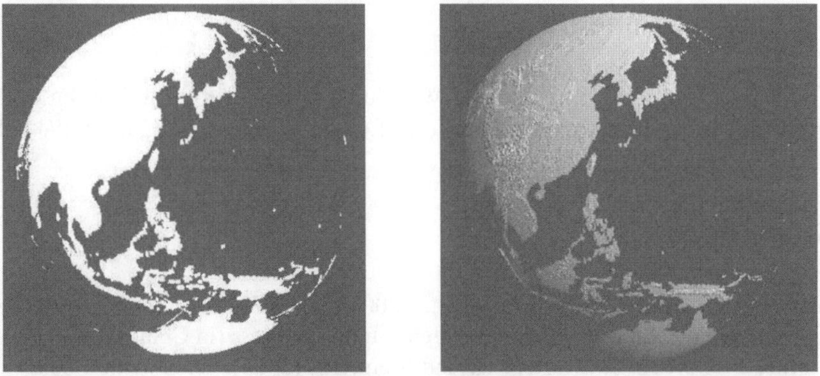

(a) τ = 0.5 *N*=99 529

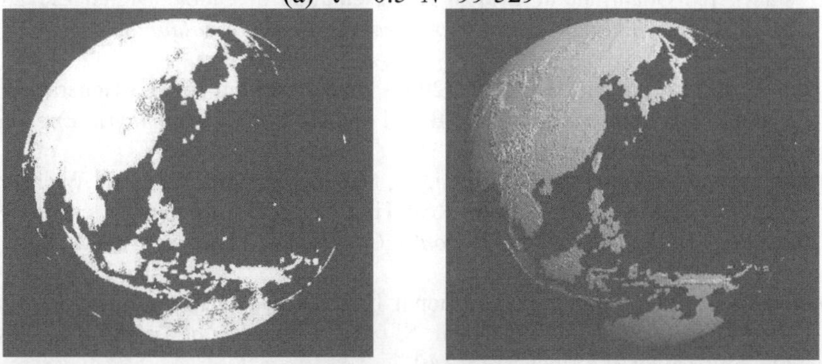

(b) τ =1; *N*=70 912

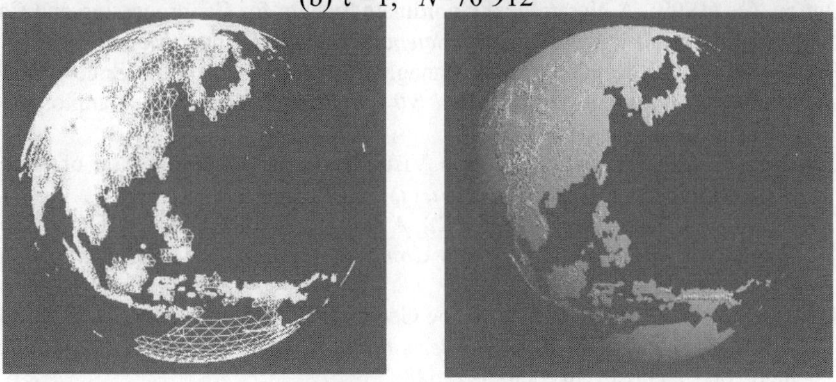

(c) τ =2; *N*=33 786

Figure 11. The visualization of global DEMs at four different error tolerances (left: triangulation grid map, right: shaded map).

Acknowledgements

The work described in this paper was substantially supported by an award from the Natural Science Foundation of China (under grant No.40471108).

References

Bian, S.F., Cai, H.Z., and Jin, J.H., (2005), *Geoid Coordination and Geoid Benchmark*, Beijing: National Defense Industry Press (in Chinese).

Bjørke, J., Grytten, J., Morten, H. and Stein, N., (2004), Examination of a Constant-Area Quadrilateral Grid in Representation of Global Digital Elevation Models, *International Journal of Geographic Information Science*, **(18)**7: 653–664.

Chen, J., Zhao, X.S. and Li, Z.L., (2003), An Algorithm for the Generation of Voronoi Diagram on the Sphere Based on QTM, *Photogrammetric Engineering & Remote Sensing*, **(69)**1: 79–90.

Duchaineau, M., Wolinsky, M., Sigeti, D., Miller, M., Aldrich, C. and Weinstein, M., (1997), ROAMing Terrain: Real-Time Optimally Adapting Meshes, *Proceedings of the IEEE Visualization97 Conference*, Phoenix Arizona, October 1997,), ACM Press: 81–88.

Dutton, G., (1989), Modeling Locational Uncertainty via Hierarchical Tessellation, In Goodchild, M and Gopal, S. (eds.): *Accuracy of Spatial Databases,* London: Taylor & Francis: 125–140.

Dutton, G., (1999), A hierarchical Coordinate System for Geoprocessing and Cartography*, Lecture Notes in Earth Sciences,* Berlin: Springer-Verlag.

Fekete, G., (1990), Rendering and Managing Spherical Data With Sphere Quadtree, *Proceedings of Visualization '90,),* October 1990,), Los Alamitos, CA: IEEE Computer Society: 176–186.

Gerstner, T., (2003), Multi-resolution Visualization and Compression of Global Topographic Data, *Geoinformatica*, **(7)**1: 7–32.

Goodchild, M.F. and Yang, S.R., 1992,), A Hierarchical Data Structure for Global Geographic Information Systems *Computer Vision and Geographic Image Processing*, **(54)**1: 31–44.

Kolar, J., (2004), Representation of The Geographic Terrain Surface Using Global Indexing *Proceeding of 12th International Conference on Geoinformatics*, June 2004,), Gavle, Sweden: 321–328.

Lee, M. and Samet, H., (2000), Navigating through Triangle Meshes Implemented as Linear Quadtree, *ACM Transactions on Graphics*, **(19)**2: 79–121.

Li, Z.L. and Zhu, Q., (2000), Digital Elevation Model, Wuhan: Wuhan Technical University of Surveying & Mapping Press (in Chinese).

Lindstrom, P., Koller, D., Hodges, L.F., Faust, N. and Turner, G., (1996), A Real-Time Continuous Level of Detail Rendering of Height Fields, *SIGGRAPH'96 Conference Proceedings*: 109–118.

Lindstrom, P. and Pascucci, V., (2002), Terrain Simplification Simplified: A General Framework for View-Dependent Out-of-Core Visualization, *IEEE Transactions on Visualization and Computer Graphics*, July–September, **(8)**3: 239–254.

Lukatela, H., (2000), *A Seamless Global Terrain Model in the Hipparchus System.* Available on-line at: http://www.ncgia.ucsb.edu/globalgrids/papers/ (accessed 15 May 2002).

Lukatela, H., (1987), Hipparchus Geopositioning Model: An Overview, *Proceedings of the Eighth International Symposium on Computer-Assisted Cartography*, 29 March – 03 April, Baltimore, Maryland: 87–96.

NIMA, (2003), *Digital Terrain Elevation Data.* Available on-line at: http://www.niama.mil/ (accessed 12 April 2005).

Pang, M. and Shi, W., (1998), Modeling Hierarchical Structure of Spatial Processes Using Voronoi Spatial Model, *8th International Symposium on Spatial Data Handling, SDH'98,),* July 1998,), Vancouver, Canada. Columbia: Taylor & Francis: 34–43.

Sahr, K., White, D. and Kimerling, A., (2003), Geodesic Discrete Global Grid Systems, *Cartography & Geographical Information Science,* **(30)**2: 121–134.

Sun, H., Du, D., Li, Z. and Zhou, Y., (2000), Study on 3D Visualization of Earth Figure *Journal of Wuhan Technical University of Survey and Mapping,* **(25)**2: 158–162 (in Chinese).

White, D., (2000), Global Grids From Recursive Diamond Subdivisions of The Surface of an Octahedron or Icosahedron, *Environmental Monitoring and Assessment*, **(4)**1: 93–103.

Section 2: Morphological Terrain Analysis

Landform Classification of the Loess Plateau Based on Slope Spectrum from Grid DEMs

TANG Guo-an and LI Fayuan

Abstract

This chapter proposed a new method for landform classification of the Loess Plateau based on grid DEM data and the GIS method. The slope spectrum, as well as corresponding indices, was first extracted from 5 m resolution DEMs. Then, 13 slope spectrum index layers were integrated into a comprehensive layer, on which were based a series of unsupervised classifications applied to landform classification. Experimental results show that the slope spectrum is appropriate for describing the spatial distribution of the loess landform and the automatic classification of loess landform types is realizable through an integrated analysis of the slope spectrum. Furthermore, the method pays more attention to the spatial distribution of the terrain at a global level and is more fitted to loess landform classification at the regional scale. It is also of great significance in extending the application of DEMs and it improves the methodology of digital terrain analysis.

Keywords: landform classification, Loess Plateau, slope spectrum, DEM.

1 Introduction

The Loess Plateau is a hot research topic because of its unique and peculiar landforms, its serious soil erosion, and its unique and specific land use. Landform classification of the Loess Plateau has attracted world-wide interest from many geomorphologists as well as geologists since the 1950s. Luo (1956), Yang *et al.* (1957), Qi and Wang (1959), Chen (1984), Chen (1956), Zhang (1986) and Wu *et al.* (1991) investigated the landform classification of the Loess Plateau based on different classification principles and field investigations, all of which proved the need for further research. However, due to the lack of an effective methodology for collecting and processing the vast amount of geo-related data necessary, these qualitative researches need to be advanced to a new and higher quantitative level based on modern GIS methods.

DEM data availability and different state-of-the-art analysis technologies have extended the use of DEMs in many ways (Richard 2000). DEMs can contain so much information on surface landforms; that is why numerous researchers have suggested that indices computed from DEMs might help define landform units by automated procedures based on computer-assisted terrain analysis (Kühni and Pfiffner 2001, Miliaresis and Argialas 2002, Argialas and Tzotsos 2003, MacMillan *et al.* 2004). Taking the Loess Plateau in north Shaanxi province as the study area, this chapter proposes a new method of landform classification based on slope spectrum.

2 Slope Spectrum

Slope is one of the most important terrain factors in representing ground surface properties and is commonly applied in geographical research and land use planning. However, the slope derived from DEMs at local grid cell level is incapable of revealing comprehensive terrain features. Although mean slope is widely accepted as a terrain index revealing the roughness of the true surface at regional level, it cannot be employed as a unique factor in the classification of terrain type. Hence, it is an obvious limitation in the classification of landforms and other more complicated analyses. In 2003, slope spectrum was first proposed, and defined as a statistical model of surface gradient composition in a certain area. Previous research on slope spectrum showed that: (1) slope spectrum can reflect relief type more effectively than any single terrain variable; (2) transition of slope composition from positive to negative skewness is governed by a dimensionless uplift number that reflects the balance between processes of relief generation and denudation, and the slope distributions provide a useful tool to explore hillslope processes (Wolinsky and Prason 2005, Montgomery 2001); (3) rock units can be classified by the shape of the slope spectrum on stable slopes (Iwahashi *et al.* 2001); and (4) slope spectrum can be used to characterize sections of non-homogeneous topography effectively (Smith and Shaw 1989).

Tang (2003) also proposed that the slope spectrum analysis method should be a new and effective methodology in revealing the macroscopical landform features via a microcosmic surface index. Any landform type can find its corresponding slope spectrum, which exists uniquely and stably only if the test area is larger than the statistical critical area (Figure 1).

Hence, slope spectrum could be used to describe corresponding landform types. Previous research on the spatial distribution of slope spectrum in the Loess Plateau showed that slope spectrum was capable of depicting the spatial variance of loess landforms quantitatively; every quantitative index of the slope spectrum can depict the landform features from different aspects. So the automatic classification of loess landform types is realizable through an integrated analysis of slope spectrum.

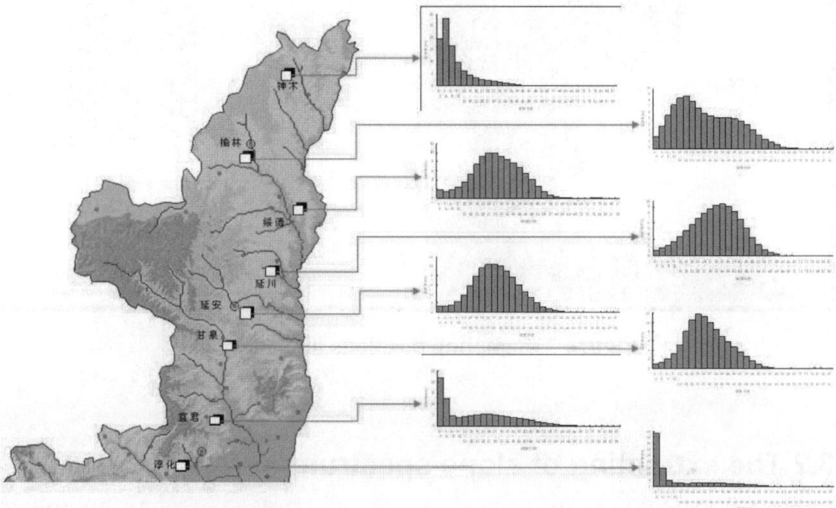

Figure 1. Slope spectrum of northern Shaanxi.

3 Method

3.1 Test area and data

1,040 set DEMs with 25 m grid size were selected as test data. The original elevations were obtained from digitized contour maps that cover the whole northern Shaanxi of the Loess Plateau (Figure 2).

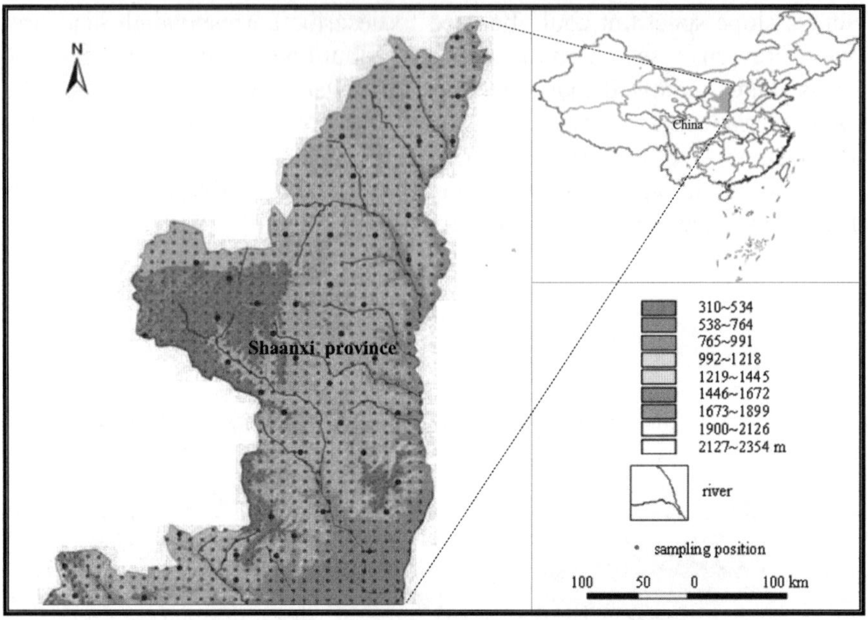

Figure 2. Sampling positions in the test area.

3.2 The extracting of slope spectrum

Figure 3 shows a procedure for extracting slope spectrum from DEMs. There follows a few key points in this process.

(1) 5 m resolution with a 3 degree gradient grading scheme proved to be suitable for describing and analysing surface features and constructing slope spectrum (Tang 2003).

(2) A key step is to determine a critical area for a stable slope spectrum. Figure 4 shows that if the test area is too small, it will be impossible to get a stable slope spectrum. Hence, an area that could guarantee the existence of a stable slope spectrum is defined as the critical area for slope spectrum in this test area.

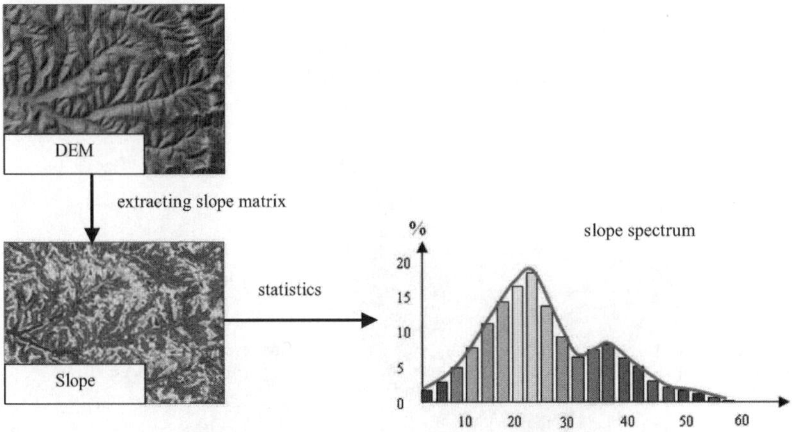

Figure 3. Procedure for extracting a slope spectrum.

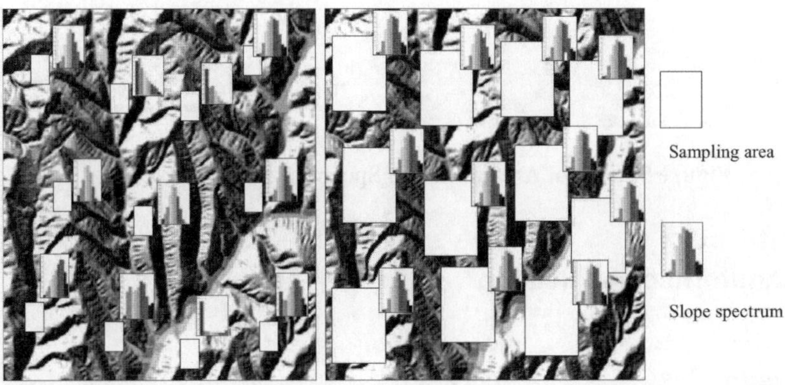

Figure 4. Slope spectrum becoming stable when the sampling area is over critical area.

The critical area for a stable slope spectrum is calculated by continuously extending the sampling windows. The area will be recorded and defined as the critical area of this terrain type when the slope histogram remains relatively unchanged when the sampling window size is increased.

The critical area is tightly correlated with landform type, sample position and the constrain condition for stable slope spectrum (Figure 5). In a certain area, a minimum critical area can be calculated when the sample area is able to represent whole geomorphic feature.

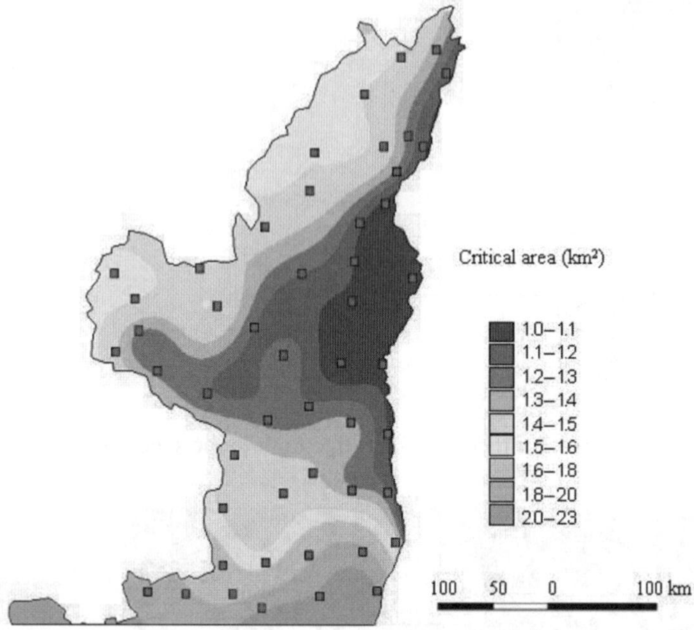

Figure 5. Critical Area for Slope Spectrum in north Shaanxi.

3.3 Landform classification

Principle

A basic principle underlying geomorphometrics is that there exists a relationship between landform and the numerical parameters used to describe it (Andrea *et al.* 2005). Since the same landform will show the greatest similarities and the smallest differences and different landforms will show the smallest similarities and the greatest differences, the integrated characteristics of the slope spectrum derived from the same landforms behave similarly, and the integrated characteristics of the slope spectrum derived from different landforms will be different. This resemblance and difference could be regarded as the rule for landform classification.

Assuming that there is a group of terrain parameters totalling *n*, and each is regarded as a single band image, the normalization value of any point (*i, j*) on the Earth's surface for every band can make up a vector X = (*x1, x2, L, x3*) T. X is the feature value of the point (*i, j*). The n-dimensional space containing X is called the feature space. Then any point on the Earth's surface can be represented by a point in the n-dimension feature space. Usually, the more similar the terrain parameters derived from the same landform, the more concentrated is the distribution of the corresponding points in the n-dimension space. Different landforms will form different point-sets. The aim of landform classification is to search for some curves or some curved surfaces in feature space and using these curves or curved surfaces to separate point-sets. The core of landform classification is then to determine a discriminant algorithm and its corresponding rules.

The above principle has proved to be effective in the classification of landforms (Bue and Stepinski 2006, Daniel *et al.* 1998, Liu *et al.* 1999, Liu and Liu 1991). However, this process takes DEM cells as a basic unit for analysis, but pays less attention to the spatial distribution of terrain at the global level. Hence, an ameliorated method is put forward in this chapter. The procedure in this experiment can be described as follows: (1) extracting slope spectrum from the corresponding DEMs; (2) extracting a series of quantitative indices based on the spectrum; (3) building continuous surfaces in the test area based on the extracted quantitative factors of the spectrum by the mathematical interpolation method; (4) treating these factor layers as a single-band image and combined them into an integrated one via the image integrating method; (5) generating classified layers after decorrelation analyses; (6) unsupervised classification; and (7) verifying the classification result (Figure 6).

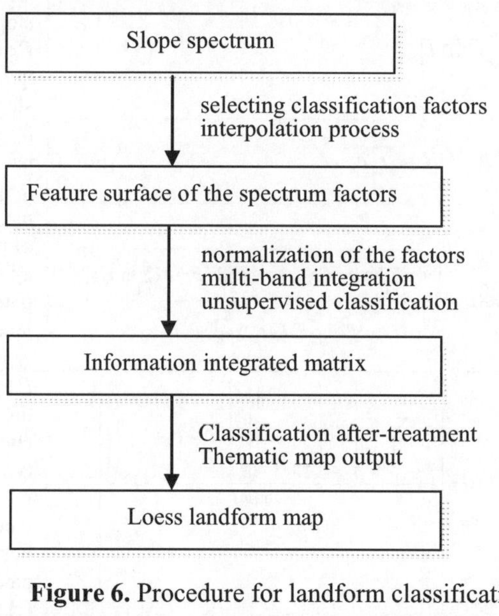

Figure 6. Procedure for landform classification.

Slope spectrum indices

Based on research by Li (2007), 13 indices from the slope spectrum are selected as basic factors for landform classification. These are slope spectrum entropy (*H*), skewness of the slope spectrum (*S*), mean elevation, hypsometric integral, slope-area integral, slope variability (*SOS*), mean plane curvature, mean profile curvature, mean patch area (*AREA_MN*), Perimeter-Area Fractal Dimension (*PAFRAC*), Contagion Index (*CONTAG*), Interspersion and Juxtaposition Index (*IJI*), Patch Cohesion Index (*COHESION*). Calculation methods for the indices are listed in Table 1.

Table 1. Algorithms for the calculation of slope spectrum indices.

Classification index	Calculation	Remark
Slope spectrum entropy (*H*)	$H = -\sum_{i=1}^{m} P_i \ln P_i$	*m*: number of slope class, P_i: denotes frequency of each slope class
Skewness of the slope spectrum (*S*)	$S = \sqrt{\frac{1}{6m}} \sum_{i=1}^{m} \left(\frac{p_i - \overline{p}}{\sigma} \right)^3$	$\overline{p}$: mean frequency *o*: standard deviation
Slope Variabiity (*SOS*)	first derivative of slope	
Slope-area integral	$-\sum_{i=1}^{m} P_i \ln P_i$	*m*: number of slope class, P_i: cumulated frequency of each slope class
Mean patch area (*AREA_MN*)	$AREA\_MN = \sum_{j=1}^{n} a_{ij} \Big/ n_i$	a_{ij}: area (m^2) of patch *ij*, n_i: patch number.
Perimeter-Area Fractal Dimesion (*PAFRAC*)	$PAFRAC = \dfrac{2}{\dfrac{\left[n_i \times \sum_{j=1}^{n}(\ln p_{ij} \times \ln a_{ij})\right] - \left[(\sum_{j=1}^{n} \ln p_{ij}) \times (\sum_{j=1}^{n} \ln a_{ij})\right]}{(n_i \times \sum_{j=1}^{n} \ln p_{ij}^2) - (\sum_{j=1}^{n} \ln p_{ij})^2}}$	p_{ij}: perimeter (m) of patch *ij*. n_i: number of patches in the landscape of patch type (class) *i*.
Contagion Index (*CONTAG*)	$CONTAG = \left[1 + \dfrac{\sum_{i=1}^{m} \sum_{k=1}^{m} \left[(P_i) \left(\dfrac{g_{ik}}{\sum_{k=1}^{m} g_{ik}} \right) \right] \times \left[\ln(P_i) \left(\dfrac{g_{ik}}{\sum_{k=1}^{m} g_{ik}} \right) \right]}{2\ln(m)} \right] \times 100$	P_i: proportion of the landscape occupied by patch type (class) *i*. g_{ik}: number of adjacencies (joins) between pixels of patch types (classes) *i* and *k*

		based on the double-count method. m: number of patch types (classes) present in the landscape, including the landscape border if present.
Interspersion and Juxtaposition Index (*IJI*)	$LJI = \left[1 + \frac{-\sum_{k=1}^{m}\left[\left(\frac{e_{ik}}{\sum_{k=1}^{m} e_{ik}}\right)\ln\left(\frac{e_{ik}}{\sum_{k=1}^{m} e_{ik}}\right)\right]}{\ln(m-1)} \right] \times 100$	e_{ik}: total length (m) of edge in landscape between patch types (classes) i and k. m: number of patch types (classes) present in the landscape, including the landscape border, if present.
Patch Cohesion Index (*COHESION*)	$COHESION = \left[1 - \frac{\sum_{j=1}^{n} p_{ij}}{\sum_{j=1}^{n} p_{ij}\sqrt{a_{ij}}}\right] \times \left[1 - \frac{1}{\sqrt{A}}\right]^{-1} \times 100$	p_{ij}: perimeter of patch ij in terms of number of cell surfaces. a_{ij}: area of patch ij in terms of number of cells. A: total number of cells in the landscape.

4 Experimental results and discussion

4.1 Spatial distribution of the slope spectrum

Based on the slope spectrum of 1,040 test areas, 13 indices are derived, which reflect the properties of the slope spectrum. Then, the spatial distribution layers of these indices are constructed with the ArcGIS geostatistical analysis tools. The process can be depicted as follows:

(1) Exploring the data distribution: if the data is not normally distributed, a transformation process should be carried out to bring it to a normal distribution pattern;

(2) Identifying global trends: the trends will be removed when modelling;
(3) Calculating distance and semi-variance between all sampling pairs and determining rational lag size;
(4) Establishing best-fit semi-variance /covariance modelling;
(5) Calculating the semi-variance of each pair of points and that between each observation and prediction value; then a matrix of these semi-variance values can be built;
(6) Calculating the weight of the measured points around the prediction point;
(7) Multiplying the weight for each measured value times the value, and adding the products together to calculate the prediction value;
(8) Applying cross-validation to evaluate prediction accuracy. The evaluation standard is as follows: Mean Error (ME) and Mean Standardized Error (MSE) should be close to zero, Average Standardized Error (ASE) should be close to the Root Mean Square Error (RMSE) and the Root Mean Square Standardized Error (RMSSE) close to unity.

Furthermore, to avoid the influence of different dimensions, all of the terrain factors should be normalized according to Equation (2).

$$x_{i,j}' = \frac{x_{i,j} - x_{\min}}{x_{\max} - x_{\min}} \times 255 \qquad (2)$$

where x is the attribute value; i j are the number of rows and columns of the interpolated surface; $x_{i,j}'$ denotes the normalization value.

Each normalized terrain derivative is treated as a single-band image and they are put into 13 channels for the integration process in the following order: slope spectrum information entropy, skewness of the slope spectrum, mean elevation, hypsometric Integral, slope-area integral, slope variability, mean plane curvature, mean profile curvature, mean patch area, Perimeter-Area Fractal Dimension, Contagion Index, Interspersion and Juxtaposition Index and Patch Cohesion Index. Then, a comprehensive multi-band image can be obtained by an integrating process of the 13 single-band images, which will be the basis for landform classification.

Actually, not every band is suitable and available for the above process. Some bands are highly correlated. Since the classification accuracy will not be affected if these bands are not used, they can be thrown out selectively. The band or band combinations used for classification are called the "features". Feature selection is the process of choosing characteristic to be used for classification and feature extraction picks up a group of new characteristics, which best reflects the different classes. There are usually two methods for feature selection: Correlation Matrix and Optimum Index Factor (OIF). The method of Correlation Matrix is used in this experiment. This method calculates the correlation coefficient matrix of every slope spectrum index, and the one with the lowest correlation will be the candidate index for the classification.

Covariance and correlation coefficients between any two band images are calculated to examine the correlatability of different terrain derivatives. The terrain derivatives with high correlatability can be put into one group. In Tables 2 and 3, 13 slope spectrum indices were classified into five groups according to their correlatability. The first group includes slope spectrum information entropy, slope-area integral, mean slope variability, mean plane curvature, mean profile curvature, Perimeter-Area Fractal Dimension and Juxtaposition Index. The second group includes skewness of the slope spectrum, Contagion Index, Interspersion, Patch Cohesion Index. Mean elevation, hypsometric integral, mean patch area are taken independently as the remaining three groups. To enhance the efficiency of the process, principal component analysis is applied to eliminate correlatability between the factors. With Principal Components Analysis (PCA), the data in the first and second groups are integrated. Finally, a multi-band image with comprehensive terrain information can be derived using an integrating process with above data sets.

4.2 Classification of landform

The ISODATA (Iterative Self-Organizing Data Analysis Technique) was applied in the landform classification. This method includes three steps: (1) clustering analysis, (2) split of assemblage, and (3) merge of assemblage. The mechanism of ISODATA is a process of automatic iterative clustering. Split and merge can run automatically; the operator can determine model parameters in the process of continuous clustering, so as to construct finally the best fitted discriminant function. It is also a process of adjustment and training of the discriminant function using statistical features of the optical spectrum.

Under the environment of ERDAS 8.5, some controlling parameters should be appointed first:

(1) Number of class. This number is often appointed twice as the ultimate number of classes.

(2) Maximum iteration: cycle number of clustering, to avoid lengthy processing or dead cycles in running the program.

Table 2. Correlation coefficient matrix of slope spectrum factors used for classification.

	H	*S*	*Mean elevation*	*Hypsometric Integral*	*Slope-area integral*	*SOS*	*Mean plan curvature*	*Mean profile curvature*	*AREA_MN*	*PAFRAC*	*CONTAG*	*IJI*	*COHESION*
H	1	-0.9365	0.3855	0.1930	0.8883	0.9497	0.8745	0.8988	-0.1781	0.6791	-0.9842	0.8820	-0.8595
S	-0.9365	1	-0.4327	-0.1732	-0.9094	-0.9408	-0.8847	-0.9101	0.1323	-0.6110	0.9577	-0.7695	0.9115
Mean elevation	0.3855	-0.4327	1	0.2216	0.4914	0.3827	0.3878	0.3224	-0.1621	0.3212	-0.4326	0.1632	-0.3687
Hypsometric Integral	0.1930	-0.1732	0.2216	1	0.1900	0.2145	0.1597	0.2136	-0.0429	0.1993	-0.2210	0.1800	-0.1869
Slope-area integral	0.8883	-0.9094	0.4914	0.1900	1	0.9445	0.9333	0.8871	-0.1221	0.5920	-0.9216	0.7560	-0.9291
SOS	0.9497	-0.9408	0.3827	0.2145	0.9445	1	0.9408	0.9659	-0.1455	0.6483	-0.9616	0.8459	-0.9577
Mean plan curvature	0.8745	-0.8847	0.3878	0.1597	0.9333	0.9408	1	0.9284	-0.1236	0.6172	-0.8919	0.7654	-0.9483
Mean profile curvature	0.8988	-0.9101	0.3224	0.2136	0.8871	0.9659	0.9284	1	-0.1345	0.6573	-0.9319	0.8203	-0.9700
AREA_MN	-0.1781	0.1323	-0.1621	-0.0429	-0.1221	-0.1455	-0.1236	-0.1345	1	-0.7427	0.1818	-0.2930	0.1141
PAFRAC	0.6791	-0.6110	0.3212	0.1993	0.5920	0.6483	0.6172	0.6573	-0.7427	1	-0.7035	0.6624	-0.6051
CONTAG	-0.9842	0.9577	-0.4326	-0.2210	-0.9216	-0.9616	-0.8919	-0.9319	0.1818	-0.7035	1	-0.8568	0.9011
IJI	0.8820	-0.7695	0.1632	0.1800	0.7560	0.8459	0.7654	0.8203	-0.2930	0.6624	-0.8568	1	-0.7631
COHESION	-0.8595	0.9115	-0.3687	-0.1869	-0.9291	-0.9577	-0.9483	-0.9700	0.1141	-0.6051	0.9011	-0.7631	1

Correlation is significant at the 0.01 level (2-tailed).

Table 3. Covariance matrix of slope spectrum factors used for classification.

	H	*S*	*Mean elevation*	*Hypsometric Integral*	*Slope-area integral*	*SOS*	*Mean plan curvature*	*Mean profile curvature*	*AREA MN*	*PAFRAC*	*CONTAG*	*IJI*	*COHESION*
H	0.7025	-1.4209	110.31	0.0169	0.1085	2.3442	16.348	14.477	-63.202	0.0320	-20.529	11.218	-14.164
S	-1.4209	3.2769	-267.38	-0.0327	-0.2398	-5.0151	-35.718	-31.658	101.37	-0.0627	43.143	-21.302	32.441
Mean elevation	110.31	-267.38	116521	7.8826	24.436	384.67	2952.3	2114.7	-23429	6.1760	-3674.7	846.70	-2474.5
Hypsometric Integral	0.0169	-0.0327	7.8826	0.0109	0.0029	0.0658	0.3712	0.4278	-1.8916	0.0012	-0.5731	0.2885	-0.3830
Slope-area integral	0.1085	-0.2398	24.436	0.0029	0.0212	0.4052	3.0326	2.4836	-7.5308	0.0049	-3.3413	1.6861	-2.6611
SOS	2.3442	-5.0151	384.67	0.0658	0.4052	8.6726	61.793	54.662	-181.41	0.1080	-70.479	38.023	-55.450
Mean plan curvature	16.348	-35.718	2952.3	0.3712	3.0326	61.793	497.47	397.93	-1167.0	0.7818	-495.06	261.39	-415.83
Mean profile curvature	14.477	-31.658	2114.7	0.4278	2.4836	54.662	397.93	369.27	-1094.2	0.7161	-445.65	240.99	-366.46
AREA_MN	-63.202	101.371	-23429	-1.8916	-7.5308	-181.41	-1167.0	-1094.2	179283	-1.7272	1916.1	-309.06	950.02
PAFRAC	0.0320	-0.0627	6.1760	0.0012	0.0049	0.1080	0.7818	0.7161	-1.7272	0.0033	-0.9836	0.5605	-0.6765
CONTAG	-20.529	43.143	-3674.7	-0.5731	-3.3413	-70.479	-495.06	-445.65	1916.0	-0.9836	619.34	-323.35	440.87
IJI	11.218	-21.302	846.70	0.2885	1.6861	38.023	261.39	240.99	-309.06	0.5605	-323.346	236.79	-229.97
COHESION	-14.164	32.441	-2474.5	-0.3830	-2.6611	-55.450	-415.83	-366.46	950.02	-0.6765	440.87	-229.97	386.52

Correlation is significant at the 0.01 level (2-tailed).

(3) Convergence threshold: maximum percentage of pixels that remains unchanged between two classifications.

In this research, the number of classifications is set at 30, maximum iterations as 24, and the convergence threshold as 0.95.

4.3 After-treatment of classification result

Because clustering analysis of ISODATA is virtually based on the optical spectrum features of the image, many small patches will be generated in the process. Hence, an after-treatment is necessary to eliminate these patches in order to obtain a desirable thematic map. This process can be realized with ERDAS 8.5 software with the following steps: (1) producing a primary classification image (Figure 7); (2) eliminating the patches with the functions of [Clump] and [Eliminate] in ERDAS 8.5; (3) determining of the attributes of the thematic map derived in the above process; and (4) recoding to the images after the after-treatment process. Within the process, the thematic attributes of each primary class were determined, based mainly on the Geomorphology Sketch Map of the Loess Plateau (Meng 1996) as well as the 1:500,000 scale Geomorphology Map of the Loess Plateau (Zhang 1986).

The primary image can be split into nine classes; they are meadow-basin, loess-low-hill, gorge-hill, middle-low-mountain, loess-hill-and-loess hill-ridge, loess-yuan-ridge, loess-platform-yuan, loss-ridge-low- mountain, and loess yuan (Figure 8).

4.4 Verification of classification result

To verify the classification, field exploration and comparison were adopted. A series of test sites was selected along a north to south profile in northern Shaanxi, which represented almost all the relief types in the Loess Plateau. Then, the results at these field investigation points were compared with the automatic-classified results to test the correctness of the classification.

The experiment showed that the correct result was obtained at 28 out of the 34 points. The classification method has proved to be credible, although more work is needed to perfect the theory and the method.

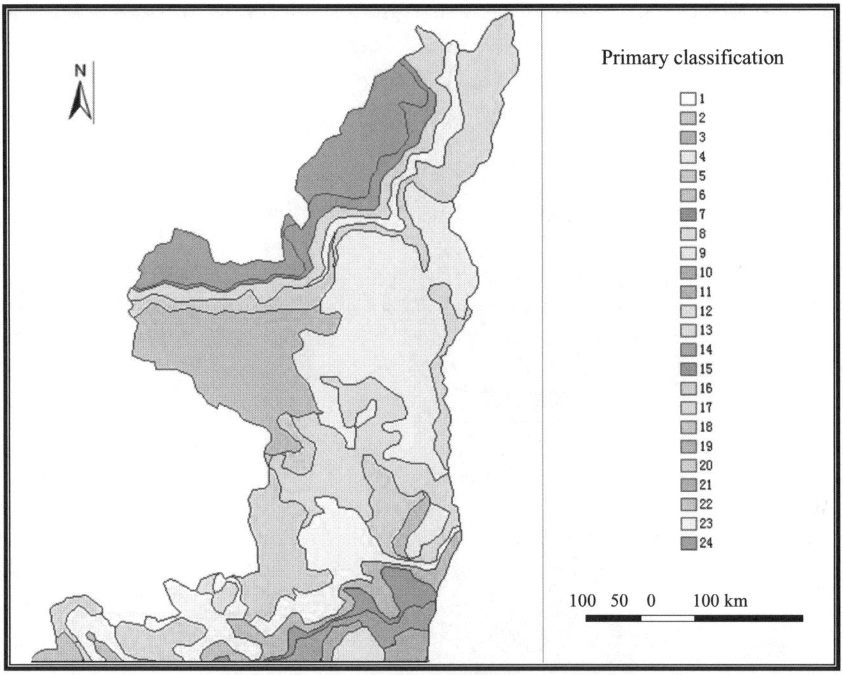

Figure 7. Map of primary classification.

5 Conclusion

The geographic significance of slope spectrum is emphasized through landform classification. Any geomorphic type can find its corresponding slope spectrum, which exists uniquely and is stable only if the test area is larger than the statistical critical area. Slope spectrum is appropriate for describing the spatial distribution of the loess landform.

The method provides a new link between remote sensing and digital terrain analysis by treating each layer of slope spectrum indices as a single-band image. Our classification process pays more attention to the spatial distribution of terrain at the global level and is more fitted to loess landform classification at the regional scale. It is also of great significance in extending the application of DEMs and it improves the methodology of digital terrain analysis.

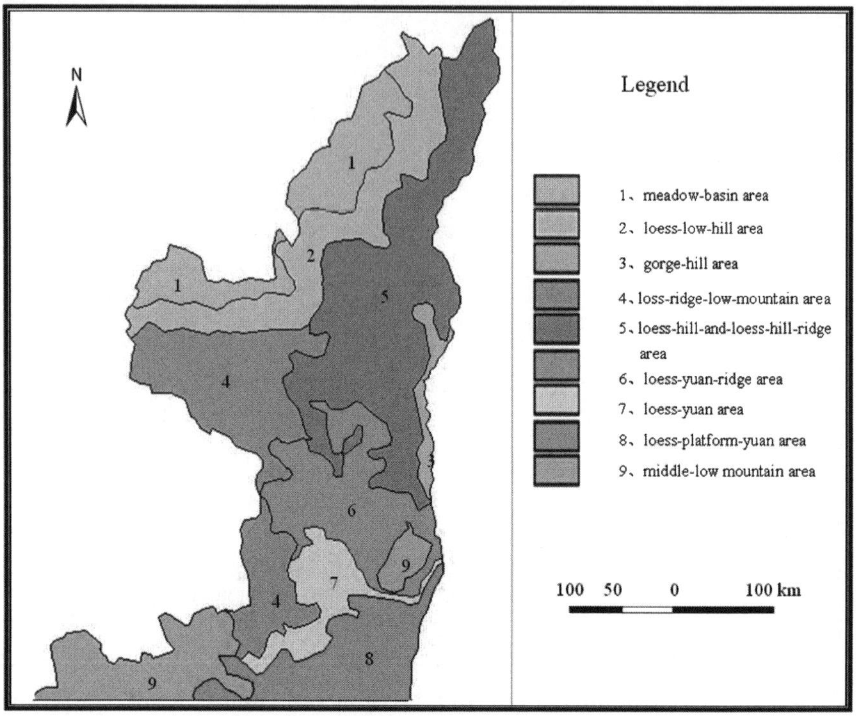

Figure 8. Map of landform classification of the Loess Plateau in north Shaanxi province.

Acknowledgement

Thanks for financial support from the National Natural Science Foundation of China (No.40571120, No.40671148), Specialized Research Fund for the Doctoral Program of Higher Education (No.20050319006).

References

Andrea, B.C., Vicente, T.R. and Valentino, S., (2005), Geomorphometric analysis for characterizing landforms in Morelos State, Mexico., *Geomorphology*, **67**: 407–422.

Argialas, D. and Tzotsos, A., (2003), Geomorphological Feature Extraction from a Digital Elevation Model Through Fuzzy Knowledge-based Classification, *Remote sensing for environmental monitoring, GIS applications, and geology II*, 516–527.

Bue, B.D. and Stepinski, T.F., (2006), Automated classification of landforms on Mars, *Computers & Geosciences*, **32**: 604–614.

Chen, C.K., (1956), Landform type and its evolution in south-eastern of east Gansu province, *Acta geographica sinica,* **22**(3): 223–231 (in Chinese).

Chen, Y.Z., (1984), Ravine type in loess hill-gully region in middle reaches of the Yellow river, *Journal of Geographical Science*, **4**(4): 321–327 (in Chinese).

Daniel, G.B, David, P.L. and Kenneth, A.D., (1998), Supervised classification of types of glaciated landscapes using digital elevation data, *Geomorphology*, **21**: 233–250.

Miliaresis, G.C. and Argialas, D.P., (2002), Quantitative representation of mountain objects extracted from the global digital elevation model (GTOPO30), *Int. J. Remote Sensing*, **23**(5): 949–964.

Iwahashi, J.S., Watanabeb, S. and Furuya, T., (2001), Landform analysis of slope movements using DEM in Higashikubiki area, Japan, *Computers & Geosciences*, **27**: 851–865.

Kühni, A. and Pfiffner, O.A., (2001), The relief of the Swiss Alps and adjacent areas and its relation to lithology and structure: topographic analysis from a 250-m DEM, *Geomorphology*, **41**: 285–307.

Li, F.Y. and Tang, G.A., (2006), DEM based research on the terrain driving force of soil erosion in the Loess Plateau, In Gong J.Y.and Zhang J.X. (eds.): *Geoinformatics 2006, Geospatial Information Science, Proc. of SPIE*, Vol. 6420, 64201W.

Li, F.Y., (2007), *Slope spectrum and its spatial distribution on the Loess Plateau,* Doctoral dissertation of Graduate University of the Chinese Academy of Sciences, Beijing: Graduate University of the Chinese Academy of Sciences (in Chinese).

Liu, Y.L. and Liu, Y.F., (1991), The Optimum Selection of Bands and It's Combinations in Remote Sensing Classification, *Journal of Geomatics*, **2**: 13–18 (in Chinese).

Liu, Y., Wang, Y.X. and Pan, B.T., (1999), A preliminary approach on the 3D presentation and quantitative analysis of planation surface, *Geographical Research*, **18**(4): 391–399.

Luo, L.X., (1956), A tentative classification of landforms in the Loess Plateau, *Acta geographica sinica*, **22**(3): 201–222 (in Chinese).

Meng, Q.M., (ed.), (1996), *Soil and water conservation on Loess Plateau*, Zhengzhou: Yellow River Hydrological Publishing (in Chinese).

Montgomery, D.R., (2001), Slope distributions, threshold hillslopes, and steady-state topography, *American Journal of Science*, **301**: 432–454.

Qi, Y.N. and Wang, Z.C., (1959), Geomorphology and new tectonic movement in the Guanzhong plain and south Loess Plateau of north Shaanxi province, *Acta geographica sinica*, **25**(4): 286–298.

MacMillan, R.A., Jones, R.K. and David, H.M., (2004), Defining a hierarchy of spatial entities for environmental analysis and modeling using digital elevation models (DEMs), *Computers, Environment and Urban Systems*, **28**: 175–200.

Richard, J.P., (2000), Geomorphometry–diversity in quantitative surface analysis, *Progress in Physical Geography*, **24**(1): 1–30.

Smith, D.K. and Shaw, P.R., (1989), Using Topographic Slope Distributions to Infer Seafloor Patterns, *IEEE Journal of Oceanic Engineering*, **14**(4): 338–347.

Tang, G.A., (2003), Research on terrain information automatic mining and slope spectrum in the Loess Plateau, Post-doctoral report of Northwest University, Xi'an: Northwest University (in Chinese).

Wang, C., (2005), The Uncertainty of Slope Spectrum Derived from DEM in the Loess Plateau of Northern Shaanxi Province, Master's paper of Northwest University, Xi'an: Northwest University (in Chinese).

Wolinsky, M.A. and Prason, L.F., (2005), Constrain on landscape evolution from slope histograms, *Geology*, **33**(6): 477–480.

Wu B.F., Chen M.R. and Chen Z.X. *et al.* (1991) The Loess Plateau of China. Shaanxi: people's publishing house of Shannxi (in Chinese).

Yang, H.R, Yu, X.J. and Han, T.C., (1957), Geomorphology of the Loess Plateau of south-western Shanxi, *Acta geographica sinica*, **23**(1): 17–53 (in Chinese).

Zhang, Z.H. (ed), (1986), *Geomorphologic map of the Loess Plateau, China (1:500,000)*, Beijing: Geological Publishing House (in Chinese).

Segmentation-based Terrain Classification

Josef STROBL

Abstract

Terrain analysis is based on a set of methods and techniques developed since the 19th Century addressing a broad range of application domains. Classification is a key scientific approach to extracting information and ultimately knowledge from data collections. Terrain classification prepares information from digital terrain models for various application scenarios. Spatial segmentation is a relatively novel approach based on classification and regionalization.

This chapter is intended as a review of current practice with an emphasis on providing the basic conceptual foundations for segmentation in terrain analysis. It starts by linking segmentation approaches to basic geographical concepts such as spatial categories vs. regions, the process of regionalization, concepts of scale, interfacing with human perception, and physical spatial processes.

Segmentation attempts to combine powerful characteristics from the continuous field view with spatial object/topographic type/landscape unit views. With an ever increasing number of sensors and terrain data acquisition technologies, terrain analysis should no longer be driven by traditional data-centric approaches, but rather by semantics-centred concepts from the respective application domains. Only by identifying and delineating elementary spatial units independently from data sources, can physical models become truly interoperable and transferable across different types of terrain data sets.

1 Introduction

Due to the controlling effect of terrain shapes and terrain configurations on many physical processes, and because of interest in terrain shape as the key indicator for landscape evolution, the description, classification, and analysis of terrain goes back to the roots of geomorphology in pre-digital times (Penck 1894). Back then researchers already attempted to identify 'typical' landforms as basic building blocks of complex landscapes, attributing certain formative processes to these landforms.

With the emergence of quantitative methodologies and digital processing, the atomic unit for analysis went from landform units to smaller regular (raster) or irregular (e.g. triangles) geometric objects. More powerful processing and data acquisition techniques (e.g. early photogrammetric stereocorrelation, see Kelly *et al.* 1977) reduced the spatial resolution and increased data volume exponentially. This led to pseudo-realistic display options, and to analytical tools exploring new levels of detail. Data acquisition and analysis in many respects converged with remote sensing based image analysis toolsets, sometimes unfortunately discarding the distinction between radiometric image (pixel) data and thematic (point, lattice) data sets.

In this process, the semantic meaning of more complex landform units was essentially lost, with combinations of simple topographic descriptors (like slope, aspect, curvature) serving as proxy indicators for process-oriented analysis. Interestingly, due to the toolset convergence between image processing and terrain analysis, the technique of image segmentation (Haralick and Shapiro 1985, Pal and Pal 1993) began to be explored independently by several groups of researchers for applications in terrain analysis (e.g. Dragut and Blaschke 2006).

Essentially, this means going back 'full circle' to much earlier concepts using landform units as building blocks for landscape-scale process models. Spatial segmentation aggregates high resolution samples into homogeneous and contiguous patches, which in turn are considered semantic objects according to the respective application domain's ontology.

While 'patches' had been considered a terrain modelling tool for decades (see Mark 1979), segments are semantically identified 2D shapes projected onto 2.5D surfaces, which can be flexibly attributed by statistical, geometrical, and topological descriptors other than the former concept of patches as mere geometrical surface approximations (e.g. as piecewise splines).

Due to the semantics-driven generation and use of spatial segments, there is a clear linkage to natural language and visual perception, and thus potential use throughout user interfaces. This altogether paves the way towards a more application-centric language, an analytical approach, a processing logic and end user communication in terrain analysis, moving from highly abstract formal indicators towards ontologically founded analysis based on functional or homogeneous terrain units.

This review chapter starts from the above outlined situation and pursues three objectives:

- Makes explicit the transition of segmentation from an image processing methodology into a thematic spatial data aggregation method, discussing constraints and limitations;
- Relates segmentation methodology to established geographical concepts and methodologies, such as numerical taxonomy and regionalisation; and
- Discusses the distinction of data-centric analysis of sampled data vs. semantics-oriented analysis of elementary spatial units based on suitable application logic.

2 Segmentation and Classification

The term 'segmentation' is widely used for a loosely defined set of methods related to classification. Taxonomic classification attempts to group objects into a set of categories ('classes') based on object properties. Classes are intended to be internally homogeneous, with maximum separation between classes. Classification implements the principle of generalization in order to focus on broader, more general insights into an empirical or conceptual domain.

Generalization is a synthetic, bottom-up approach and can essentially be considered an abstraction operation, focused on what features have in common. As an example in a terrain context, an elongated set of low elevation points surrounded by higher elevation points on two or more sides is generally considered a 'valley'. Individual valleys come in different shapes and are distinctly different from each other, but share common 2.5D characteristics, helping us to recognize and identify them as some kind of 'valley'.

Aggregative classification in a geospatial context starts from sets of individual observations such as points or pixels, looks at their characteristics (attribute values) and attempts to group them into classes based on similarity. (In contrast, divisive techniques follow a top-down logic only mentioned here for completeness, but not further considered in this chapter.) Resulting classes can either be organized in a flat schema or as a hierarchy where broader, more general classes contain multiple specialized classes more similar to each other than to sub-classes from another super-class.

Classification can be based on a single attribute (elevation zones, slope categories) or simultaneously use multiple attributes, leading to multivariate classification. The latter are well established in all sciences using taxonomies (biology, anthropology, etc.) and have been brought into the geospatial sciences through image processing and analytical techniques in regional sciences. Based on multivariate statistical techniques, a broad range of classification techniques has been developed over several decades and is implemented in readily available software tools.

The simplest, uni-variate classification technique is thresholding at either regular or distribution-dependent irregular intervals (e.g. at histogram minima). This basic approach to categorization is widely used in cartography as well as in many other applications where simplification is desired.

Segmentation as a trivial concept, e.g. in marketing (Steenkamp and Ter Hofstede 2002), demographics, or simplistic image processing, follows exactly the same strategy of thresholding along one or over several metric dimensions in order to break a continuous set of elements into discrete groups. Such distinct subsets of the entire domain of study are supposed to display similar behaviour and homogeneous characteristics.

Frequently, segmentation of a domain (like a population or an image of an area's land cover) aims at identifying 'types'. Types are considered named categories or classes based on common characteristics (see above) and are understood to exhibit unique behaviour. Types like other generalizations are an important concept in scientific research, and 'terrain types' will be discussed below.

Obviously, as a broad concept, the idea of segmentation cannot be strictly distinguished from classification. In some disciplines and for certain techniques, these terms are used interchangeably. For further discussion, it therefore is critical to constrain terminological semantics to 'spatial segmentation', with the understanding that spatial position and/or relations (like distance) are always considered. As a spatial analytical technique, a purely statistical, non-spatial approach to segmentation would not generate any added value; it thus is imperative to discuss segmentation from an explicitly spatial perspective.

Image segmentation is mostly implemented (see Appendix A) as techniques considering similarities in pixel values as well as proximity between pixels. Resulting aggregates therefore are not only classes (when mapped to a spatial view class members can be scattered all over the study area), but are spatially contiguous clusters of pixels. Translating this concept to terrain segmentation, the results are patches of similar surface characteristics.

Following the general, higher level objectives of classification, the generalization of individual objects or measurements like pixels or terrain elevation points into classes or spatial segments is an indispensable tool for scientific enquiry, needed for conceptualization, understanding, and communication of the respective study matter. Studies of terrain stand to benefit substantially from this kind of otherwise mostly continuous phenomena, but are frequently carried out without a solid conceptual underpinning. Discussing the latter is the main objective of this chapter.

3 The Concept of Regionalization

Focusing on spatial segmentation as outlined above, segmentation actually is not only rooted in classification, but just as much in regionalization. As a general geographical concept, regionalization refers to the process and techniques of delineating contiguous regions on the Earth surface, either by aggregating elementary units or by dividing larger entities. Regions are either identified based on homogeneity, or on relations or flows (functional region). A uniformly plane slope would be a homogeneous region; a watershed is considered a functional region.

Classification of spatial data and regionalization therefore are closely related techniques. Classification will lead to spatial types or zones according to map algebra terminology (Tomlin 1990), with a zone consisting of many spatially disjoint patches belonging to the same class. Precisely the same classification technique could then be used for regionalization, requiring an additional spatial constraint: objects are only combined into one class if they meet similarity criteria AND if they are adjacent.

Unfortunately, few software tools offering multi-variate statistical classification support a regionalization option (i.e. an added contiguity constraint) as well. One of the few exceptions is the Clustan package (Wishart 1987) controlling cluster analysis with a contiguity matrix to join operational taxonomic units only when their areas are spatially adjacent. Resulting regions are comparable to results from current segmentation processors, even though fewer control parameters regarding region shape, hierarchies, etc. are available.

As attribute and spatial constraints are employed to delineate homogeneous regions, it might be of interest to consider briefly spatial-constraint-only techniques. Essentially, these would result in something like Thiessen polygons based on proximity only (Gold 1989). From a terrain analysis perspective, this is of particular interest as Thiessen polygons can be considered a dual structure to Triangulated Irregular Networks (TINs), which are frequently used for terrain representation and surface analysis. Spatial segmentation therefore covers the middle ground between purely attribute-based classification and an exclusively spatially oriented Thiessen tessellation.

A tessellation or tiling of any space is considered a plan structuring of a study area without gaps or overlaps. Frequently a regionalization results in a tessellation, i.e. leads to an exhaustive and mutually exclusive spatial pattern (like a puzzle game 'tiling' a study area). A spatial segmentation therefore will in most cases deliver a tessellation of the study area.

Another core concept of classification and regionalization, and thus segmentation as well, is working with hierarchies. Again we need to distinguish between categorical (subclass – superclass) and spatial hierarchies (subregion – superregion). Hierarchies are closely linked to region size and thus scale. Region growing algorithms typically stop even before meeting other regions' boundaries when reaching thresholds of internal heterogeneity or size. A size or total-number-of-regions factor therefore terminates the aggregation process just like the statistical generation of clusters. By combining several adjacent, relatively similar regions into larger entities, hierarchical structures of higher level objects are constructed; multi-level or multi-scale segmentation is a frequently implemented tool to achieve this aim.

Spatial hierarchies are not necessarily based on similarity, but frequently occur when elementary objects 'belong together' and combine into an ontological feature (see above, 'functional regions'). This could be various slopes forming a river catchment, facets combined into a building on a digital surface model (Miliaresis and Kokkas 2007), or simply the brightly lit plus shaded sides of a hill. Building up these types of feature representations is beyond segmentation and requires semantics-rich spatial processing in order to bridge the gap between segments and ontological features.

Based on the concepts discussed above, a well-founded approach to segmentation-based terrain analysis can now be pursued. Even though segmentation is generally considered to have evolved out of computer vision and image processing (e.g. Shapiro and Stockman 2001, Shi and Malik 1997) it is important to recognize that the fundamental concepts of spatial classification and regionalization provide critical insights into the use of segmentation for thematic geospatial datasets like terrain surfaces.

4 Segmentation and Terrain

Terrain surfaces are a particular type of spatial entity, where it is important to understand the key pertinent characteristics before selecting analytical approaches and methods:

- Terrain is generally considered a spatially continuous phenomenon, requiring discretization and adequate data modelling (Kemp 1997 a, b).

- Continuous surfaces are treated as single-valued surfaces, often referred to as 2.5D entities.
- When using a raster/grid model to represent terrain, it is important to consider this as a regular point sample ('lattice') and not a cell grid; this is a key distinction for image data types where segmentation has originated.
- The presence, shape, and characteristics of terrain features depend heavily on resolution and scale; terrain models from different acquisition techniques and resolutions should not be mixed.
- Terrain characteristics are highly spatially autocorrelated; this is an essential prerequisite for spatial segmentation.
- Although terrain is generally modelled as 2.5D, most measures are treated as planar attributes. For example, slopes are represented as two independent attributes (slope angle, azimuth), and it is not yet well understood how this might affect multi-variate analyses.
- Most terrain representations are not built on original elevation measurements, but rather on interpolated or resampled values. Again, this affects subsequent processes, e.g. by systematic smoothing of edges.
- True vertical faces (or even more so, overhangs) are not properly represented in 2.5D data structures. This increasingly is becoming an issue with very high density data acquisition over built-up areas and for work with digital surface models.
- Generally speaking, terrain segmentation aims at transforming a quasi-continuous terrain representation (like a regular lattice) into a discrete and adaptive representation of terrain units.

There is an extensive body of literature about delineating terrain units and using them as a basis for physical models as well as for visualization, both looking at homogeneous units (like uniform slope) and functional units (basic hydrological entities) - see Lane *et al.* (1998), Wilson and Gallant (2000), Bue and Stepinski (2006).

Terrain and other surface representations only rather recently were based on the data- and computation-intensive very dense sampling facilitated by automated, remote sensing supported acquisition. Before that, most analytical work could not adequately deal with terrain as a continuum, but rather had to use 'terrain units' as a basic entity. Due to this long tradition, the concept of terrain units is firmly established across several disciplines. It was therefore considered a boon to ultimately – via segmentation – be able to replace the laborious, subjective process of delineating terrain units with sets of automatic algorithms.

4.1 Issues with terrain segmentation

It needs to be stressed, though, that while terrain segmentation holds promise and potential to abstract terrain analysis from the vagaries of different source data and more tightly connect modelling and simulation with application domains' ontologies, there are numerous unresolved conceptual as well as algorithmic issues that at least need to be acknowledged.

First, segmentation processes tend to exhibit unstable behaviour; segmentation results can vary widely with only little change to the underlying data sets or even by just shifting the extent of a study area. It should be expected that similar segments are derived from different data sets representing the same terrain, as long as parameters controlling the processing are stable.

This generally is not the case, due to several reasons. Even though open source implementations are available, algorithms for segmentation are not fully published, they tend to obviously vary considerably, and thus lead to variable results. Beyond that even the control (input) parameters differ, with multiple approaches controlling attribute homogeneity, size, shape, and uniformity. As opposed to, for example, multi-variate statistical methods, segmentation frequently is implemented as a black box where users need to resort to learning from experimentation and experience.

Segmentation has been related above to cluster analysis. As in aggregative clustering, the issue of where to start the clustering process, how to initiate 'seeding' of a cluster, is a core methodological issue having a huge impact on outcomes. Implementations range from random seeding to using centroids of highly homogeneous neighbourhoods to stochastic optimization. In some cases it might be desirable to control these starting locations, which generally is not foreseen in software implementations. One example might be the use of particular points in a hydrographic network (pour points, confluence points) to make sure that such key locations control the segmentation process.

Overall, spatial segmentation today is still far removed from a mature methodology with implementations reaching from trivial thresholding to proprietary, copyrighted and non-disclosed algorithms. Only when the methodology becomes fully documented and a consensus converges on types of implementations will segmentation techniques find their way into the standard toolset for the analytical spatial sciences.

Finally, it needs to be pointed out that segmentation does not necessarily aim to be a one-step process leading from dense sampling data to semantic objects (Argialas and Miliaresis 2000). As briefly pointed out above, segments are homogeneous and/or functional units, and quite frequently objects are built from different, potentially very dissimilar segments. Therefore 'object building' (sometimes misleadingly termed 'classification') is used as a second processing phase. In this stage, segment characteristics based on attribute statistics, geometry, topology, and hierarchy are used by crisp or fuzzy rule sets to construct features representing real world entities from the basic segmentation results.

5 Why Terrain Segmentation?

Before going any deeper, this question should be addressed briefly. We are moving to ever more detailed and accurate representations of terrain and other physical surfaces; why do we want to 'return' to characterizing coarse patches of terrain? Why look at larger area entities, when we can easily inspect every bit of detail? Being able to dynamically change scales, to look at detail without losing the wider context, to analyse the big picture and still be able to drill down to the local scale is essential for most analysts.

This issue on first sight is closely related to a discussion around 'multi-resolution terrain models' in the late 1990s (see e.g. de Floriani *et al.* 2000). Still, the focus at that time was on supporting dynamic visualization across many different scales, and multi-resolution storage like pyramids and other hierarchical structures were considered essential to achieving the desired visual performance.

The above stated question is aiming beyond real-time visualization, though. And there is likely more than one answer, and not only one single motivation to consider segmentation:

Interpretation and identification of landforms is inherently multi-scalar. A small local minimum might be an artefact, or 'surface roughness', or part of a much larger depression. This is about seeing the forest for the trees, about interpreting local surface attributes in a wider context. Larger units provide this kind of indispensable context.

Standard topographic attributes are by default determined in a local neighbourhood, like a 3 by 3 kernel or within a search radius of 12 points. With changing data sources and increasing resolutions, these neighbourhood definitions are not stable, but change scale as well. In order to allow stable parameterization, larger units related to process modelling scale instead of data acquisition scale need to be identified.

Many physical processes operate at a different order of magnitude, and depend on a wider framework than local landform. Those higher orders of magnitude cannot easily be derived from local detail. Wind-exposed ridges and low wetlands are features that require the identification of landforms at scales related to the respective processes.

From a broader perspective, we aim at distinguishing measurement and sampling from semantics-driven analysis and modelling. Currently, most analyses are directly tied to the respective data model, which in turn in most cases is (for good reason) driven by a data acquisition technique. As elevation data today are derived from many different sources and technologies, a direct link between data sources and processing methodology would compromise the transferability and general applicability of models: e.g. a landslide susceptibility model would work well with the NEXTMap radar elevation data sets it has been developed for, but would not yield useful results with Lidar data, photogrammetric elevation data, etc.

Facing various and continuously changing data acquisition technologies, different sampling resolutions and levels of accuracy, there is a clear need for achieving interoperability between, for example, process models and terrain elevation data sets. Introducing terrain units as a semantics-based interface, built on the domain ontology of a given research discipline, might provide a future-proof solution to this problem. Terrain segmentation then will be the toolset to translate sampling data into domain information by building semantics into the terrain representation.

6 Towards a Unified Terrain Representation (?)

Digital terrain modelling uses a rather stable set of conceptual data models (Mark 1979), only recently extended to accommodate high volume elevation mass points as generated through large-area Lidar campaigns in commercially available geodatabase models.

Over several decades technologies for terrain data acquisition and the resulting quantities and qualities of elevation data have changed considerably, though. From arc-seconds and deka-metre resolutions, frequently derived as secondary data from contour maps, technology has progresses to high resolution elevation data acquisition based on remote sensing methods like photogrammetry, radar interferometry, and laser scanning. Spatial resolutions frequently exceed one point per square metre.

Although resolution therefore has improved by several orders of magnitude, the set of methods for terrain analysis have by and large remained stable. As already indicated above, neighbourhood-based topographic attributes are inherently unstable with changing resolution. In order to proceed from the traditional data-driven approaches to analysis towards a more stable semantics-oriented definition of terrain units, segmentation has the potential to help with defining semantically defined objects, essentially de-coupling terrain-based physical process or evaluation models from changing data sources. This in turn is a key step towards interoperability and moving away from stovepipe-style models, where terrain data feed directly into the application logic.

This is a fundamental change in the processing logic of terrain analysis. And, it needs to be recognized that there is no simple or singular way to identify 'natural' terrain units. Therefore, just like with other approaches to classification, it is not feasible to start with delineating topographic features '*per se*' from a terrain data set, but as different application domains will have very different views and demands, the starting point needs to be the respective application logic.

Most terrain modelling applications are based on physical spatial processes, and different types of processes interact very differently with terrain. Some major process categories are:

- Radiative, line-of-sight processes: solar irradiation, intervisibility, etc.
- Advective or convective flow processes: e.g. wind field simulation, exposure.
- Gravity-induced processes: surface runoff, fluvial erosion, potential energy.
- Centrifugal diffusion: like accessibility, contamination, fire, etc.

Due to the large volume of published work, special emphasis is frequently put on geomorphological, actually geomorphogenetical, analysis. While this set of processes of course is primarily controlled by gravitational forces as well, it serves the purpose of explaining the genesis of landscapes *per se*.

Geomorphological landscape research (Lane *et al.* 1998) therefore has produced typologies of landforms and slopes (headslope, toeslope, noseslope, sideslope, backslope … and many more) primarily defined by dominance of different geomorphological processes along a gravity-controlled sequence and described by typical topographic landform parameters (like e.g. curvature). Geomorphology therefore has for a long time seriously worked towards defining 'landscape units', but it needs to be acknowledged that these 'natural units' help to explain the genetical aspects for today's morphology, but are largely irrelevant for current processes on a much shorter temporal scale.

As indicated above, it therefore is indispensable to first study the process-oriented application logic of a physical process to be analysed in a given application context. Depending on whether researchers are interested in predicting soil erosion, analysing surface runoff patterns, studying vegetation distribution according to solar input and wind exposure, model snowmelt patterns, or estimate hydropower potentials, very different sets of 'landform units' will be suitable as a basis for the process models.

This in turn means that different terrain ontologies are required for different application domains (Miliaresis *et al.* 2005, or Miliaresis and Kokkas 2007). In one early paper, Kemp and Vckovski (1998) explored general ontologies of 'fields'. Subsequently, starting from a wider framework of ontology research, Mark and Smith (2002) developed a framework for a terrain ontology primarily based on topographic mapping (i.e. mid-scale general visualization purposes). Beyond that, little work has been done in the area of explicit application-specific terrain ontologies (e.g. for soil erosion, visible exposure, wind impact, etc.), identifying this clearly as an area of high research demand and potential.

Finally, it shall be pointed out that terrain units, depending on application logics, are not necessarily expected to be exhaustive, mutually exclusive tessellations of space. Rather, terrain units can just as well overlap and leave gaps, therefore not always fitting into simple, topologic polygonal data models. This will apply especially whenever terrain units are linked to natural language communication, where the identification of landform elements is not at all considered a neat spatial subdivision of a larger landscape.

7 Outlook

Looking back at the initially outlined objectives, it has been demonstrated that the transition of a method originally developed for multispectral image analysis into the multi-faceted thematic domains of terrain analysis is beyond the trivial and involves explicit referencing with established concepts for spatial modelling and analysis.

The application of segmentation techniques for thematic data is not yet fully understood and requires additional research in several directions. It has become clear, though, that the potential for more semantically oriented processing and a closer connection with the domain logic in various application areas hold considerable promise and make further research definitely worthwhile.

With higher and higher resolution sensors, data volumes continue to expand exponentially; this increases the attractiveness of data reduction techniques. Essentially, segmentation can be considered a data mining and information extraction tool, translating data into domain specific information.

It is anticipated that segments/patches/regions will play an important role in future terrain analysis. In order to leverage this potential, several requirements have to be met: (a) application domains need to define explicit terrain ontologies; (b) segmentation techniques need to develop from a loosely defined set of related approaches into a sound methodology with results reproducible across implementations, and (c) clear conceptual linkage with geographical concepts like regions, hierarchies, scale, and process modelling.

Appendix 1: Software references

Statements and assessments regarding the state of technology in spatial segmentation methods made in this review chapter are based on the author's experience with particular software products. Other implementations might yield different insights and conclusions; therefore the 'implicit background' software products are mentioned below in alphabetical order:

Definiens professional – Understanding Images - Munich, Germany http://www.definiens.com/
GeoAIDA (Geo Automatic Image Data Analyser) Hannover University http://www.tnt.uni-hannover.de/project/geoaida/
SAGA – System for Automated Geographical Analysis – Göttingen University http://www.saga-gis.uni-goettingen.de/
TNTmips The Map and Image Processing System – Lincoln, Nebraska http://www.microimages.com/

References

Argialas, D.P., and Miliaresis, G.CH., (2000), Physiographic Region Interpretation: Formalization With Rule Based Structures and Object Hierarchies, *Int. Archives of Photogrammetry & Remote Sensing (ISPRS)*, Vol. XXXIII, Part B4: 91–98.

Baatz, M., and Schaepe, A., (2000), Multi-resolution segmentation. An optimization approach for high-quality multiscale image segmentation, In Strobl, J., Blaschke, T. and Griesebner, G. (eds.): *Angewandte Geographische Informationsverarbeitung XII*, Heidelberg: Wichmann Verlag: 12–23.

Blaschke, T., Burnett, C. and Pekkarinen, A., (2004), New contextual approaches using image segmentation for object-based classification, In De Meer, F. and de Jong, S. (eds.): *Remote Sensing Image Analysis: Including the spatial domain*, Dordrecht: Kluver Academic Publishers: 211–236.

Bue, B.D. and Stepinski, T.F., (2006), Automated classification of landforms on Mars, *Computers and Geosciences*, **32**: 604–614.

Burrough, P.A., van Gaans, P.F. and MacMillan, R.A., (2000), High resolution landform classification using fuzzy k-means, *Fuzzy Sets Systems,* **113**, 37–52.

Buttenfield, B. P., and McMaster, R.B. (eds.), (1991), *Map Generalization: making rules for knowledge representation*, New York: John Wiley and Sons.

Campbell, J., (2001), *Map Use and Analysis* (4th ed.), New York: McGraw Hill.

Dragut, L. and Blaschke, T., (2006), Automated classification of landform elements using object-based image analysis, *Geomorphology*, **81**: 330–344.

de Floriani, L., Magillo, P. and Puppo, E., (2000), A library for multiresolution modeling of field data in GIS, *Int. Workshop on Emerging Technologies for Geo-Based Applications*, Lausanne: Swiss Federal Institute of Technology: 133–151.

Gold, C., (1989), Voronoi diagrams and spatial adjacency, *Proceedings G.I.S. - challenge for the 1990s*, Ottawa, Canada: 1309–1316.

Haralick, R. and Shapiro, L., (1985), Survey: image segmentation techniques, *Computer Vision, Graphics, and Image Processing*, **29**: 100–132.

Irvin, B.J., Ventura, S.J. and Slater, B.K., (1997), Fuzzy and isodata classification of landform elements from digital terrain data in Pleasant Valley, Wisconsin, *Geoderma*, **77**: 137–154.

Kelly, R.E., McConnell, P.R. and Mildenberger, S.J., (1977), The Gestalt photomapping system, *Photogrammetric Engineering and Remote Sensing*, **(43)**11: 1407–1417.

Kemp, K.K., (1997a), Fields as a framework for integrating GIS and environmental process models. Part one: Representing spatial continuity, *Transactions in GIS*, **(1)**3: 219–234 and **(1)**4: 335.

Kemp, K.K., (1997b), Fields as a framework for integrating GIS and environmental process models. Part two: Specifying field variables, *Transactions in GIS* **(1)**3: 235–246.

Kemp, K.K. and Vckovski, A., (1998), Towards an ontology of fields, In *Proceedings of the Third International Conference on GeoComputation*, Bristol, UK, (CD-ROM).

Lane, S., Richards, K. and Chandler, J., (1998), *Landform Monitoring, Modelling and Analysis*, John Wiley & Sons.

Mark, D.M., (1979), Phenomenon-based data-structuring and digital terrain modelling, *Geo-Processing*, **1**: 27–36.

Mark, D.M., and Smith, B., (2002), Do mountains exist? Ontology of landforms and topography, In *Environment & Planning B*, From: <http://wings.buffalo.edu/philosophy/faculty/smith/articles/Mountains.htm>.

Miliaresis, G., Sabatakakis, N. and Koukis, G., (2005), Terrain pattern recognition and spatial decision making for regional slope stability studies, *Natural Resources Research*, **14**: 91–100.

Miliaresis, G. and Kokkas, N., (2007), Segmentation & object based classification for the extraction of building class from LIDAR DEMs, *Computers and Geosciences*, **33**: 1076–1087.

Pal, R. and Pal, K., (1993), A review on image segmentation techniques, *Pattern Recognition,* **26**, 1277–1294.

Penck, A., (1894), *Orometrie und Morphologie der Erdoberfläche, Vol. 2*, Stuttgart: J. Engelhorn: 339–343 (in German).

Pike, R.J., (1988), The geometric signature - Quantifying landslide - terrain types from digital elevation models, *Mathematical Geology*, **20**: 491–511.

Senin, N., Ziliotti, M. and Groppetti, R., (2007), Three-dimensional surface topography segmentation through clustering, *Wear*, **262**: 395–410.

Shapiro, L. and. Stockman, G., (2001), *Computer Vision*, New Jersey: Prentice-Hall.

Shi, J. and Malik, J., (1997), Normalized Cuts and Image Segmentation, *IEEE Conference on Computer Vision and Pattern Recognition*: 731–737.

Steenkamp and Ter Hofstede (2002), International market segmentation: issues and perspectives, *Intern. J. of Market Research,* **19**: 185–213.

Tang, G., (2000), A research on the accuracy of digital elevation models, Beijing, New York: Science Press.

Tomlin, C.D., (1990), *Geographic Information Systems and Cartographic Modeling*, Englewood Cliff, NJ: Prentice Hall.

Wishart, D., (1987), *Clustan User Manual*, University of Edinburgh.

Wilson, J. and Gallant J., (2000), *Terrain Analysis, Principles and Applications*, John Wiley & Sons.

Terrain Segmentation and Classification using SRTM Data

Lucian D. DRĂGUȚ and Thomas BLASCHKE

Abstract

The main objective of this chapter is to segment and classify Shuttle Radar Topography Mission (SRTM) data into specific landforms. Based on the results of previous research (Drăguț and Blaschke 2006), a classification system of landform elements was improved and adapted for SRTM 3 arc second data. Terrain derivatives such as elevation, slope gradient, slope aspect, profile curvature, and plane curvatures were classified in a multi-resolution object-oriented approach comprising four different scale levels. We carried out object-based image analysis, using a software program called eCognition Professional 4.0, to segment terrain derivatives into relatively homogeneous objects, which were further classified using fuzzy logic rule sets. Special emphasis was put on the accuracy assessment of the results as well as on the transferability of the procedure between study areas. We classified two SRTM datasets comprising a rolling hill landscape, which covers small areas of the states of Arkansas, Missouri and Oklahoma, USA, and a high mountain area of 50 square km around Hochkalter Peak, Berchtesgaden National Park, Germany. Results were visually compared and accuracy assessments using fuzzy classification options and an error matrix were performed. The classification system proved to be transferable between hilly and high mountain areas, its outcomes being satisfactorily accurate.

Keywords: terrain segmentation; SRTM; accuracy assessment; fuzzy logic; geomorphometry.

1 Introduction

The ability to recognize landforms and to infer processes from them is an important skill in geomorphology (Goudie 2002). Although some reluctance to abandon the qualitative approach to landform description was noticed (Pike 1995), landform quantification and classification have been gaining increasing interest due to practical needs. Topography of various types and scales can be fingerprinted by computer analysis of DTMs (Pike 1988), thus increasing tremendously the versatility, speed, and accuracy of geomorphic applications. As mapping and assessing landforms and erosion is essential in order to understand landscape denudation and complex feedback mechanisms, the development and evaluation of new approaches in remote sensing and geomorphometry is required (Bishop *et al.* 2003).

Many planning applications and a majority of research topics concerning environmental processes and environmental management do need a dissection of the continuous surface of the Earth into 'tangible objects'. Such 'landscape units', 'facets', or 'patches' are certainly imperfect generalizations of the reality but they are needed in daily life. Any discrete zoning decision may be better expressed by gradients of functions but when DTM information supports spatially explicit decisions we need to discretize continuous surfaces or pseudo-continuous representations, respectively, into relatively homogeneous zones. As landscape ecology theory tells us, there will never be the one and only dissection of the real world into pieces. This is a fundamental and well known discussion in cartography. In geomorphology, we hypothesize that we need such a dissection of the real world or its representation (Blaschke and Strobl 2001). As well as applications in Earth sciences, new requirements from industrial engineering and planetary sciences stimulate research on terrain classification. Techniques from geosciences are adapted to the partitioning of the three-dimensional micro and nano topography of engineered surfaces (Senin *et al.* 2007) or to the classification of landforms on Mars (Bue and Stepinski 2006). Thus, numerous systems for automated classification of topography from DTMs have been proposed during the last two decades (see Iwahashi and Pike (2007) for a review). Latest technical developments in this field include land surface segmentation, which has recently started being consolidated theoretically (Minár and Evans 2007).

An important question related to the classification of spatial entities is how reliable the outputs are. While accuracy assessment is standard in classification of remotely sensed data, 'objective' methods for validating terrain classification beyond visual inspection have not yet been proposed. Within this chapter, we aim at exploring innovative ways of plausibility checks and accuracy assessment. This is especially important in the light of increased data availability and the increased transferability of the landscape classification procedures as introduced by Drăguţ and Blaschke (2006).

With the advent of the SRTM data, which covers almost 80% of the Earth surface, new avenues are open to digital terrain modelling applications. In spite of the relatively coarse spatial resolution (90 m), the availability of this data set fosters both development of (semi)-automated and transferable applications. Moreover, the results of DTM modelling applications are more comparable between various geographic areas than in the past. Among the most promising applications are recognition and quantification of landforms, which are some of the main tasks of geomorphometry (Rasemann *et al.* 2004). In the future, we will see many more applications dealing with landform classification and quantification for large regions.

This work forms part of a project called 'Complex Landscape Units for Environmental Assessment and Modelling' (CLUE), which aims at developing and testing a new methodology for the integration of landform elements within the landscape definition, using geographic information systems (GIS), object-based image analysis, digital terrain models (DTM), and satellite imagery. In this approach, terrain information would bring the third dimension in landscape structure assessment as an alternative to current two-dimensional methods developed in landscape ecology (Blaschke and Drăguţ 2003).This chapter reports on the first step of the CLUE approach, namely the classification of landform elements. Based on the methodology developed by Drăguţ and Blaschke (2006), a classification system of landform elements was improved and adapted for SRTM 3 arc second data. Although developed independently, this classification system is in compliance with the requirements of maximizing internal homogeneity and external differences (Minár and Evans 2007). Within this chapter, special emphasis is placed on the accuracy assessment of the results.

2 Methods

The workflow includes two main steps, namely terrain analysis and object-based image analysis. We carried out object-based image analysis using a software program called eCognition Professional 4.0 (Definiens Imaging, GmbH, München, Germany). For the first step, several data layers were produced from SRTM data: elevation, profile curvature, plan curvature, slope aspect, and slope gradient. As basic attributes of the land surface at a point, these derivatives 'form a coherent system for its description and analysis' (Evans 1998). Second, relatively homogenous objects were delineated at several levels through an image segmentation technique. Image segmentation is not new (see Haralick and Shapiro 1985). Several comprehensive reviews exist, e.g. Pal and Pal (1993). Until some years ago, segmentation techniques were mainly used to produce image objects, which were then either extracted or classified. In recent years, segmentation is increasingly used as a first step in image analysis and the results serve as inputs for modelling and complex object based image analysis (OBIA); see Hay & Castilla (2006).

For a general discussion on segmentation approaches applied to geographical data – as opposed to industrial image processing where similar algorithms are often used but the application contexts are very different – see Blaschke *et al.* (2004). In brief, one main obstacle with remote sensing and DTM derived raster information is its multi-scale nature. Industrial image processing has a clear level and image scale. For instance, the verification of bank notes, the recognition of human faces or fingerprints, and similar applications are clearly defined in terms of typical object sizes. For the delineation of landforms, this is less intrinsically defined. Depending on the application, one can be interested in delineating 'mountainous' areas from a world wide data set. Another application may aim for micro-relief in a terrace-dominated fluvial landscape. Some years ago, these extremes had their very different data sources. No high-resolution data sets were available over large areas. Therefore, the described micro-relief studies were only undertaken for specifically derived local data sets. Only recently have high resolution DTMs been created for large areas, e.g. through airborne laser scanning. Methodologically challenging is then the fact that one can be interested in both fine scale objects and large, overview-like or overarching forms within the same data set. This is one of the reasons why we build on a multi-scale object-based image analysis methodology as suggested by Burnett and Blaschke (2003).

In the methodology developed by Drăguţ and Blaschke (2006), object primitives were classified as landform elements using a relative classification model. The model builds both on the surface shape (Dikau 1989) and on the altitudinal position of objects. As described before, it is acknowledged that one resolution dependent parameterization may not be sufficient to describe surface form (Wood 1996), at least not application-independent. Our classification system is built on four hierarchical levels, the lowest one being designed for representation at the landscape scale (cf. Phillips 2005). The classification at this level was preset to nine classes: peaks and toe slopes (defined by the altitudinal position or the degree of dominance), steep slopes and flat/gentle slopes (defined by slope gradients), shoulders and footslopes (defined by profile curvatures), head slopes, side slopes and nose slopes (defined by plan curvatures). Classes are defined using flexible fuzzy membership functions (Drăguţ and Blaschke 2006). Fuzzy logic seems to be more suitable than crisp classification methods in the attempts to delineate terrain units that have vague spatial extents (Irvin *et al.* 1997, Lagacherie *et al.* 1997, Burrough *et al.* 2000, Fisher *et al.* 2004, 2005). This property arises 'from the continuity of the classified objects in space, i.e. the land surface forms a continuous field' (Minár and Evans 2007). While one can recognize the occurrence of a specific landform, e.g. peak, its spatial extent is not so obvious (see Smith and Mark 2003). Thus, we can imagine a model of 'core' landforms separated by 'transitional' units, which allows for a complete, exhaustive classification of a study area into a finite number of classes. Considering the preset classes as ideal landform elements, we can express the degree of affiliation of an object primitive to these classes through fuzzy membership values. Therefore, a high degree of membership will indicate that an object represents a 'core' landform, while a low degree of membership means that the object belongs to a transition area. A 'transitional' object typically has multiple membership values, carrying the degrees of membership to different classes. Fuzzy logic also allows for the integration of a wide spectrum of different characteristics, such as spectral values, form, texture (Bolongaro-Crevenna *et al.* 2005), and neighbourhood relationships.

Including slope aspect into the classification system gives rise to some problems related to the circular particularity of the data (e.g. 0 equals 360 degrees) (Evans 2006), especially when processed in an object based image analysis (OBIA, see Hay and Castilla 2006) environment. Since the segmentation algorithm of the software used (eCognition) relies on the "spectral" distance among pixel values (Baatz and Schäpe 2000), the difference between 1 and 359 degrees is a natural threshold in the process of object generation. No other derivative layer shows such a high difference between neighbour pixels so that the threshold above mentioned will prevail over all other break lines in terrain attribute data. As a result, North-facing slopes are systematically divided. Another problem is segmentation along centrelines of valleys and ridges, as adjacent slopes have opposite aspect values. To overcome these drawbacks, we reclassified the slope aspect data on eight directions and used this layer to perform a second segmentation, this time classification based. Following the first run of the classification, all classes apart from 'Peak' and 'Toeslope' were grouped and a second segmentation was performed on objects belonging to the group. Consequently, the delineation of objects based on aspect values was applied only at the level of slopes. In this way, classes 'Peak' and 'Toeslope' remain untouched in the second segmentation process. Other solutions, like cos/sin disaggregation were not used in our methodology since they would increase the number of layers without avoiding segmentation along centrelines of valleys and ridges.

It has to be emphasised–again–that slope aspect was not used in the definition of the classification system in order to create a better visualization of landform classification. Nevertheless, adding slope aspect into the segmentation leads to more realistic shapes of the emerging objects. The resulting mean object values of this parameter additionally reflect changes in slope aspect, which may be of importance in various ecological applications.

We ran the new classification system on two very different areas: first, it was applied on a SRTM dataset that covers small areas of the states of Arkansas, Missouri, and Oklahoma (Figure 1). The second test run was upon an area of approximately 50 sq km around the Hochkalter Peak, in the National Park Berchtesgaden, Germany (Figure 2). The aim was to assess the applicability of the method to two different areas in terms of terrain characteristics: a rolling landscape and a high mountain area, respectively. The accuracy assessment and classification interpretation were conducted at the smallest or 'operational' level. Maps of the results were prepared with ArcGIS version 9.2 software.

3 Accuracy Assessment

The greatest problem at the moment is the lack of quantitative methods for assessing the accuracy of DTM based classifications. As long as this is not solved, many geomorphologists and planners will be sceptical about the relevance of DTM classification results in general, including those of our segmentation based classification approach.

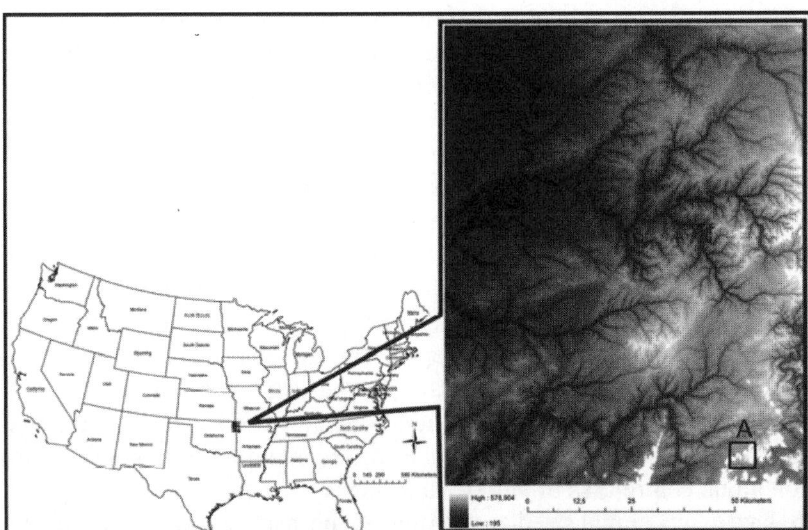

Figure 1. Location of the American study area. A indicates the subset used for visualization and interpretation of the results.

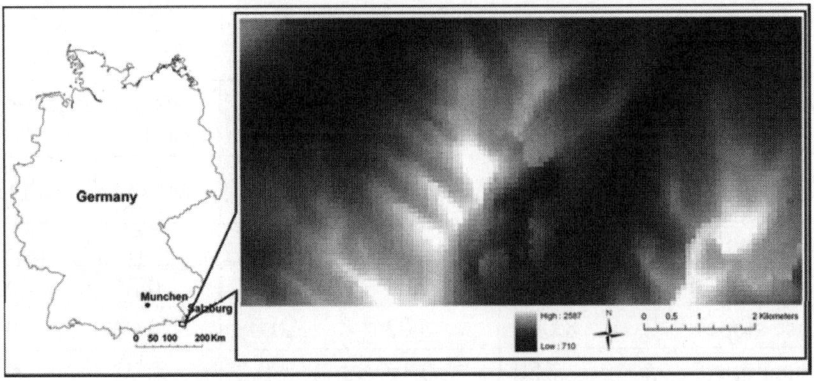

Figure 2. Location of the Hochkalter area.

Therefore, a significant amount of work was dedicated to (a) the identification of potential methods and (b) empirical work on the two case studies. The accuracy assessment was performed using fuzzy classification tools and (for the Hochkalter area only) a 'classic' error matrix. For the first approach, two of eCognition's tools, namely 'Classification stability' and 'Best classification result', were applied. Classification stability evaluates the differences in degrees of membership between the best and the second best class assignments of each object. The smaller the value of an object/class, the more ambiguous its classification is. 'Best classification result' assesses how the objects of a class fulfil the class description. Results in both study areas are very close to each other (Table 1). It is arguable whether these methods really fall under 'accuracy assessment' or should be called 'consistency checks' since they do not use independent external data sets.

These two assessments show that classification is ambiguous, particularly for the classes 'Head slope' and 'Nose slope' (Table 1). Mean and maximum values reveal a systematic tendency of objects belonging to these classes to overlap the definition of other classes. This is particularly true for confusions between these two classes and the classes 'Footslope' and 'Shoulder', respectively. This may relate to the way one variable, profile curvature, is used in both their definitions. While the classes 'Footslope' and 'Shoulder' were defined by profile curvature only (negative and positive, respectively), the classes 'Head slope' and 'Nose slope' additionally include plan curvature in their definition. To avoid arbitrary assignments of objects among these classes, we set up priority rules in the fuzzy classification system.

Table 1. Statistics of 'Classification stability' and 'Best classification result' assessment.

		'Classification stability'		'Best classification result'	
Class	Objects	Mean	StdDev	Mean	StdDev
Fayetteville					
Head slope	3,347	0.04	0.12	0.83	0.17
Nose slope	8,001	0.02	0.08	0.93	0.12
Side slope	350	0.16	0.2	0.72	0.14
Footslope	6,037	0.33	0.39	0.88	0.14
Shoulder	6,966	0.27	0.35	0.93	0.12
Flat	6,126	0.18	0.18	0.85	0.13
Peak	10,658	0.8	0.28	0.89	0.14
Toeslope	13,589	0.73	0.29	0.85	0.14
Hochkalter area					
Head slope	53	0.014	0.10	0.83	0.17
Nose slope	80	0.004	0.04	0.97	0.08
Footslope	61	0.362	0.46	0.87	0.15
Shoulder	31	0.385	0.46	0.95	0.09
Steep slope	140	0.678	0.38	0.94	0.13
Peak	35	0.664	0.45	1.00	0.00
Toeslope	46	0.614	0.35	0.84	0.17

The assessment 'Best classification result' indicates that the objects fulfil their class descriptions to high degrees (Table 1). This is particularly suggested by high mean values of the membership degree across all objects of a particular class.

As Fisher *et al.* (2005) pointed out "landscape morphometric units have a vague spatial extent, which may be modelled by fuzzy sets" (p. 209). In such a fuzzy approach, objects may have membership degrees in more than one class. Thus, modelling with fuzzy logic implies finding a trade-off between classification stability and assigning all objects a membership to (at least) one class. The results of the accuracy assessment illustrate that our classification rules are flexible enough to allow all objects to be classified, although the classification is ambiguous for some classes. Reducing the ambiguity would mean making the classification rules more specific. But in this case some objects would hardly fulfil a class definition, thus being left unclassified, which is undesirable. Rather, our goal is a complete 'wall to wall' classification without gaps or holes in the sense of a 'tesselation'.

However, these two assessment tools represent relative qualities. They do not indicate how much the classification outputs match the real landscape. Hence, thematic accuracy (Rossiter 2001) is not evaluated. For this reason, a 'classic' accuracy assessment (Congalton 1991) was applied for the Hochkalter area. The resulting error matrix scores an overall accuracy of 75.8 % (Table 2). Producer's accuracy ranges between 60 % (class 'Footslope') and 100 % (class 'Toeslope'). But user's accuracy, which allows for identifying confusions between classes (Congalton 1991, Lillesand *et al.* 2004), is more relevant. The lowest user's accuracy is assigned to the class 'Toeslope' (55.1 %), mostly due to the confusion with 'Footslope', while the class 'Steep slope' scored the highest percentage (87.8 %).

Nevertheless, these values should be viewed critically, since the application of the error matrix on DTM based classification revealed some associated shortcomings. Apart from typical and well known sources of confusion in remotely sensed data (Congalton and Green 1993), we identified two main limitations in our attempt to transfer the error matrix method to terrain data: (1) suitability of the terrain itself for collecting reference data; and (2) the subjective nature of reference data acquisition.

The first limitation is related to error propagation during DTM production. As a result, DTMs and real surfaces might be different, consequently leading to biased results compared with reference data collected on the ground. This problem is particularly valid for SRTM data in high mountain areas because of the large gaps in the original data. The third version has these gaps filled in to provide continuous elevation surfaces (CGIAR-CSI 2004) but important mismatches between interpolated surfaces and the real terrain were still identified. Figure 3 depicts such a situation. Contour lines derived from the SRTM data were draped on a 5 m DTM, which is displayed using a hillshading effect. One of the most obvious mismatches between the two surfaces is visible inside the white box on the top-right graphic. The area corresponding to the upper part of a glacial valley (depicted by the hillshaded DTM in the background) is represented on the SRTM data as a complex surface including a saddle and a nose slope (depicted by the contour lines). This limitation, together with limited accessibility in the study area, forced us to assign values to reference points through visual interpretation of aerial photographs draped on the DTM in a

3D view. 50 samples per class were randomly generated and then interpreted.

Table 2. Error matrix comparing classified landform elements to the photo interpreted reference data sites in Hochkalter area, Berchtesgaden National Park, Germany.

		Photo interpreted reference data sites						
		Head	Foot-slope	Nose	Peak	Shoulder	Steep	Toe-slope
Classified landform elements	Head	38	6	4	1	0	1	0
	Footslope	2	39	7	0	0	2	0
	Nose	2	1	41	0	0	5	0
	Peak	3	0	1	34	6	6	0
	Shoulder	0	7	0	1	41	1	0
	Steep	3	0	3	0	0	43	0
	Toeslope	5	12	4	0	0	1	27
	Total	53	65	60	36	47	59	27

User's accuracy		Producer's accuracy		
Head	= 76.00%	Head	= 71.70%	
Negative	= 78.00%	Negative	= 60.00%	
Nose	= 83.67%	Nose	= 68.33%	Overall
Peaks	= 68.00%	Peaks	= 94.44%	accuracy =
Shoulder	= 82.00%	Shoulder	= 87.23%	75.79%
Steep	= 87.76%	Steep	= 72.88%	
Toeslope	= 55.10%	Toeslope	= 100.00%	

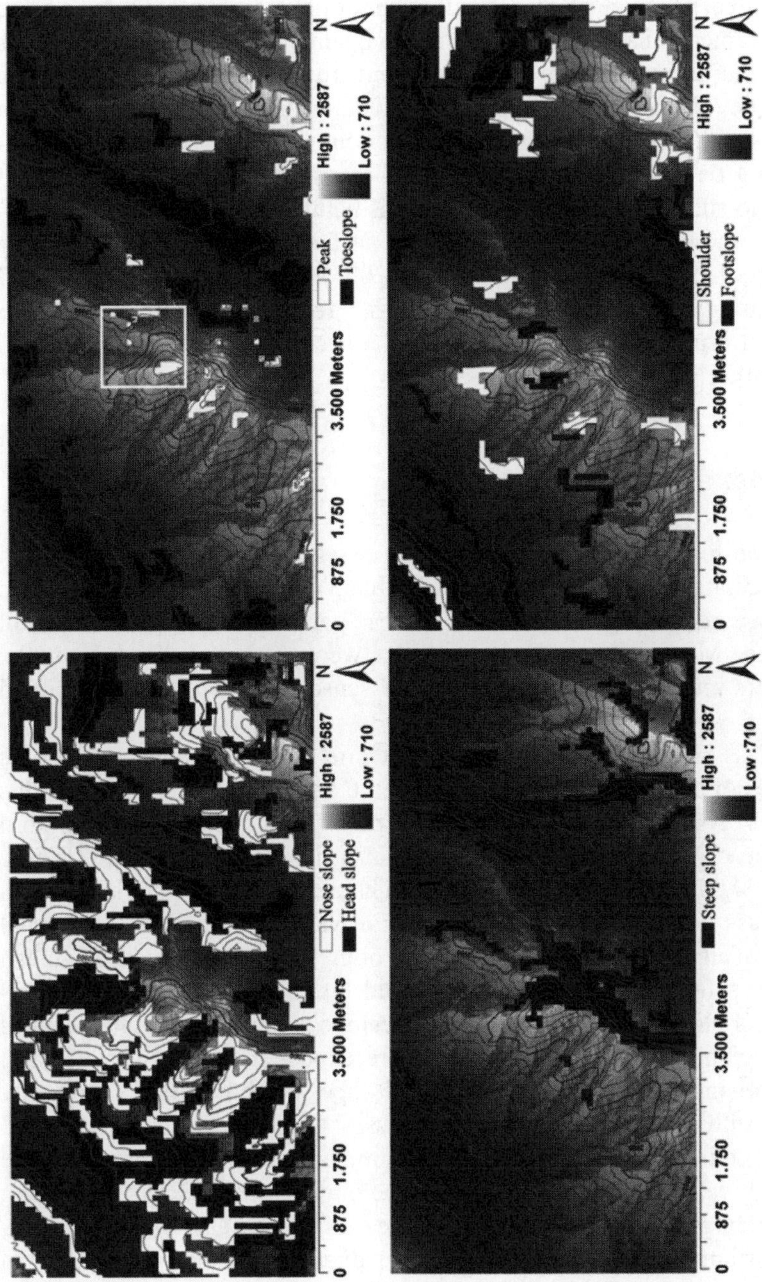

Figure 3. Classification results for Hochkalter area, National Park Berchtesgaden, Germany. The white box on the top-right graphic indicates a mismatch between the SRTM data and the real surface.

The second limitation is due to difficulties of interpreting point locations on continuous surfaces (the terrain itself or its model, e.g. the DTM). In a mathematical sense, a point basically cannot have a geomorphology since it is one-dimensional. Even the attempt to quantify individual parameters is problematic because of the lack of ground truth data for geomorphologic features beyond altitude and slope gradient. Assigning a class value to a sample is still more difficult under these circumstances. Reference data obtained in a field survey or through visual interpretation of the model itself is prone to subjective errors, as transitions between 'core' classes are typically soft and points from the continuous surfaces obviously do not belong to a single landform class (Fisher *et al.* 2005). Moreover, the accuracy of interpretation depends on the operator's 'feeling of scale'. All these sources of errors might have impacted the accuracy assessment of terrain classification in the Hochkalter area.

4 Results and Discussions

In addition to the results of the accuracy assessment, visual interpretation (Figures 3 and 4) and class statistics (Table 3) also indicate that landform elements, as defined by their morphometric parameters, are generally well delineated. Mean values of slope aspect were excluded from the table, since their interpretation does not make sense, as Evans (2006) pointed out.

The class 'Steep slope' is not represented in the American study area, as the values of slope gradient range between 0 and 26°. On the contrary, this class is very representative for the German study area, which in turn misses the class 'Flat/gentle slopes' and also the class 'Side slope'. These differences emphasize particular terrain characteristics. Mean class values fit the class definitions and mean elevation values agree with the normal position of classes in the landscape for both areas (e.g. maximum mean elevation is for the class 'Peak', followed by 'Shoulder', etc., and minimum mean elevation is for the class 'Toeslope'). While the highest values of mean plan and profile curvatures were expected for the class 'Peak', they surprisingly occur for the class 'Toeslope' (with negative signs). Associated with the confusion between classes 'Toeslopes' and 'Footslope' (Table 2), this situation leads to the preliminary conclusion that the scale parameter was too high for the objects belonging to the class 'Toeslope', thus including portions of lower slope areas. This can also explain the unusual distribution of mean slope gradient values between the two classes in the Hochkalter area (Table 3). Tackling this problem is relatively easy and does not require changing the classification system. It requires differential segmentation parameters for the class ‘Toeslope’ only, e.g. a finer segmentation followed by the reconstruction of meaningful objects, which is possible with the new version of Definiens Professional 5 (Tiede and Hoffmann 2006). Mean slope gradient is distributed according to our expectations, with highest values attached to slope categories, excepting the class ‘Flat/gentle slopes’ that has the minimum value.

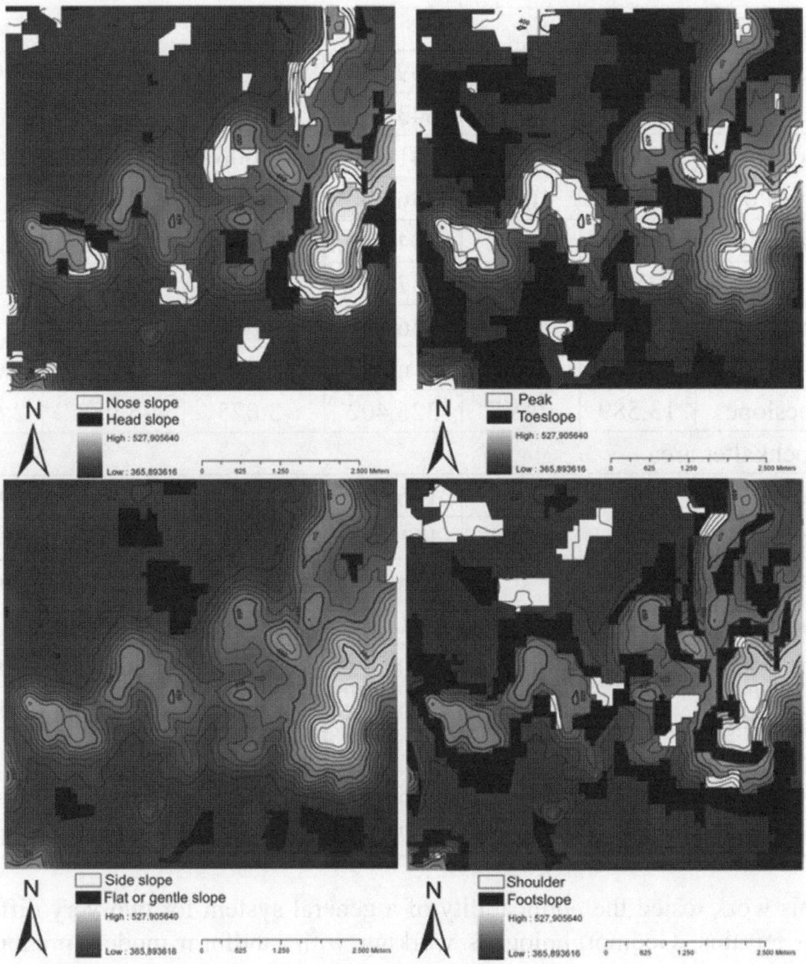

Figure 4. Classification results on a small subset within the town of Fayetteville, Arkansas, USA.

As expected, the method reaches its limits in flat regions, at least at this level of representation. In the North-Western corners of the maps in Figure 3, classification does not match the DTM very well. However, at higher levels, this area is classified as flat, since in a multi-scale approach each location has multiple feature attributes (Wood 1996).

Table 3. Class statistics.

Class	Objects	Objects (%)	Mean elevation (m)	Mean plan curvature	Mean profile curvature	Mean slope (^{0})
Fayetteville						
Head	3,347	6.08	339.570	2.759	-3.470	3.994
Nose	8,001	14.53	343.362	3.145	3.321	4.212
Side	350	0.64	341.329	-0.006	-0.003	2.289
Footslope	6,037	10.96	336.602	-0.199	-3.369	4.159
Shoulder	6,966	12.65	344.819	0.844	2.656	3.289
Flat	6,126	11.12	337.733	0.016	-0.009	1.063
Peak	10,658	19.35	346.568	3.149	4.463	2.748
Steep	0	0.00	n/a	n/a	n/a	n/a
Toeslope	13,589	24.67	326.402	-3.025	-5.083	2.655
Hochkalter area						
Head	53	11.88	1467.78	-13.37	-22.66	29.07
Nose	80	17.94	1618.44	28.03	35.18	30.36
Side	0	0	n/a	n/a	n/a	n/a
Footslope	61	13.68	1419.17	-9.61	-20.59	27.43
Shoulder	31	6.95	1700.91	7.86	20.71	30.69
Flat	0	0	n/a	n/a	n/a	n/a
Peak	35	7.85	1999.19	161.41	219.27	36.46
Steep	140	31.39	1686.23	-0.55	-18.78	51.78
Toeslope	46	10.31	1373.62	-32.36	-72.04	28.59

This work tested the applicability of a general system for two very different terrains. Geomorphologists working with landform models predominantly seek to obtain better and better approximations of physical reality. In the research described in this chapter, we are rather aiming for a method to be able to map landforms based on SRTM data (semi)automatically. The near worldwide availability of such data allows researchers to obtain geometric characteristics, including numerical descriptions of topographic forms of the Earth, at several scales. In this context, breadth of applicability could be defined spatially (e.g. how suitable a method is across different terrain characteristics) or in terms of application purposes (e.g. how suitable a method is for modelling similar phenomena). Here we discuss the first part of the definition. In this respect, extent of applicability can be increased through a system built on morphometric parameters rather than on morphogenetic description. Since values of morphometric parameters (apart of elevation) fall within expected ranges (e.g. slope gradient and aspect) or signs (e.g. curvatures), they allow for the creation of logical rules to separate surfaces with expected attributes, which potentially describe any terrain. This represents a possible way of transferring geomorphometric parameters from points (as information carried by cells of a DTM, including elevation derivatives) to objects (Schmidt and Dikau 1999). Object-based image analysis is an appropriate framework to accomplish this goal.

As Minár and Evans (2007) pointed out ‘the basic geomorphological goal of land surface segmentation should be to distinguish segments (elements) that are homogeneous genetically and therefore also morphologically’. However, as the authors explained, the link between form and processes that created it is not always straightforward. Among other reasons, this relationship is problematic because of equifinality, e.g. ‘similar landforms might arise as a result of quite different sets of processes and histories’ (Beven 2006). The problem is even more complicated through scale issues. Thus, the traditional concept of ‘landform’ is difficult to transfer fully to automatic classification approaches. While landform geometry can be classified straightforwardly using computer techniques, as in our approach, landform morphogenesis and morphology are rather in the realm of human interpretation. Therefore, the attempts at building an automatic and transferable system for landform classification should limit the concept ‘landform’ to its morphometric dimension, or restrict the approach to the recognition of specific forms, e.g. terraces (see Demoulin *et al.* 2007).

As stated in Section 3, one main remaining task is the assessment of how well the landforms match ‘reality’. Since there is not one specific measure for landform assessment, methodologies vary. The concept of landform is essentially an idealized one and every human interpreter will end up with different results. We therefore started recently to develop a methodology for this accuracy assessment. First results of the accuracy assessment are promising, particularly since the classification system is not site specific. We are currently working toward an integration of the resulting shapes of the delineated two-dimensional landform objects in the accuracy assessment process.

Acknowledgments

This research was supported by a Fulbright Senior Fellowship, and a Marie Curie Fellowship awarded to Dr. Drăguţ (**MEIF-CT-2005-009532**). We are grateful to Dr. Ian S. Evans for his suggestions, which were useful in improving this paper.

References

Baatz, M. and Schäpe, A., (2000), Multi-resolution segmentation. An optimization approach for high-quality multiscale image segmentation, *Angewandte Geographische Informationsverarbeitung XII,* Heidelberg: Wichmann Verlag: 12–23.

Beven, K., (2006), A manifesto for the equifinality thesis, *Journal of Hydrology*, **320**: 18–36.

Bishop, M., Shroder, J. and Colby, J., (2003), Remote sensing and geomorphometry for studying relief production in high mountains, *Geomorphology*, **55**: 345–361.

Blaschke, T. and Strobl, J., (2001), What's wrong with pixels? Some recent developments interfacing remote sensing and GIS, *GIS – Zeitschrift für Geoinformationssysteme*, **14**: 12–17.

Blaschke, T. and Drăguţ, L., (2003), Integration of GIS and object-based image analysis to model and visualize landscapes. In *ISPRS workshop "Challenges in Geospatial Analysis, Integration and Visualization II"*, 8–9 September, 2003, Stuttgart, Germany, pp. 18-23.

Blaschke, T., Burnett, C. and Pekkarinen, A., (2004), New contextual approaches using image segmentation for object-based classification, In de Meer, F. and. de Jong, S.: *Remote Sensing Image Analysis: Including the spatial domain*, Dordrecht: Kluver Academic Publishers: 211–236.

Bolongaro-Crevenna, A., Torres-Rodriguez, V., Sorani, V., Frame, D. and Ortiz, M.A., (2005), Geomorphometric analysis for characterizing landforms in Morelos State, Mexico, *Geomorphology*, **67**: 407–422.

Bue, B.D. and Stepinski, T.F., (2006), Automated classification of landforms on Mars, *Computers & Geosciences*, 32: 604- 614.

Burnett, C. and Blaschke, T., (2003), A multi-scale segmentation/object relationship modelling methodology for landscape analysis, *Ecological Modelling*, **168**: 233–249.

Burrough, P.A., van Gaans, P.F. and MacMillan, R.A., (2000), High resolution landform classification using fuzzy k-means, *Fuzzy Sets Syst.*, **113**: 37–52.

CGIAR-CSI (2004), *SRTM Data Processing Methodology*, Available: on-line: http://srtm.csi.cgiar.org/SRTMdataProcessingMethodology.asp.

Congalton, R.G., (1991), A review of assessing the accuracy of classification of remotely sensed data, *Remote. Sens. Environ.*, **37**: 35–46.

Congalton, R.G. and Green, K., (1993), A practical look at the sources of confusion in error matrix generation, *Photogrammetric Engineering & Remote Sensing*, **59**: 641–644.

Demoulin, A., Bovy, B., Rixhon, G. and Cornet, Y., (2007), An automated method to extract fluvial terraces from digital elevation models: The Vesdre valley, a case study in eastern Belgium, *Geomorphology*, **91**: 51–64.

Dikau, R., (1989), The application of a digital relief model to landform analysis, In Raper, J.F. (ed.): *Three dimensional applications in Geographical Information Systems,* London: Taylor and Francis: 51–77.

Drăguț, L. and Blaschke, T., (2006), Automated classification of landform elements using object-based image analysis, *Geomorphology*, **81**: 330–344.

Evans, I.S., (1998), What do terrain statistics really mean? In Lane, S., Richards, K. and Chandler, J.: *Landform Monitoring, Modelling and Analysis*, Chichester: Wiley: 119–138.

Evans, I.S., (2006), Local aspect asymmetry of mountain glaciation: a global survey of consistency of favoured directions for glacier numbers and altitudes, *Geomorphology*, **73**: 166–184.

Fisher, P., Wood, J. and Cheng, T., (2004), Where is Helvellyn? Fuzziness of multi-scale landscape morphometry, *Trans Inst Br Geogr*, **29**: 106–128.

Fisher, P., Wood, J. and Cheng, T., (2005), Fuzziness and ambiguity in multi-scale analysis of landscape morphometry, In Petry, F.E., Robinson, V.B. and Cobb, M.A.: *Fuzzy Modeling with Spatial Information for Geographic Problems,* Berlin, Heidelberg, New York: Springer: 209–232.

Goudie, A.S., (2002), Aesthetics and relevance in geomorphological outreach, *Geomorphology*, **47**: 245–249.

Haralick, R. and Shapiro, L., (1985), Survey: image segmentation techniques, *Computer Vision, Graphics, and Image Processing*, **29:** 100–132.

Hay, G.J. and Castilla, G., (2006), Object-based image analysis: strengths, weaknesses, opportunities and threats (SWOT), *International Archives of Photogrammetry, Remote Sensing and Spatial Information Sciences vol. XXXVI-4/C42*, (CD ROM).

Irvin, B.J., Ventura, S.J. and Slater, B.K., (1997), Fuzzy and isodata classification of landform elements from digital terrain data in Pleasant Valley, Wisconsin, *Geoderma*, **77**: 137–154.

Iwahashi, J. and Pike, R.J., (2007), Automated classifications of topography from DEMs by an unsupervised nested-means algorithm and a three-part geometric signature, *Geomorphology*, **86**: 409–440.

Lagacherie, P., Cazemier, D.R., van Gaans, P.F.M. and Burrough, P.A., (1997), Fuzzy k-means clustering of fields in an elementary catchment and extrapolation to a larger area, *Geoderma*, **77**: 197–216.

Lillesand, T.M., Kiefer, R.W. and Chipman, J.W., (2004), *Remote sensing and image interpretation*, John Wiley & Sons.

Minár, J. and Evans, I.S., (2007), Elementary forms for land surface segmentation: the theoretical basis of terrain analysis and geomorphological mapping, *Geomorphology*, doi: 10.1016/j.geomorph.(2007.),06.003.

Pal, R. and Pal, K., (1993), A review on image segmentation techniques, *Pattern Recognition*, **26**: 1277–1294.

Phillips, J.D., (2005), Weathering instability and landscape evolution, *Geomorphology*, **67**: 255–272.

Pike, R.J., (1988), The geometric signature - Quantifying landslide - terrain types from digital elevation models, *Mathematical Geology*, **20**: 491–511.

Pike, R.J., (1995), Geomorphometry - progress, practice, and prospect, *Zeitschrift fur Geomorphologie Supplementband*, **101**: 221–238.

Pike, R.J., (2000), Geomorphometry: diversity in quantitative surface analysis, *Prog. Phys. Geogr.*, **24**: 1–20.

Rasemann, S., Schmidt, J., Schrott, L. and Dikau, R., (2004), Geomorphometry in mountain terrain, In Bishop, M. and Shroder, J.F.: *GIS in mountain geomorphology*, Berlin: Springer: 101–146.

Rossiter, D.J., (2001), *Assessing the thematic accuracy of area-class soil maps.* [On-line]. Available: http://www.itc.nl/~rossiter/docs/AccuracySoilMaps_Preprint31Aug2001,pdf

Schmidt, J. and Dikau, R., (1999), Extracting geomorphometric attributes and objects from digital elevation models- semantics, methods, future needs. In Dikau, R. and Saurer, H.: *GIS for Earth Surface Systems: Analysis and Modelling of the Natural Environment,* Berlin, Stuttgart: Borntraeger: 153–173.

Senin, N., Ziliotti, M. and Groppetti, R., (2007), Three-dimensional surface topography segmentation through clustering, *Wear*, **262**: 395–410.

Smith, B. and Mark, D.M., (2003), Do mountain exist? Towards an ontology of landforms, *Environment and Planning B. Planning and Design,* **(30)**3: 411–427.

Tiede, D. and Hoffmann, C., (2006), Process oriented object-based algorithms for single tree detection using laser scanning data, *EARSeL-Proceedings of the Workshop on 3D Remote Sensing in Forestry*, 14th-15th Feb 2006, Vienna.

Wood, J., (1996), *The Geomorphological Characterisation of Digital Elevation Models,* Ph.D. thesis [On-line]. Available: http://www.soi.city.ac.uk/~jwo/phd/

Modelling Terrain Complexity

LU Huaxing

Abstract

Terrain complexity is an important terrain feature in digital terrain analysis; however, unlike aspect or slope, terrain complexity is an ambiguous terrain feature that until now has had no optimal index to quantify it. The traditional terrain complexity definitions can be classified as statistical, geometrical, and semantic indices. These indices evaluate terrain complexity only from one perspective of geomorphometry, and will cause more or less prejudice when modelling the real world. This chapter wants to seeks an optimal terrain complexity index (TCI) based grid DEM. Firstly, we select four traditional indices (total curvature, rugosity, local relief, local standard deviation) that can easily be evaluated by a local kernel window, then deduce the compound terrain complexity index (CTCI) using the normalization factor. In order to validate the CTCI, four study areas with typical terrain characteristics of plane, gully, hill and hybrid landforms are selected for experimentation. The results show CTCI to be a sound terrain parameter to evaluate terrain complexity. Terrain complexity is a regional feature, while CTCI is a local index, so the statistics (Mean CTCI, Maximum CTCI, and SD CTCI) are proper indicators to statistically evaluate terrain complexity.

Keywords: DEM, terrain complexity, terrain complexity index.

1 Introduction

Digital terrain data are useful for all kinds of applications in digital terrain analysis (DTA). Recently, terrain feature extraction methods have generally been based on grid DEMs because most terrain data are organized in a raster format. Terrain complexity, which describes turbulence and complexity of the terrain surface, is not only an important terrain parameter in digital terrain analysis, but also widely applied in the fields of reduction of topographic data, terrain classification and visualization, mapping and surveying, landuse, soil erosion, surface turbulence and biological richness assessment, and DEM accuracy modelling.

Terrain complexity is involved to varying extents in many studies. For example, Chou *et al.* (1999) discuss the reduction of topographic data based on terrain complexity. Gao (1998) studied the sampling intervals on the reliability of topographic variables from DEMs with terrain complexity of valleys, peaks, and ridges. Jie *et al.* (2003) studied thc accuracy of the digital elevation model in terms of topographic complexity. Parth and Mukunda (2005) analysed the relationship of the biological richness with terrain complexity in the eastern Himalayas. Cary *et al.* (2006) studied the sensitivity of areas burned to variations of landform (flat, undulating and mountainous). Fesquet *et al.* (2006) studied the impact of terrain heterogeneity on near-surface turbulence. In these studies, terrain complexity was only regarded as an influencing factor or analysis condition. Only in the research by Hsu (2002) was the indicator of the terrain complexity discussed quantitatively. Unfortunately, the indicator of terrain complexity is still a statistical model (single value).

So far, there exist many semantic ideas about terrain complexity, such as roughness, relief and rugosity. These concepts may be prejudiced and often misapplied. What the terrain complexity really is and how to describe it quantitatively is a key task in surveying and mapping, and in geomorphology fields, especially in GIS and terrain analysis based on DEM.

The objective of this study is to:

- Discuss the definition of terrain complexity from the viewpoint of ontology, and
- Model a quantitative index of terrain complexity based on grid DEM.

2 Terrain Complexity: Definition and Indices

Terrain complexity has a widely semantic implication and is often used to describe the variability of the terrain surface. Up to now, the existing terrain indices could be classified into three types: statistical sense, geometrical sense, and semantic sense.

In the statistical sense, the terrain complexity is described by the statistics of one or more terrain parameters such as elevation, slope, aspect, and curvature. For example, variance or standard deviation of elevations, terrain relief, and contour density within a unit area are used to describe the complexity of valleys, peaks, and ridge (Gao 1998).The commonly used statistical indices of terrain complexity are:

- Variance or standard deviation of elevations (Hsu 2002, Zhang *et al.* 1999);
- Autocorrelation of elevation (Li and Zhu 2003);
- Relief (Gao 1998): arithmetic disparity between the maximum and minimum elevations: H^{max}- H_{min};
- Contour density (Byers 1992): the ratio of contour line length to planar area: Lc/Ap, where Lc is total length of contour and Ap is the projected area within a unit area;
- Mean slope or mean aspect (Tang 2000); and
- Drainage density (Tucker and Bras 1998).

Many studies consider the terrain surface in the real world as a constant surface. Hu *et al.* (2004) thought of the Earth surface as being a complex sphere, but still a constant surface, not a stochastic field. Terrain surfaces simulated by specific mathematical surfaces are also involved in many studies; Liu (2002) assessed the accuracy of different slope and aspect algorithms based on the Gauss synthesized curve, inversed ellipsoid, elliptic dome, saddle surface, and Tsai (1994) and Yanalak (2003) used trigonometric function surfaces to assess the accuracy of interpolation algorithms. Holmgern (1994) thought the land surface should be simulated by complicated mathematical surfaces rather than simple ones when evaluating flow algorithms. So terrain complexity in the geometrical sense is to depict the shape of the landform surface or cross section, such as:

- Rugosity (Hobson 1967, 1972): the ratio of surface area to planar area: As/Ap, here As and Ap are terrain surface and planer area, respectively;
- Shape complexity index (Hengl *et al.* 2003): which is defined as how oval/dissected is a contour line: P/sqrt(A/3.1415)) where P is the perimeter of contour line, A is the area of the contour line circle;
- Curvatures (Shary *et al.* 2002);
- Fractal dimension: fractal box dimension (Zhou and Long 2006);
- Slope or slope change (Jie *et al.* 2003, Tang 2000); and
- Aspect or aspect change (Shary 2006, Tang 2000).

Terrain complexity can be described as “The terrain is very complex”, “How rough the landform surface is!”, “This is a mountainous area, not a gully area”, and so on. These are qualitative sentences and often reflect the image in the human mind; we call these the semantic sense, and this sense is the first-hand criteria to evaluate terrain surfaces.

According to the indices mentioned above, terrain complexity is basically depicted as the topographic turbulence and roughness. For example, curvature is the derivative of specific points on the mathematical curve, used to depict how fast the angle of a tangent that moves over a given arc to the length of the arc. In fact, aspect or aspect change, and the shape complexity index mentioned above, are equivalent to the plan curvature, and slope and slope change are equivalent to the profile curvature, the only difference between these geometrical indices is direction, while rugosity is a special type of roughness measure describing how wrinkled a surface is.

Homoplastically, variance, relief, standard deviation or autocorrelation of elevation represent the scope or skewness of elevations, while mean slope, mean aspect, contour or drainage density are trends to give a single value for the whole map, which is relatively difficult to evaluate for a hybrid landform that has spatially-distributed terrain complexity.

The terrain complexity index (TCI) based on curvature has been proposed. The slope change (Jie *et al.* 2003), the aspect change, the slope of slope (SOS) and the slope of aspect (SOA) (Tang 2000), contour curvature, and so on are indices only describing the curvature from the profile direction or plan direction, which cannot evaluate the anisotropy terrain. Shary (2006) proved that mean curvature, maximal curvature, minimal curvature and total Gaussian curvature are irrelevant to coordinate systems, meaning that values of these curvatures keep constant no matter how the x axis or y axis rotate. But these curvatures may generate negative or positive values, depending on whether the terrain is concave or convex, so they are not suitable for evaluating terrain complexity.

Fractal dimension is not a traditional spatial morphological index; it lacks theoretical principle, cannot be validated, and it can only give a single value for a specific area. Terrain simulations based on fractals find it difficult to depict the terrain geomorphological structure and features (Hsu 2002). What's more, Mendicino and Sole (1997) utilized the entropy theory for the estimation of topographic index; information content here can be considered as a terrain complexity index.

The question is whether terrain complexity exists; what is the essence of terrain complexity? The sense of terrain complexity may vary with different principles: geology may be interested in the change of geological structure, hydrography may be interested in the drainage density, ecology may require the slope or aspect change. It may be that for many specific purposes, a different number is likely to be required. Evans (1990) thought one number could not suffice to summarize surface roughness; he considered the terrain profile as a wave, and tried to depict the wave from five perspectives: amplitude, wavelength (Figure 1.a), massiveness, homogeneity, and asymmetry (Figure 1.b). He also considered absolute altitude above sea level as important in geomorphology, because, in a constant amplitude and wavelength, a greater proportion of the profile can be concentrated either at high or at low heights.

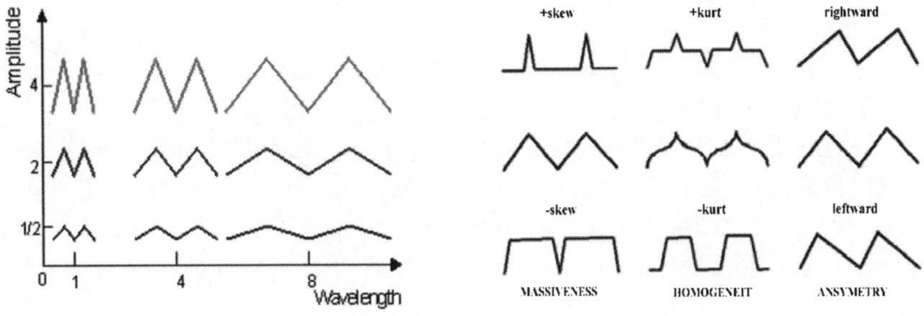

(a) (b)

Figure 1. Variables of terrain profile (Evans1990). (a) Amplitude and wavelength. (b) Massiveness, homogeneity and asymmetry.

Shary (2006, personal communication) thought that experts in a given discipline try to describe terrain complexity for their own purposes. One example is the sum of gully lengths per square kilometre to evaluate or map water erosion (note that they should have a definition of a gully to automate the approach in digital terrain analysis). So there may be numerous approaches to the same problem of deciding what terrain complexity is. A researcher already working in a given discipline (e.g. in water erosion) is free to introduce measures or indices that are specific for that discipline.

But there might be a general study of the complexity that is not related to any specific discipline. Such a study may concentrate on general properties of complexity, and try to sub-divide it into a few components that may or may not be present in specific approaches. In other words, such a study might introduce concepts that appear mutual for all specific approaches, and may therefore serve as a basis for many other approaches. The relationship between general and specific approaches is common for knowledge differentiation; compare this sub-division of geomorphometry with general and specific (Evans 1972): "General geomorphometry as a whole provides a basis for the quantitative comparison even of qualitatively different landscapes, and it can adapt methods of surface analysis used outside geomorphology. Specific geomorphometry is more limited; it involves more arbitrary decisions, and leaves more room for subjectivity in the quantification of its concepts".

Shary thought one general study of terrain complexity might be based on profile (Figure 2a, Figure 2b) and contour line (Figure 2c, Figure 2d) on land surfaces, representing horizontal and vertical dimensions, respectively (Figure 2), and these two dimensions are independent of each other.

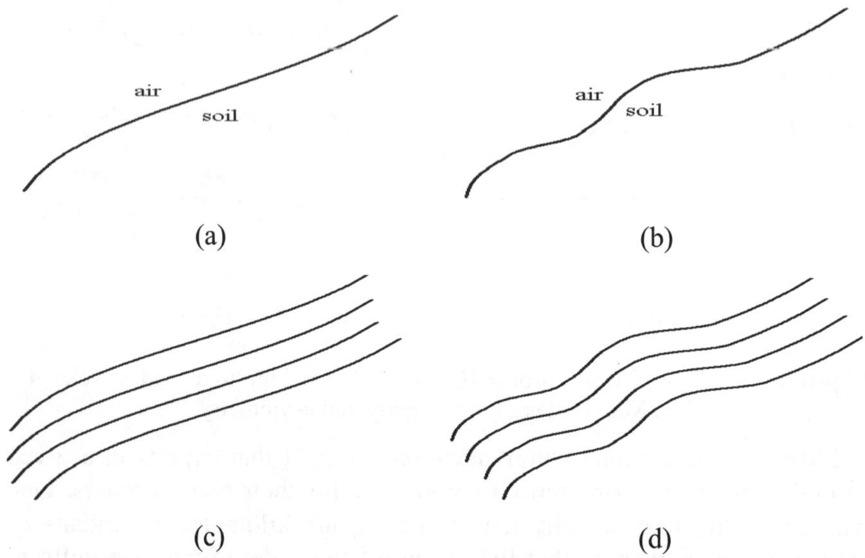

Figure 2. Terrain complexity in two dimentions (Shary 2006, personal communication). (a) Simple profile line. (b) Complex profile line. (c) Simple contour lines. (d) Complex contour lines.

Terrain complexity evaluated only from a profile (Evans 1990) and contour line (Shary 2006) still strongly depends on direction, which is somewhat difficult to evaluate in anisotropy terrain, so we need to combine all indices listed above as a compound terrain complexity index (CTCI). But how to select it and what are the criteria? Firstly, according to the Shary's viewpoint, the candidate indices must be independent of any other, that is one is irrelevant to all others. Secondly, the candidate indices must be very easy to calculate when quantifying the local terrain complexity.

3 Modelling Terrain Complexity

Although the indices listed above can roughly represent terrain complexity, they are somewhat unreasonable since they only describe one aspect of the landform shape and always fail to estimate terrain complexity; for the real world is difficult to depict by a simple mathematical formula. For example, suppose there are two kinds of landform (Figure 3), one represents a plain slope (Figure 3a) and the other represents a pleated slope (Figure 3b). They may generate the same value according to the formulae above, but actually they are of totally different shapes (Zhou and Long 2006).

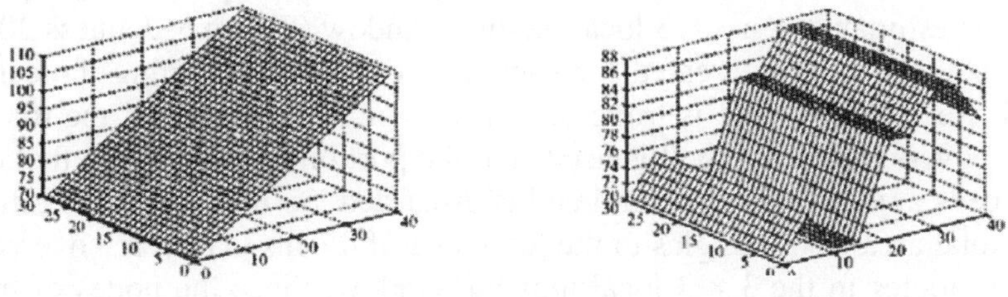

(a) (b)

Figure 3. Different terrain complexity with the same index. (a) Plane slope. (b) Pleated slope

Since different indices depict terrain complexity from different aspects, their synthesis may be the optimal solution. We investigate the candidate indices that were easily evaluated by a local moving window (Figure 4), then we find the total curvature (Wilson *et al.* 2000), rugosity, local relief, and local standard deviation that coincide with the criterion. Because total curvature will also generate a negative value, it needs to change to be a constant positive index.

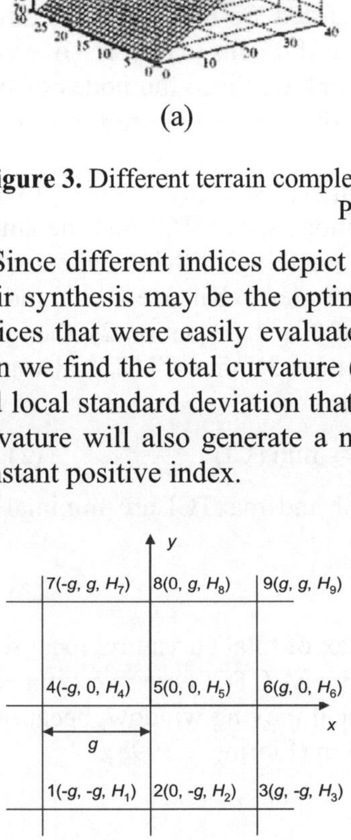

Figure 4. 3 × 3 local moving window.

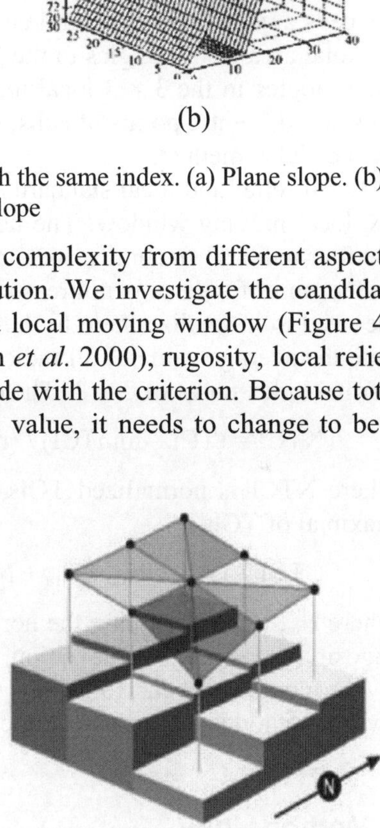

Figure 5. Rugosity estimation.

The positive form of total curvature is expressed as:

$$c = \left(r^2 + 2s^2 + t^2\right)^{1/2} \tag{1}$$

where:

$$\frac{\partial^2 f}{\partial x^2} = f_{xx} = r\,; \quad \frac{\partial^2 f}{\partial x \partial y} = f_{xy} = s\,; \quad \frac{\partial^2 f}{\partial y^2} = f_{yy} = t\,.$$

Rugosity is the ratio of surface area to planar area: As/Ap, where As and Ap are the areas of terrain surface and plan cells, respectively. Rugosity can be estimated in a 3 × 3 local moving window (Figure 5, Jenness 2002). A triangular area is created for each of the cells surrounding the central cell. Next, the triangle is truncated so that only the portion of the triangle that covers the central cell is used, and the portion of the triangle that covers cells adjacent to the central cell is discarded. So rugosity is the ratio of the total area of 8 triangles to the total area of 4 cells that were covered by the triangles in the 3 × 3 local moving window. Since the nodes of triangles are the centre points of cells, the area of each triangle can be calculate by the Helen method.

Local relief and local standard deviation can be easily calculated within the local moving window. The next question is: Is CTCI just the sum of the four indices? Generally, local relief is much higher than total curvature, so the sum of indices is not reasonable. In order to balance all the indices, we need to normalize them to the range of 0 to 1 (Equation 2), then compute the average value of all indices (Equation 3); thus CTCI is also in the normal index range of 0 to 1. The equations are:

$$\mathrm{NTCI} = (\mathrm{TCI} - \mathrm{minTCI}) \,/\, (\mathrm{maxTCI} - \mathrm{minTCI}) \tag{2}$$

where NTCI is normalized TCIs, minTCI and maxTCI are minimal and maximal of TCIs.

$$\mathrm{CTCI} = (\,\mathrm{N_{tc}} + \mathrm{N_{re}} + \mathrm{N_{ru}} + \mathrm{N_{sd}}\,) \,/\, 4 \tag{3}$$

where N_{tc}, N_{re}, N_{ru}, N_{sd} are the normal index of total curvature, local relief, rugosity, local standard deviation, respectively. CTCI here is estimated by Evans methods (Evans 1980) in a 3 × 3 local moving window, because the Evans method is the most accurate algorithm (Florinsky 1998).

4 Methodology

CTCI weights four indices that depict terrain complexity from different perspectives; it is a positive variable. Utilizing modular GIS analysis, we can easily derive the CTCI.

4.1 Data

The study area is one of the most fluctuating and fractured landforms with elevations ranging from 233 metres to 3546 metres, located in the Shaanxi Loess Plateau, North-West China (Figure 6). Three point dataset DEMs (randomly distributed elevation datasets, RDED) that were digitized from topographic maps at a scale of 1:50,000 were selected as the test area. The test data are typical landforms: Shenmu (gully landform, Figure 6a) with elevations ranging from 1,232.0 metres to 1,740.8 metres, Xi'an (plain landform, Figure 6b) with elevations ranging from 378.9 to 559.6 metres, and Qinling (hill landform, Figure 6c) with elevations ranging from 1,042.2 metres to 2,652.7 metres.

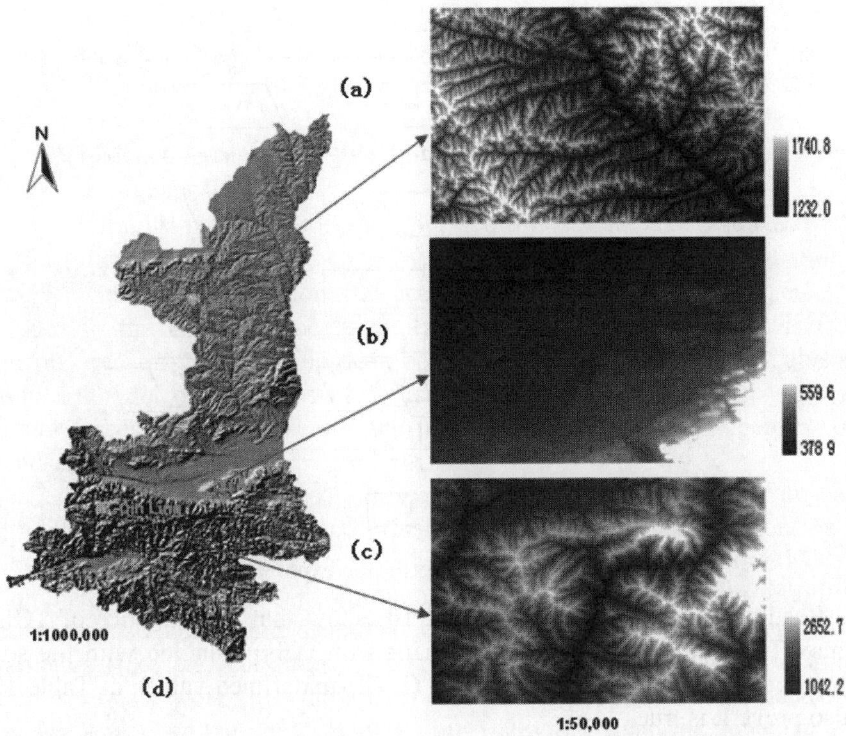

Figure 6. The three typical experiment areas in the Loess Plateau in Shaanxi, China. (a) Gully area; (b) Plain area; (c) Hill area; (d) Background map, Shannxi.

The reason that these test areas were selected followed semantic sense or common sense: the plain area represents low TCI, the hill area represents high TCI, and the gully area is in the middle.

4.2 Extract CTCI and statistics

Utilizing the tools of ESRI ArcGIS, we can convert RDED DEM to TIN (triangular irregular network), then derive grid DEMs with 25 m grid cell size utilizing the TIN to Raster tool. Then we can derive the different TCIs (total curvature, local relief, local standard deviation, rugosity) for each grid DEM, and calculate the normal TCIs (N_{tc}, N_{re}, N_{ru}, N_{sd}). Lastly, wc obtain the statistics of each TCI, as well as the CTCI, and analyse the statistics of TCIs (Figure 7). Considering the boundary influence, we need to remove the boundary cells from the result data when executing the kernel window analysis.

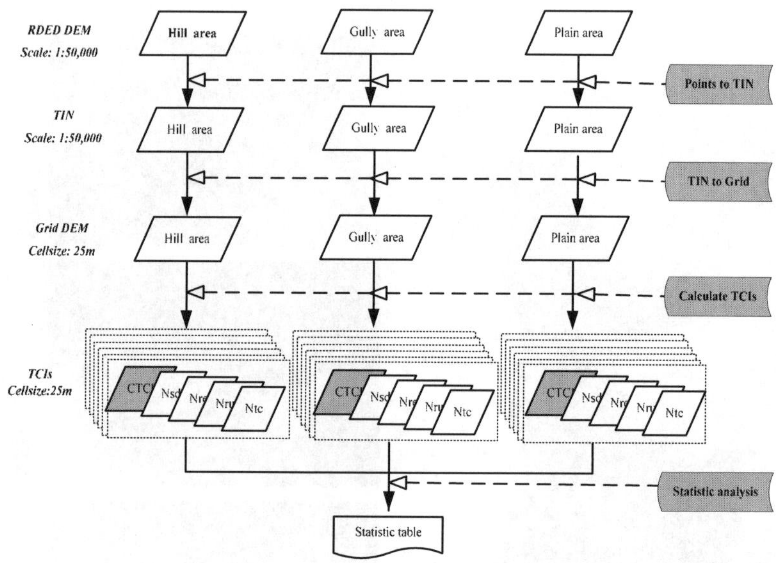

Figure 7. Flow chart for the experiment.

Results (Table 1, Figure 8, 9 and 10) show that the statistics of TCIs (max TCI, mean TCI, standard deviation of TCIs) coincide with the semantic sense or common sense. The CTCI (underlined values in Table 1) also prove it is true.

5 Model Validation

CTCI is a synthetical index modelled by the four different indices; its feasibility needs to be validated. The method for validation is based on the semantic sense: comparing the CTCI with the hillshade of the original

DEM, because we can visually judge where it is complex and where it is simple in the hillshade map. If the CTCI is relatively high in mountainous areas compared with that in plain areas, it means that the CTCI is a sound index to quantify terrain complexity. Although this visualization comparison is not so rigid, it coincides with the definition from the semantic sense.

Table 1. Statistics of TCIs (cell size: 25 m).

	Maximum	*Minimum*	*Mean*	*SD*
Xi'an				
Rugosity	1.083949	1	1.005267	0.005017
Total Curvature	0.024886	0	0.000181	0.000417
Local SD	7.908564	0	0.187605	0.320762
Local relief	21.09998	0	0.564413	0.986082
CTCI	0.113561	0	0.002409	0.004194
Qinling				
Rugosity	2.503333	1	1.232323	0.134085
Total Curvature	0.117671	0	0.013202	0.008533
Local SD	45.64762	0	13.75478	4.899303
Local relief	146.1	0	42.81402	15.42477
CTCI	0.788611	0	0.216059	0.078247
Shenmu				
Rugosity	1.948476	1	1.124266	0.082245
Total curvature	0.06425	0	0.011144	0.0072
Local SD	32.26643	0	9.590529	3.812486
Local relief	100.5	0	29.39271	11.63341
CTCI	0.588669	0	0.147977	0.056732

An equivalent method for validating CTCI is by comparing the contours and drainage density of the topographic map with the CTCI, because again it is common sense: where there is high CTCI, there must be high contour density and drainage density.

Unlike mean density, local contour density and local drainage density are somewhat difficult to quantify. The spatial analysis module in ArcGIS provides a tool to calculate the contour density. The principle is similar to the calculation of point density. The contour density is strongly dependent on the size of the kernel window and data quality, namely noise (see discussion). The drainage density extraction is involved in a series of procedures, namely pit removing and parameter input, which will introduce prejudice to the model validation. So here we validate the CTCI by comparing the CTCI and the original contour line attached hillshade map as background.

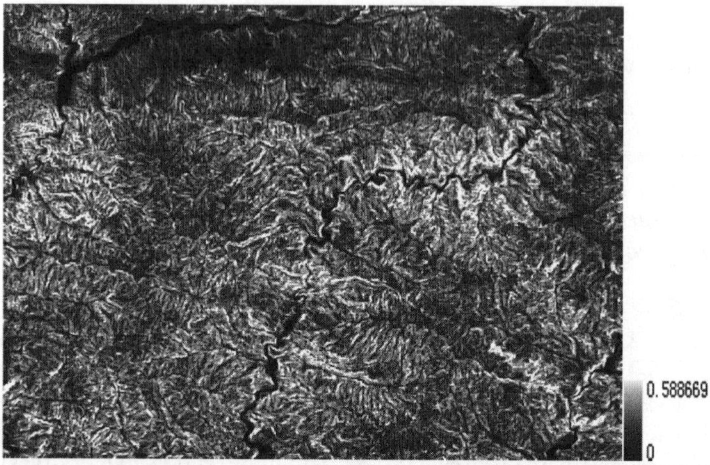

Figure 8. CTCI of Qinling area (cell size = 25 m, rows = 761, cols = 943).

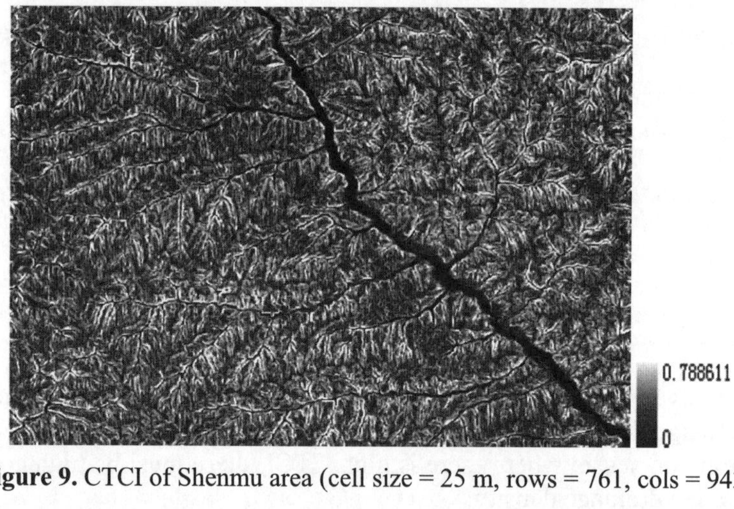

Figure 9. CTCI of Shenmu area (cell size = 25 m, rows = 761, cols = 943).

The test area is located in ShanXi, China, a typical hybrid landform (Figure 11a). The data is produced by the Provincial Bureau of Surveying and Mapping; the data format is a grid DEM that was derived from topographic maps at a scale of 1:10,000; the cell size is 5 metres.

We extract the contour line base on the grid DEM, with the interval of 10 metres in elevation. CTCI is extracted according to the flow chart

(Figure 7). We classify the CTCI into 3 classes (Figure 11b, legend): red areas represent high terrain complexity and transparent (grey colour) represents low, and the blue area is in the middle. Lastly, we overlay contour lines and the CTCI on the hillshade map (Figure 11b).

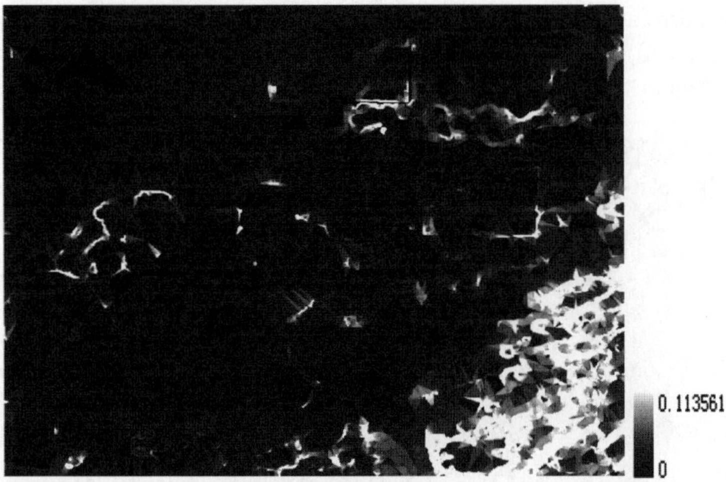

Figure 10. CTCI of Xi'an area (cell size = 25 m, rows = 765, cols = 913).

Judging from the three layers (Figure 11b), the CTCI can roughly differentiate the contour lines where there is cliff and dense contours, and where they are smooth and sparse, just as beforehand, where the terrain is complex and where it is simple (Figure 2). That also proves that the CTCI covers the function of contour density in evaluating terrain complexity; that is, the index of contour density is not independent of the CTCI.

The quantitative validation may investigate the correlation between the CTCI and contour (or drainage) density utilizing the regression model.

6 Discussion

6.1 CTCI and scale effect

Terrain complexity is an inherent attribute of terrain surface; like other terrain variables, it will change with different scales, which was known as scale effect. But relatively little literature involves the relationship between DEM scales and TCIs. Different DEM variables have different effects, and scale effects based on a grid DEM also have two types: the first is analysis scale, also the size of the local moving window, as it will definitely influence variables such as local relief, local standard deviation, and the choice of analysis scale depends on variables and applications. The second is DEM resolution. We can generate DEMs with different cell sizes utilizing the GIS tools. Lu *et al.* (2007) found the statistics of total curvature have strong correlation (negative correlation by a power function, $y = axm$, where $m < 0$) to the DEM resolution; the correlation coefficient is commonly higher than 0.96.

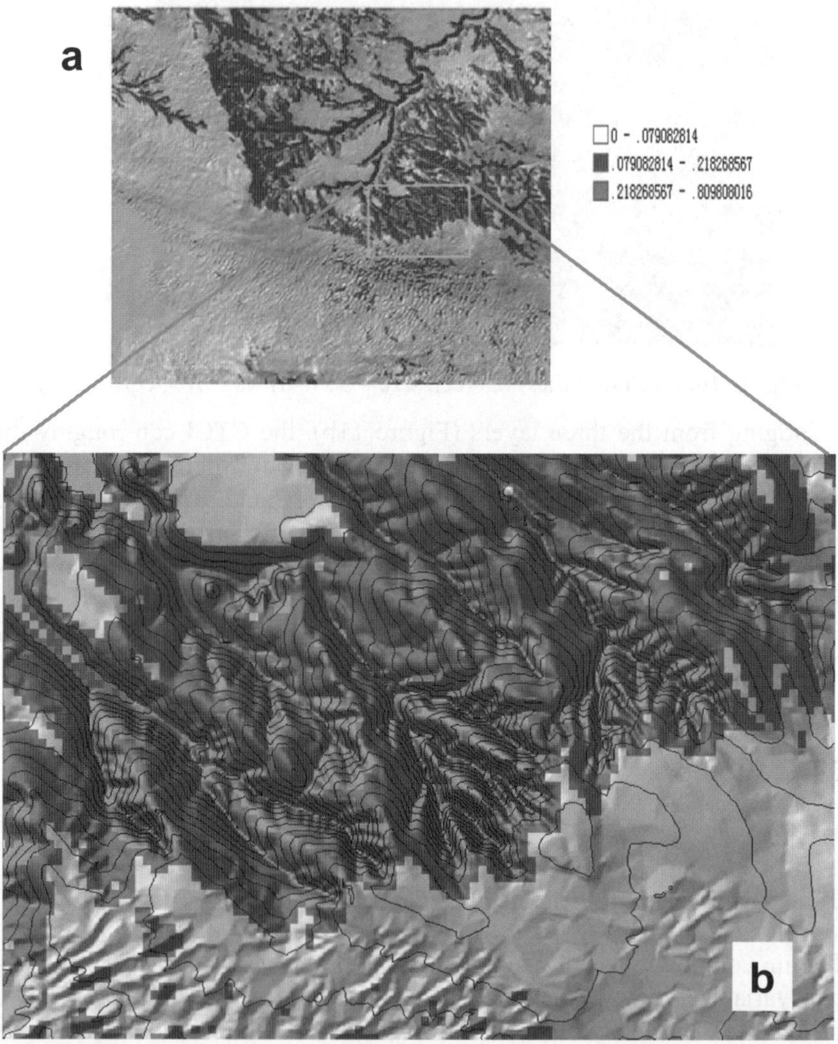

Figure 11. CTCI and contour line (1:10,000). (a) Test area; (b) CTCI and contour line (local view).

6.2 CTCI and DEM local accuracy

DEM accuracy has a strong correlation with local morphometric variables. Fisher (1998) found significant correlation between DEM errors and slope angle. Gao (1997) observed that DEM errors seem lower in less complex terrain. Hunter and Goodchild (1997) stated that DEM error is probably related to slope steepness. Ehlschlaeger and Shortridge (1997) reported that empirical studies have shown DEM error to be related to gradient and propose that it may also be related to other elevation derivatives. Kyriakidis *et al.* (1999) found that DEM error is correlated with terrain ruggedness, with the correlation coefficient being up to 0.64. Guth (1992) found DEM error to be highly correlated with gradient, aspect and satellite image reflectance values. Tang (2000) proposed a representation error and deduced the empirical linear formulae for different landforms.

According to the above-mentioned literature, higher DEM errors always occur in the more complex landforms, so terrain complexity can be assumed as some combination of terrain variables that have strong correlation with DEM error. Carlisle (2005) modelled spatial-distributed DEM error using multi-variable regression, finding an empirical formula that had strong correlation with DEM error. For this application, CTCI will serve as a useful index in modelling local DEM accuracy. But DEM accuracy may not rely on terrain complexity for the pixels nearby the original points that are used for interpolating the grid DEM, so this factor must be taken into account when modelling DEM accuracy.

As to applications, CTCI can be applied to mapping of terrain classifications and can also be used as a criterion in data generalization to indicate where the redundant data should be removed and where it should be retained.

7 Conclusion

Unlike slope and aspect, terrain complexity is ambiguous in itself, at least in the sense that it can be considered in the vertical and horizontal directions: local relief and local standard deviation are characteristics of vertical complexity, while aspect change and contour shape index are of horizontal ones. These characteristics are independent; that is, total curvature may be large even if local relief is small. So their synthesis is a reasonable method.

Terrain complexity is a regional feature, although we calculate with a local window and model it spatially, so the statistics (max, min, standard deviation) should be taken into account. The local variables are always too sensitive to DEM noise and scale effects. The statistics of the CTCI are the reasonable solution for evaluating terrain complexity for the whole map. Different indices have different value ranges (rugosity is always higher than total curvature) but the solution to this problem is to weight each of the indices by a normalized transform.

Under the same DEM resolution, plain areas have a very low Max CTCI, Mean CTCI and SD of CTCI compared with hilly areas, and the gully areas are in the middle. Utilizing visualization, we can validate that the CTCI is a sound model, but further experimentation is needed to test whether it is universal to any landform, and the issues for ontology of terrain complexity still need further research.

Acknowledgement

We particularly want to acknowledge Peter A Shary, Tang Guo'an, Qiming Zhou, Zhilin Li, John Gallant and Ventura for their guidance and suggestions. This work is financially supported by the National Natural Science Foundation of China (No. 40571120).

References

Byers, J.A., (1992), Grid Cell Contour Mapping of Point Densities: Bark Beetle Attacks, Fallen Pine Shoots, and Infested Trees, *Oikos,* **63**: 233–243.

Carlisle, B.H., (2005), Modelling the Spatial Distribution of DEM Error, *Transactions in GIS,* **(9)**4: 521–540.

Cary, G.J., Keane, R.E., Gardner, R.H., Lavorel, S., Flannigan, M.D., Davies, I.D., Li, C., Lenihan, J.M., Rupp, T.S. and Mouillot F., (2006), Comparison of the Sensitivity of Landscape-fire-succession Models to Variation in Terrain, Fuel pattern, Climate and Weather, *Landscape ecology*, **(21)**1: 121–137.

Chou, Y., Liu, P. and Dezzani, R.J., (1999), Terrain Complexity and Reduction of Topographic Data, *Journal of Geographical Systems*, **(1)**2: 179–198.

Ehlschlaeger, C.R. and Shortridge, A., (1997), Modelling Elevation Uncertainty in Geographical Analyses, In Kraak, M.J. and Molenaar, M., (eds): *Advances in GIS Research: Proceedings of the Seventh International Symposium on Spatial Data Handling*, London: Taylor and Francis: 585–95.

Evans, I.S., (1972), General Geomorphometry, Derivatives of Altitude and Descriptive Statistics, In Chorley, R.J. (ed.): *Spatial Analysis in Geomorphology*, London: Methuen and Co.: 17–90.

Evans, I.S., (1980), An Integrated System of Terrain Analysis and Slope Mapping, *Zeitschrift Fiir Geomorphologie. Supplement Band*, **36**: 274–95.

Evans, I.S., (1990), General geomorphometry, In Goudie, A., (ed.): *Geomophological techniques*, London: Allen and Unwin.

Fesquet, C., Barthlott, C., Drobinski, P., Dubos, T., Pietras, C. and Haeffelin, M., (2006), *Impact of Terrain Heterogeneity on Near-surface Turbulence: Long Term Investigation at Sirta Observation,* URL: http://ams.confex.com/ams/BLTAgFBioA/techprogram/paper_111039.

Fisher, P., (1998), Improve Modelling of Elevation Error with Geostatistics, *Geoinformatica*, **(2)**3: 215–233.

Florinsky, I.V., (1998), Accuracy of Local Topographic Variables Derived From Digital Elevation Models, *International Journal of Geographical Information Science*, **12**: 47–61.

Gao, J., (1997), Resolution and Accuracy of Terrain Representation by Grid DEMs at a Micro-scale, *International Journal of Geographic Information Science*, **11**: 199–212.

Gao, J., (1998), Impact of Sampling Intervals on the Reliability of Topographic Variables Mapped from Grid DEMs at a Micro-scale, Geographical Information Science, **12**(8): 875–890.

Guth, P., (1992), Spatial analysis of DEM error, *Proceedings of the ASPRS/ACSM Annual Meeting, Washington D.C.*: 187–96.

Hengl, T., Gruber, S. and Shrestha, D.P., (2003), *Digital Terrain Analysis in ILWIS*, Netherlands: ITC.

Hobson, R.D., (1967), FORTRAN IV programs to determine the surface roughness in topography for the CDC 3400 computer, *Computer Contribution State Geol. Survey Kansas*, **14**: 1–28.

Hobson, R.D., (1972), Surface roughness in topography: quantitative approach, In Chorley, R.J., (ed.): *Spatial analysis in geomorphology*: 225–245.

Holmgren, P., (1994), Multiple flow direction algorithm for runoff modelling in grid based elevation models: an empirical evaluation, *Hydrology Processes*, **8**: 327–334.

Hsu Chiu-Ling, (2002), *An Indicator Research of the Terrain Complexity - A Classification of Gully Scale Based on DEM,* Taiwan: Taiwan University, Master Dissertation.

Hu Peng, Wu Yanlan and Hu Hai, (2005), A New Research on Fundamental Theory of Assessing the Accuracy of DEMs, *Geo-Information Science*, **7**: 28–33 (in Chinese).

Hunter, G.J. and Goodchild, M.F., (1997), Modelling the Uncertainty of Slope and Aspect Estimates Derived From Spatial Database, *Geographical Analysis*, **29**: 35–49.

Jenness, J. (2002), *Surface Areas and Ratios from Elevation Grid (surfgrids.avx) Extension for ArcView3.x, v.1.2,* Jenness Enterprises, URL: http://www.jennessent.com/arcview/surface_areas.htm .

Jie Shan, Muhammad Zaheer and Ejaz Hussain, (2003), Study on Accuracy of 1-Degree DEM Versus Topographic Complexity Using GIS Zonal Analysis, *Journal of Surveying Engineering*, **(129)**2: 85–89.

Kyriakidis, P.C., Shortridge, A.M. and Goodchild, M.F., (1999), Geostatistics for Conflation and Accuracy Assessment of Digital Elevation Models, *International Journal of Geographical Information Science*, **13**: 677–707.

Li, Z. and Zhu, Q., (2003), *Digital Elevation Model*, Beijing: Science Press (in Chinese).

Liu Xuejun, (2002), On the Accuracy of the Algorithms for Interpreting Grid-based Digital Terrain Model, Ph.D. Dissertation (in Chinese).

Lu Huaxing, Liu Xuejun and Bian Lu, (2007), Terrain Complexity: Definition, Index and DEM Resolution, Geoinformatics 2007: Geospatial Information Science, Jingming Chen and Yingxia Pu (eds.), *Proc. of SPIE*, Vol. 6753 675323-1.

Mendicino, G. and Sole, A., (1997), The Information Content Theory for the Estimation of the Topographic Index Distribution Used in Topmodel, *Hydrological Processes*, **11**: 1099–1114.

Parth Sarathi Roy and Mukunda Dev Behera Arunachal Pradesh, (2005), Assessment of biological richness in different altitudinal zones in the Eastern Himalayas, *Current Science*, **(88)**2: 250–257.

Shary, P.A., Sharaya, L.S. and Mitusov, A.V., (2002), Fundamental Quantitative Methods of Land Surface Analysis, *Geoderma*, **107**: 1–32.

Shary, P.A., (2006), Predictable and terrain –Specific Landform Classification. *Proceeding for International Symposium on Terrain Analysis and Digital Terrain Modelling*, CD

Tang Guo'an, (2000), *A Research on the Accuracy of DEMs*, Beijing, New York: Science Press.

Tsai Tsung-Hsiun, (1994), *A Morphometric Study of Digital Elevation Models*, Taiwan University, Masters Dissertation.

Tucker, G.E. and Bras, R.L., (1998), Hillslope Processes, Drainage Density, and Landscape Morphology, *Water Resources Research*, (**34**)10: 2751–2764.

Wilson, J.P., Repetto, P.L. and Snyder, R.D., (2000), Effect of Data Source, Grid Resolution, and Flow-routing Method on Computed Topographic Attributes, In Wilson, J.P. and Gallant, J.C. (eds): *Terrain Analysis: Principles and Applications*, New York: John Wiley & Sons: 133–161.

Yanalak, M., (2003), Effect of Gridding Method on Digital Terrain Model Profile Data Based on Scattered Data, *Journal of Computing in Civil Engineering*, **(17)**1: 58–68.

Zhang Chao, Chen Bingxian and Wu Lun, (1999), *Geographic Information Systems*, Beijing: High Education Press (in Chinese).

Zhou Qiming and Liu Xuejun, (2006), *Digital Terrain Analysis*, Beijing: Science Press (in Chinese).

Zhou Tong and Long Yi, (2006), A Fractal Method to Describe the Terrain Complexity Reflected by the Raster DEM, *Geography and Geo-Information Science*, **(22)**1: 26–30 (in Chinese).

DEM-based Analysis of Local Relief

LIU Aili

Abstract

Local relief (LR) mainly describes the complexity of the Earth's surface at large spatial scales and reflects the degree to which the Earth's surface is incised by exogenic agents. This chapter probes the methodology of extracting LR from a 1-km DEM and discusses the large-scale geomorphological characteristics of China. It proves that the optimum unit for measuring LR is $42.25km^2$. Analyses of the LR map show that most areas in China have mountainous relief (200 < LR < 600m), and the higher the LR, the lower the proportion of the national territory. From space, LR shows obvious differences from East to West and from NE to SW. Moreover, the spatial distribution of LR exhibits three wavy levels from North to South and four wavy levels from SE to NW. Compared with the morphological classes of a published geomorphologic map, LR exhibits similarly in Highest mountain, High mid-mountain and High mountain areas. Additionally, the LR values in the Loess Plateau change between values in Low-middle mountain and values in Low mountain. This study shows that the geomorphologic characteristic of the whole of China could be well represented by the LR derived from the DEM. It carries sufficient physiographic information and can provide an important basis for further quantitative classification of relief forms.

Keywords: Local relief, DEM, China.

1 Introduction

Local relief (LR) is an important quantitative topographical parameter and was introduced from the International Organization of Geomorphologic Mapping in the 1970s as a means of giving expression to the new concept of relief forms (Chen *et al.* 1993). This parameter is also called amplitude of relief (Demek and Embleton 1989, Tu and Liu 1990, 1991, Centamore *et al.* 1996, Hurtrez and Lucazeau 1999, Kühni and Pfiffner 2001) or relative relief (Chai 1983, Vogt *et al.* 2003, Sreedevi *et al.* 2005). It is similar to but different from the former concept of relative height. The relative height represents an elevation difference between points on the ground and local datum, while the local relief does not have a necessary relationship with relief form evolution and is usually calculated in a smaller area than relative height (Li 1982). As for the definition of LR, we do not have a uniform one at present. In Mapping Criteria and Standard of Geomorphologic Maps of 1:1,000,000 in China, LR is defined as the elevation difference from a ridge (or a peak) to the place of intersection of the closest river (conflux area is greater than 500 km^2) or the closest wider plain or mesa down slope (Institute of Geography, CSA 1987). In the Geomorphological map of China and its Adjacent Areas (1:4,000,000), it is defined as the maximal elevation difference of the datum plane from the vertex of a ridge to the first infall of the downstream closest river (Chen *et al.* 1993). Additionally, Kühni and Pfiffner (2001) regarded LR as the elevation difference from the peak of the highest mountain to the valley bottom in a region. Montgomery and Brandon (2002) defined the parameter as the difference between the elevation of one point and the mean elevation in a given area. Tu and Liu (1991) believed that not only relief amplitude but also relative height should be considered in the definition of LR, since one feels relief fluctuating both looking up and down. Despite all these definitions, LR is usually measured as an elevation difference on some length scale (Summerfield 1991).

LR describes the complexity of the Earth's surface at large spatial scale and reflects the degree to which the Earth's surface is incised by exogenic agents, for instance, wind or river. Hurtrez and Lucazeau (1999) calculated LR for three basins and studied the relationship between lithologies and LR. Kühni and Pfiffner (2001) analysed the dependence of Alpine topography on bedrock lithology by means of numerical analysis of the morphometry, such as LR, mean elevation, etc. Montgomery and Brandon (2002) investigated the global relation between LR and erosion rate for tectonically active mountain ranges using the 30 arc-second GTOPO30 digital elevation model. In addition, LR also plays an important role in describing regional relief characteristics (Bishop *et al.* 2003) and classifying relief forms (Brown *et al.* 1998, Miliaresis and Argialas 2002, Gallant *et al.* 2005). However, as a snapshot of the constant competition between tectonics and erosion through space and time, LR is usually difficult to explain (Kühni and Pfiffner 2001).

In the past, limited topographic data and technical conditions made the acquisition of LR difficult. But with the rapidly increasing availability of digital elevation data and GIS-assisted processing of DEMs, geomorphology has been promoted and in recent years some attempts have been made to develop automated mapping routines for LR from DEMs. One method of numerical analysis uses a moving n × n window (or n-diameter circle window) for calculating LR by determining the maximum change in elevation within the window. Ahnert (1984) showed a non-linear relationship between the window and LR by examining a variety of mountain belts. But a unit area of 16 km^2 was still used as an experiential value for LR in the International Geomorphological Map of Europe (1:2,500,000) (Demek and Embleton 1989). Furthermore, the window can also be obtained by fuzzy mathematics (Tu and Liu 1990). Apparently, LR is mainly dependent on the window over which elevation is measured. Another method for calculating LR is by subtracting base-level map elevations from summit-level map elevations (Kühni and Pfiffner 2001). The base-level map is interpolated by a standard Kriging method with all the lowest points in a unit area. The summit-level map is derived from all the highest points. Unfortunately, no reports have been seen on the comparative analysis of the two methods. These two methods are applicable to large-scale DEMs, but other approaches (Zhu 2003, Sreedevi *et al.* 2005) have measured LR in a drainage basin (the least area of the catchment basin usually being larger than 500 km^2) mainly for small-scale DEMs.

China has one of the most complicated landforms in the world, exhibiting great relief amplitude differences in each of the three terrace planes. According to statistical analysis, the average altitude above sea level of China's entire territory is about 1.8 times the world average. In addition, China possesses a higher than world average 'Relief degree of land surface' (Niu and Harris 1996). Since LR is an expression of incision by rivers and glaciers and is often used to describe the characteristics of geomorphological forms, it is significant that LR can reveal geomorphologic features and regional relief distribution and variation patterns in China. In the early 1990s in China, Liu *et al.* (1990) first used the DEM from the State Bureau of Surveying and Mapping of China for preliminary quantitative analysis and mapping of LR of China. Later, LR was applied in soil and water loss assessment based on DEMs by Liu *et al.* (2001). Compared with research on the micro-scale, geomorphological analysis is still poor on the macro-scale.

The aim of this chapter is to probe the methodology of extracting LR from DEMs, and to analyse geomorphologic features and regional relief distribution and variation patterns in China based on the results.

2 Materials and Methods

The data used in this study was a 1-km DEM of China. From it, we calculated LR by determining the maximum change in elevation within a moving rectangular window. Here the window size is very important and, according to previous research, two principles should be followed. The window size should meet the integrality of the mountains and adapt to any relief form. So we sampled 800 points randomly on mountains and, based on window size-LR broken-line graphs, 42.25 km^2 was found to be the optimum unit for deriving LR from the 1-km DEM.

2.1 Data

The DEM of China with a horizontal grid spacing of 1km (Figure 1) is available from the National Geomatics Center of China at a global scale, providing a digital representation of Chinese relief at a sampling interval of 28.125" (longitude difference) × 18.750" (latitude difference). It is based on elevation data derived from 8,740 1:50,000 scale topographic map sheets and 3,861 1:100,000 scale topographic map sheets. The vertical error of the data is 10–20m. Mean square errors (MSE) of any point interpolated with the data model are showed in Table 1 and the accuracy of interpolation is in conformity to the standard of the 1:1,000,000 scale topographic maps.`

Table 1. Mean square errors of the 1-km DEM of China.

Landform name	*High mountain*	*Low mid mountain*	*Hill*	*Plain*
MSE (m)	70	41	20	1

At the same time, various digital processes are applied in order to automate the quantitative analysis and interpretation of terrain features, i.e. the DEM is resampled to 1,000m in both the x and y directions, converted to a conical equal-area projection, and finally rectified (Table 2). Because of the high sampling accuracy of the DEM, landform characteristics of the whole of China could be well represented by the DEM at a large scale. Moreover, the DEM can be used as a basic spatial information source for research into macro relief characteristics, regional planning, protection and development of the ecologic environment, and so on. One of the objectives of this study is to investigate the availability of this DEM, which has been an issue of concern in China for many years.

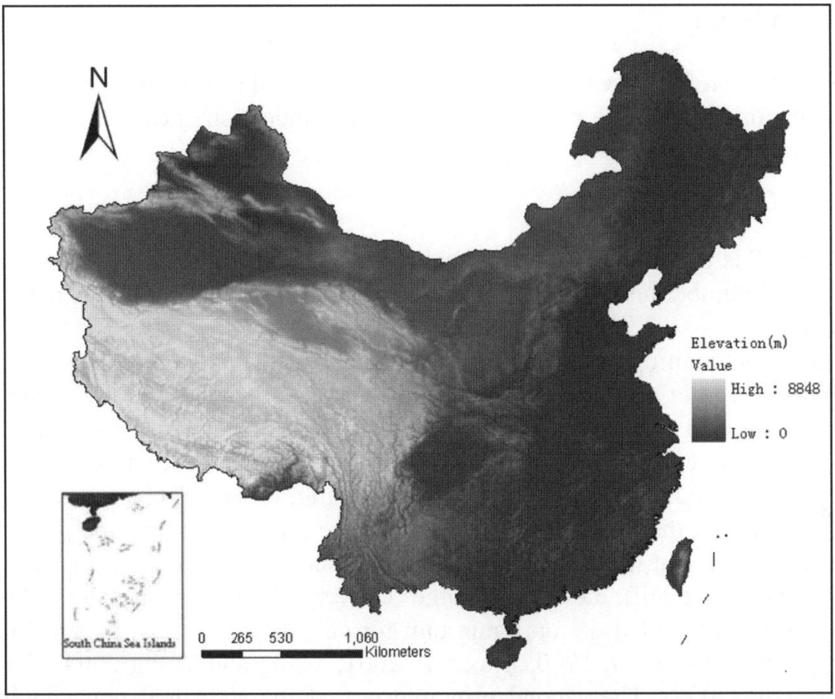

Figure 1. Digital Elevation Model of China with a horizontal grid spacing of 1 km (size: 6,835 columns and 4,681 rows). The elevation values (minimum 0m, maximum 8,848m) were rescaled to the interval 0–255 (the lighter a pixel, the higher its elevation).

Table 2. Basic parameters of the 1-km DEM of China.

Scale	1: 1,000,000	Min	0m
Projection	conical equal-area projection	Max	8,848.0m
Horizontal grid dimensions	1 × 1km	Mean	1,832.0m
Number of rows/columns	4,681/6,835	Std. Dev.	1,741.3m

2.2 Methods

By using ArcGIS® (version 9.0), a commercial GIS software, the DEM was parameterized in a given statistical unit (window size) calculated with Equation (1):

$$LR_u = Z_{u\,max} - Z_{u\,min} \tag{1}$$

where *LR* is the abbreviation for local relief; *Z* stands for the elevation; *u* is a natural number and represents the statistical unit over which LR is measured.

Obviously, an optimum unit must be determined first. According to the reports of Liu *et al.* (1990), two principles should be upheld in the experiment. The first is to keep the integrality of the mountain in the statistical unit. Former researchers have shown that with the statistical unit (which is centred on a point on any mountain) increasing, elevation changes in the unit enlarge at the beginning. At this time, the unit just covers one part of the mountain. But when the statistical unit reaches a certain area (just covering the whole hill), elevation changes converge to a constant value. This area is usually called the optimum unit for calculation of LR (Tu and Liu 1990, 1991, Liu *et al.* 1990, Liu *et al.* 2001, Wang and Wang 2004, Gallant *et al.* 2005). The second principle is that the statistical unit should adapt to any relief form.

To determine the optimum unit, we took the 'Geomorphologic Map of China and Its Adjacent Areas' (1:4,000,000) as reference, and sampled 100 points randomly on each relief form's mountaintops. These relief forms included highest mountain (elevation > 5,500m), high mountain (elevation 3,000–5,500m), high mid-mountain (elevation 2,000–3,000m), low mid-mountain (elevation 800–2,000m), low mountain (elevation 500–800m and LR > 150m), hill (elevation < 500m; elevation > 500m and LR 20–150m), the Loess Plateau and the Sichuan basin of China. For each sampled point, we calculated 25 pairs of window size-LR data, and plotted broken-line graphs in which the x-axis denotes the window size and the y-axis denotes the corresponding elevation changes (LR). From the graphs, we could pick up one window size in which LR tended to become constant. And that window will be regarded as the best window for the point. Thus, for all sampled points, the best window, which appeared most frequently, is the optimum analysis window adapted to all landforms. It is worth noting that if we sampled at the hillside, LR will tend to be stable earlier, before the statistical unit covers the whole mountain. Therefore, sampling at the hillside will not meet the principle of mountain integrity. Additionally, in order to avoid the analysis windows increasing rapidly, the DEM was previously resampled to 500m in both x and y directions and rectangular windows were used in the experiment. Finally, a 13 × 13 window size appeared most frequently and the area of $S = 2\ (13 \times 0.5) = 42.25\ \text{km}^2$ is regarded as the optimum unit (Figure 2). The results for LR are shown in Figure 3.

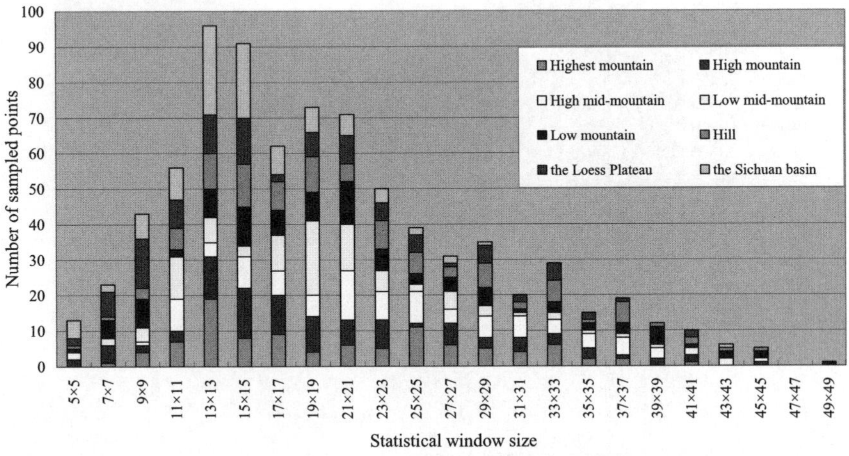

Figure 2. Number of samples in a series of windows in which LR is nearly constant. These samples are located in different relief forms.

3 Discussion

3.1 Statistical characteristics of the LR in the whole of China

Figure 4 shows that in China the higher the LR, the lower is the proportion of national territory. The proportions of LR to the national territory are enumerated as: 0 < LR < 100m, 30.71%; 0 < LR < 200m, 46.18%; 1,000m < LR, 7.34%. Furthermore, most areas in China are mountainous. By adopting the 5-class system of the published 1:4,000,000 scale Geomorphologic Map of China, the proportions of LR classes to the national territory are enumerated as: faint relief (0 < LR < 20m), 10.04%; low relief (20 < LR < 75m), 15.34%; moderate relief (75 < LR < 200m), 21.09%; mountainous relief (200 < LR < 600m), 33.22% and high-mountain relief (600m < LR), 20.31%. All of the above show that China possesses a high degree of relief in its land surface. And the highest LR appears in the Hengduan Mountains, located in the South West of China.

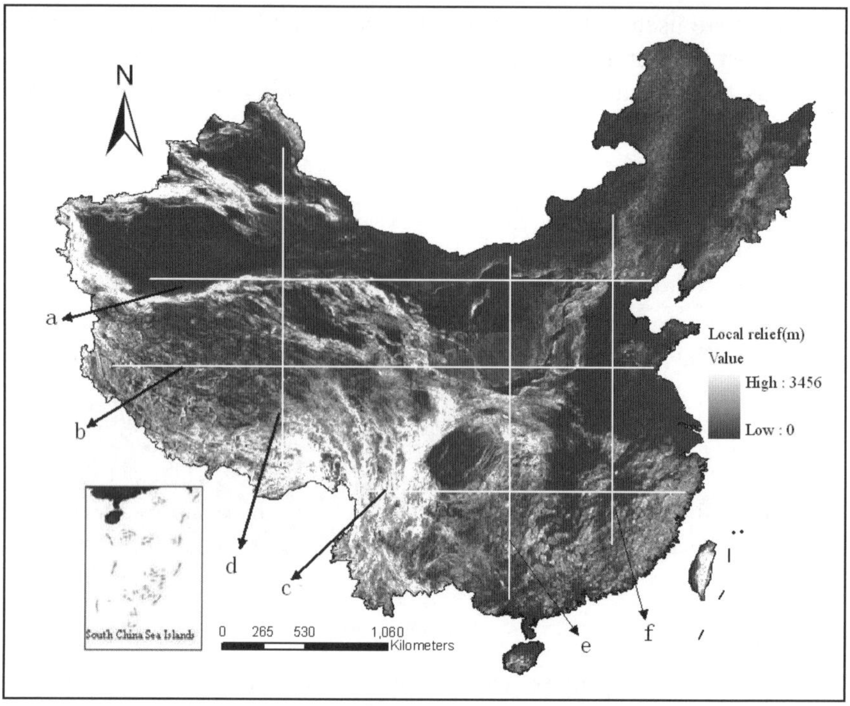

Figure 3. Map of LR of China (minimum 0m, maximum 3,456m, mean 356.88m, Std. Dev. 380.66m). The LR value is rescaled to the interval 0–255 with an equalization stretch applied to the histogram (the lighter a pixel, the higher its value). Straight solid lines are the traces of six cross-sections.

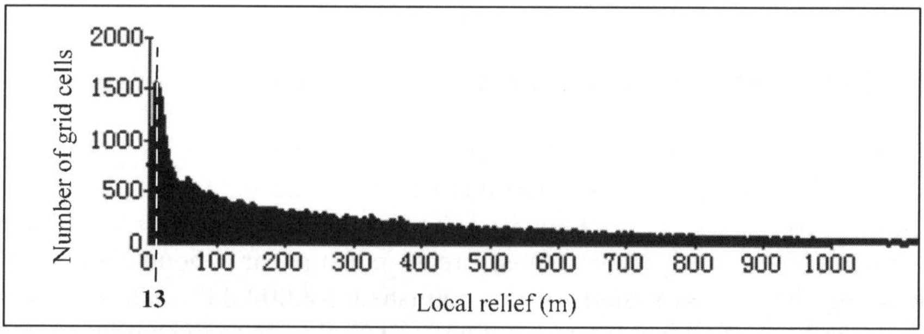

Figure 4. Frequency distribution of LR.

3.2 Spatial distribution characteristics of the LR in the whole of China

From space, the LR over the whole of China exhibits significant differences.

- LR spatial distribution shows obvious differences in the West to East and NE to SW orientation (Figure 3, Figure 5). Lower LR mainly occurs in the east, while higher LR occurs in the southwest. Lower LR mainly occurs in the north, while higher LR mainly occurs in the south.
- From the west to the east, LR shows a descending trend with two terrace plains (Figure 5a). The division belt extends through the middle of China, mainly in the Helan mountain of Inner Mongolia, the Liupan mountain, Sichuan basin and the Yunnan-guizhou plateau. The same also occurs from SW to NE, with the division belt in the Yinshan mountain, Loess Plateau, Wudang mountain, the middle-lower reach of the Yangtze River Plateau, and Wuyi mountain (Figure 5b).
- Moreover, from space, LR from the north to the south exhibits three wavy levels (Figure 5a). The peak values of LR in the first level are distributed in areas of the Tianshan mountain, Kunlun mountain, and Altun mountain; the peaks value in the second level is in the area of the Hengduan mountain; and the peak values in the third level is found in Hainan island. In the SE-NW direction, values of LR show four wavy levels (Figure 5b). From SE to NW, the peak values of LR in the first level locates in Tianshan mountain and Kunlun mountain. The highest LR in the second level is in Hengduan mountain. The highest LR in the third stands in the Wuyi mountain and the Nanling area. The highest LR of the fourth level lies in Taiwan Island. As a whole, LR shows a descending trend both from the West to the East and from the SW to NE. LR exhibits wavy spatial distributions both in the N–S orientation and in the NW–SE orientation.

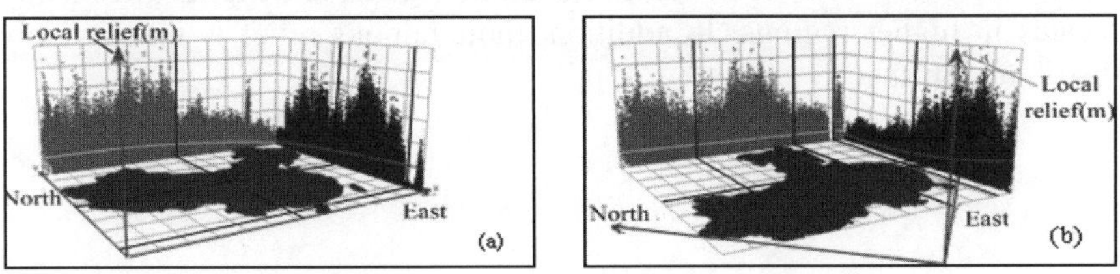

Figure 5. Trend analysis of LR in China.

3.3 Characteristics of LR compared with the elevation

To further investigate the LR features in China, this factor is also compared to the elevation from six traverses. The locations of the traverses are given in Figure 3. Traverse (a), (b) and (c) are in the east-west direction and traverse (d), (e) and (f) are in north-south direction.

Figure 6 shows the results using the profile method. Each black line cuts out of the LR map along each profile. Then, elevation data (grey lines) are projected along these profiles onto the black lines. As can be seen, most of the LR profiles are spatially rugged. From the traverse (a) to (c) and from the traverse (f) to (d), LR tends to fluctuate strongly. The cross-section (a) spans the smallest difference, while the cross-section (c) spans the largest difference. This observation is due to the Hengduan Mountains coexisting with the Sichun basin and the middle-lower reach of the Yangtze River plain along the traverse (c).

Inflexions of LR profiles are found with higher values in boundaries of different landforms, especially for mountains and plains. As compared to elevation lines, these LR traverses indicate that higher elevation does not necessarily mean higher LR. For example, LR in the Qinghai-Tibet plateau and Loess Plateau in Shaaxi province show small differences.

In Figure 7, the elevation data is divided into six classes, which basically correspond to seven relief forms: plain and hill, low mountain, low mid-mountain, high mid-mountain, high mountain and highest mountain. Then, in each elevation belt, the frequency of LR is calculated with the 5-class system just like before. As can be seen, the proportions of the former four LR classes (i.e. faint relief, low relief, moderate relief and mountainous relief) in the 0–500m elevation belt are dominant and very close. In the 500–800m elevation belt, mountainous relief is dominant. In the 800–2,000m elevation belt, moderate relief and mountainous relief are found dominant, while in the three behind mountainous relief and high-mountain relief are found absolutely dominant. Generally speaking, statistically lower LR mainly locates in lower elevation regions and higher LR mainly locates in higher regions. In addition, mountainous relief is most widely distributed in space.

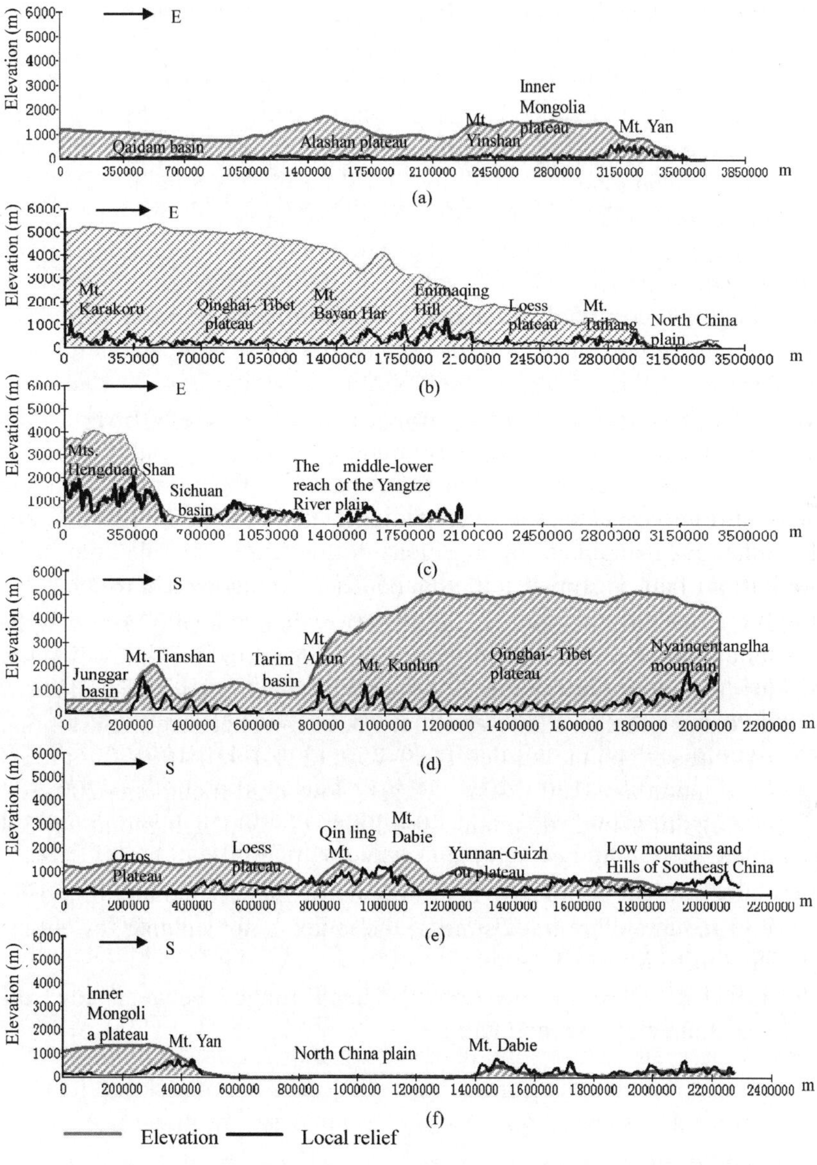

Figure 6. Comparison between profiles of LR and elevation in China.

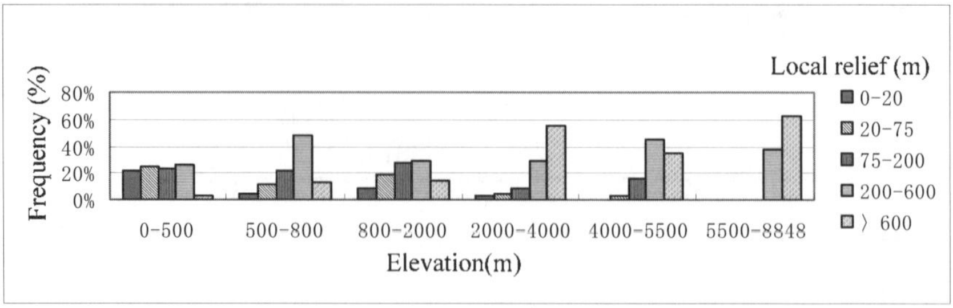

Figure 7. Frequency distribution of LR in different elevation belts.

3.4 Characteristics of LR in different landforms of China

We analysed LR in different relief forms based on the morphological classifications of the published geomorphologic map of China. Figure 8 shows the results.

- LR in plains is similar to that in deserts, ranging from 0 to 75m.
- LR wave crests move to a smaller value gradually from highest mountain, high mountain, high mid-mountain, low mid-mountain, low mountain, hill, plain to desert. That decreasing trend shows coherence with elevation variation. For each relief form, the dominant LR class is: plain and desert 20–75m (33%), hill 100–200m (33%), low mountain 100–200m (32%), low mid-mountain 200–300m (19%), high mid-mountain 400–500m (11%), high mountain 300–400m (10%), highest mountain 500–600m (11%).
- Since the LR frequency distributions of highest mountain and high mid-mountain are most similar, the index is not suitable for classifying these two relief forms.
- The LR value in the Loess Plateau ranges between low mid-mountain and low mountain.

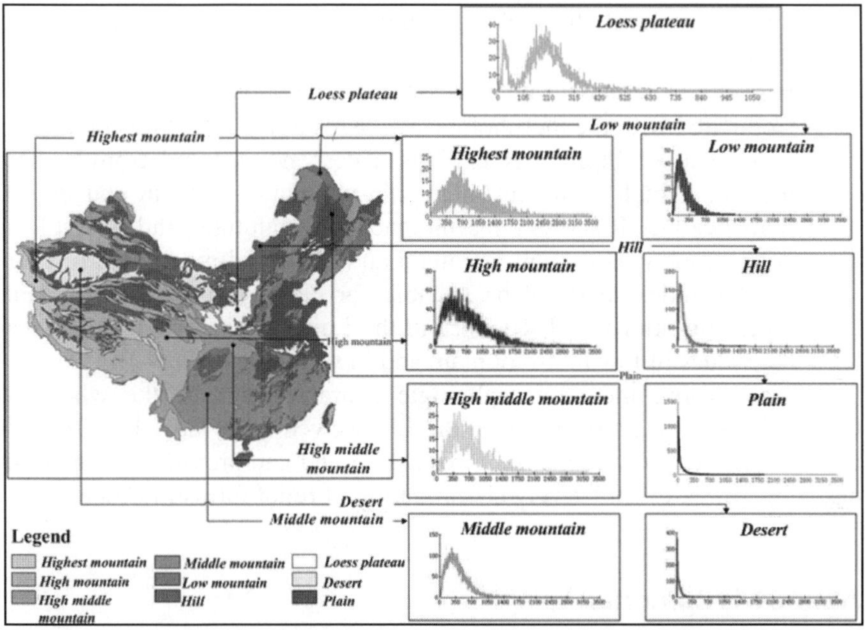

Figure 8. Frequency distribution of LR in different geomorphological areas of China (x-axis denotes the LR (m) and y-axis denotes numbers of raster cells).

4 Conclusions

In this paper, we probe the method of extracting LR from a 1-km DEM of China. Then, based on numerical analysis of the LR map, the large-scale geomorphological features of China are discussed. The results of this study show that:

(1) The 1-km DEM of China is of practical significance and potential in macroscopic geomorphological feature analysis.

(2) The experiment shows that the optimum unit for measuring LR in the 1-km DEM is 42.25km^2, based on theories of geomorphological development and numerical analysis.

(3) The LR map in our experiment is not absolutely accurate. It is influenced by the location of the optimum unit. But it still could reveal the large-scale characteristics of relief forms in China. Analysing the LR map shows that:

- Most areas belong to mountainous relief (200<LR<600m) in China.
- From space, LR shows obvious differences from East to West and from NE to SW. Moreover, the spatial distribution of LR exhibits

three wavy levels from North to South and four wavy levels from SE to NW.

- Higher elevation does not necessarily mean higher LR.
- Compared with the morphological classes of a published geomorphologic map of China, LR exhibits a descending trend from highest mountain, high mountain, high mid-mountain, to plain and desert. Statistically, LR shows a resemblance to the altitude, and since LR in Highest mountain, High mountain and High mid-mountain are similar, the index may not be suitable for classifying these three land types. Additionally, LR values in the Loess Plateau change between the values of low mid-mountain and low mountain.

In conclusion, the geomorphologic characteristics of the whole of China can be well represented by the LR derived from the 1-km DEM. It carries sufficient physiographic information and can provide an important basis for further quantitatively classification of relief forms in China.

Acknowledgments

Thanks for financial support are due to the National Natural Science Foundation of China (No.40571120, No.40671148) and the National High Technology Research and Development Program of China (No. 2006AA12Z212).

References

Ahnert, F., (1984), Local relief and the height limits of mountain ranges, *Am. J. Sci*, 2**84**: 1035–1055.

Bishop, M.P., Shroder Jr., J.F. and Colby, J.D., (2003), Remote sensing and geomorphometry for studying relief production in high mountains, *Geomorphology*, **55**: 345–361.

Brown, D.G., Lusch, D.P. and Duda, K.A., (1998), Supervised classification of types of glaciated landscapes using digital elevation data, *Geomorphology*, **21**: 233–250.

Centamore E., Ciccacci S., Del Monte, M., Fredi, P. and Palmieri, E.L., (1996), Morphological and morphometric approach to the study of the structural arrangement of northeastern Abruzzo (central Italy), *Geomorphology*, 1**6**: 127–137.

Chai, Z., (1983), The suggestion of using relative altitude to divide the geomorphologic forms, In *Geographical Society of China*, *Theses of Geomorphology*: 90–97, Beijing: Science Press.

Chen, Z. *et al*., (1993), An Outline of China's Geomorphology [Explanations of Geomorphological Map of China and its Adjacent Areas (1: 4,000,000)], China Cartographic Publishing House.

Demek, J. and Embleton, C., (1989), *International Geomorphological Map of Europe (1: 2,500,000),* Cartography, Lithography and Praba, S.P.

Gallant, A.L., Brown, D.D. and Hoffer, R.M., (2005), Automated mapping of Hammond's landforms, *IEEE Geoscience and Remote Sensing Letters*, 2 **(4)** : 384–388.

Hurtrez, J-.E. and Lucazeau, F. (1999), Lithological control on relief and hypsometry in the Hérault drainage basin (France), *Earth & Planetary Sciences*, **328**: 687–694.

Institute of Geography, CSA, (1987), Mapping Criteria and Standard of Geomorphologic Maps of 1: 1,000,000 in China, Beijing: Science Press.

Kühni, A. and Pfiffner, O.A., (2001), The relief of the Swiss Alps and adjacent areas and its relation to lithology and structure: topographic analysis from a 250-m DEM, *Geomorphology*, **41**: 285–307.

Li, J., (1982), A primary study on the quantitative index of basic types of geomorphologic forms in China, *Acta Geographica Sinica*, **37** (1) : 327–335.

Liu, X., Yang, Q. and Tang, G., (2001), Extraction and application of relief of China based on DEM and GIS method, *Bulletin of Soil and Water Conservation*, (2**1)** 1: 57–62.

Liu, Z., Sun, Y. and Tu, H., (1990), Preliminary study on compilation of relief amplitude map in small scale with DTM data, *Acta Geodaetica et Cartographica Sinica*, **(19)** 1: 57–62.

Miliaresis, G.Ch. and Argialas, D.P., (2002), Quantitative representation of mountain objects extracted from the global digital elevation model (GTOPO30), *Int.J.Remote Sensing*, (2**3)** 5: 949–964.

Montgomery, D.R. and Brandon, M.T., (2002), Topographic controls on erosion rates in tectonically active mountain ranges, *Earth and Planetary Science Letters*, **201**: 481–489.

Niu, W. and Harris, W.M., (1996), China: the forecast of its environmental situation in the 21st century, *Journal of Environmental Management*, **47**: 101–114.

Sreedevi, P.D., Subrahmanyam, K. and Ahmed, S., (2005), The significance of morphometric analysis for obtaining groundwater potential zones in a structurally controlled terrain, *Environmental Geology International Journal of Geosciences*, **(47)** 3: 412–420.

Summerfield, M.A., (1991), Sub-aerial denudation of passive margins: regional elevation versus local relief models, *Earth Planet. Scilett*, 1**02**: 406–469.

Tu, H. and Liu, Z., (1990), Demonstrating on optimum statistic unit of relief amplitude in China, *Journal of Hubei University (Natural Science),* **(12)** 3: 266–271.

Tu, H. and Liu, Z., (1991), Study on relief amplitude in China, *Acta Geodaetica et Cartographica Sinica*, **(20)** 4: 311–319.

Vogt, J.V., Colomboa, R. and Bertolob, F., (2003), Deriving drainage networks and catchment boundaries: a new methodology combining digital elevation data and environmental characteristics, *Geomorphology*, **53**: 281–298.

Wang, H. and Wang, J., (2004), Preliminary study on specification of basic terrain-unit dataset. *Science of Surveying and Mapping*, **(29)** 3: 22–25.

Zhu, H., (2003), A research on spatial data mining of digital elevation models: a case study in the loess plateau of North Shaanxi province, Master thesis, Northwest University.

Re-Scaling Lower Resolution Slope by Histogram Matching

YANG Qinke, David JUPP, LI Rui and LIANG Wei

Abstract

Slope is a key parameter for regional scale hydrology, soil erosion, and relevant land surface process modelling. Slope information is usually extracted from moderate and lower resolution DEMs because of time and financial limitations. But research has shown that the slope will become flatter as the resolution of the DEMs becomes coarser. That is, *slope reduction* (SR) has occurred. Due to the reduction of slope as the DEM resolution becomes coarser, the slopes from coarser DEMs have to be re-scaled, firstly in order to represent terrain accurately and secondly to directly support the environmental modelling mentioned above. A transform approach by slope histogram matching is suggested in this chapter. A case study in a loess hill area shows that a coarser resolution slope histogram (derived from a 1:250,000 topographic based DEM), can be successfully re-scaled into the slope histogram at a finer resolution (using a 1:100,000 topographic based DEM). The transformed slope surface has been improved as an estimate of the finer resolution DEM slope in comparison with the coarser resolution DEM, the statistical results are similar to those from the finer resolution DEM in the loess hill area, and little new distortion has been introduced.

Keywords: Loess plateau, DEM, change in scale, histogram matching.

1 Introduction

Slope is a key parameter for environmental modelling, especially for modelling hydrology and soil erosion processes. At the field scale of soil and soil erosion survey and soil conservation planning, slope can be obtained by measuring in the field, or by mapping based on large scale topographic maps, e.g. 1:10,000 in China (Tang *et al.* 2002), or 1:24,000 in USA (Martin *et al.* 2002). At the regional scale, the only practical way to obtain slope variables is based on existing moderate to coarser resolution DEMs

(Martin *et al.* 2002, Armstrong and Martz 2003). But research has shown that the slope will become flatter as the resolution of the DEMs become coarser. That is, slope reduction (SR) will occur (Gao 1997, Yin and Wang 1999, David and McCabe 2000, Thompson *et al.* 2001, Armstrong and Martz 2003, Tang *et al.* 2003, Wu *et al.* 2003, Hao *et al.* 2005). Because the reduced slope does not represent the relief sufficiently, it is therefore difficult to extract the terrain variables for hydrological and regional soil erosion models (Thompson *et al.* 2001, Hao *et al.* 2005), limiting the accuracy of modelling (Zhang *et al.* 1999, Wu *et al.* 2003, Cochrance and Flanagan 2004, Ren *et al.* 2004, Liu 2007).

To meet the needs for slope maps in regional soil erosion modelling, a number of approaches have been explored by researchers. One is to develop an alternative, or substitute, parameter, e.g. elevation difference (ED) in specified analysis windows. This has been done by Liu *et al.* (2001) and Zhao (2007). Another strategy is to transform the coarser slope distribution. Tang *et al.* (2001) and Chen *et al.* (2004) have attempted to transform coarser resolution slope by analysing the spectrum of slope (similar to a histogram) and constructing a transform table (TT). Finally, Zhang *et al.* (1999) suggested a method based on fractal analysis (Zhang *et al.* 1999). As it is difficult to define a relationship between ED and real slope, because TT does not represent the spatial pattern of slope surface in the current stage and because of limitations to the general application of Zhang's approach that were recognized by the authors, the subject of coarser resolution slope (CRS) re-scaling or transform is still not operational and is therefore an important topic for research.

In this chapter, an approach of re-scaling the coarser resolution slope by histogram matching is introduced aimed at practical support for hydrology, erosion, vegetation suitability mapping at regional scale, and to help understand issues of SCALE in geography, as well as the use of coarser resolution DEM resources.

2 Methodology Description

2.1 Principles of CRS in geomorphology and cartography

The capacity to depict the higher spatial frequencies (or fine scale patterns) of terrain in a topographic map becomes weaker as the cartographic scale becomes smaller due to the map generalization (e.g. from 1:10,000 to 1:50,000) (Yang *et al.* 2006). At the same time, specific lines/points of landforms (e.g. stream networks, ridges, saddles, etc.) remain and are sometimes enforced during the map generalization process according to the topographic cartography (Wang 1993). Meanwhile, the elevation will not be generalized; therefore the overall elevation and relief will not be generalized, so that despite the slope reduction created by the reduction in higher spatial frequencies, a significant amount of geomorphologic structure, or pattern, still remains in moderate or smaller scale topographic maps, or in coarser resolution DEMs that are based on smaller scale topographic maps (Yang *et al.* 2006). This characteristic is displayed clearly in Figure 1. In Figure 1a, a DEM based on a 1:100,000 scale map (SLP_{10}) has been colour coded according to the slope ranges shown below. In Figure 1b, the same table is applied to slopes extracted from a DEM based on a 1:250,000 scale map (SLP_{25}). The slope reduction is clearly visible. However, by comparing the most common values in the reduced slope image corresponding to each of the ranges in the DEM based on the small scale map, it is possible to redefine the ranges with the result that much of the colour classification of the original image is retrieved. In Figure 1c, SLP_{25} is displayed with the reclassified slope categories and looks steeper than the slope gradient in Figure 1b. This illustration suggests that if a look-up table (LUT) for the slope transformation of the original and reclassified cases can be built, the slope reduction can theoretically be resolved.

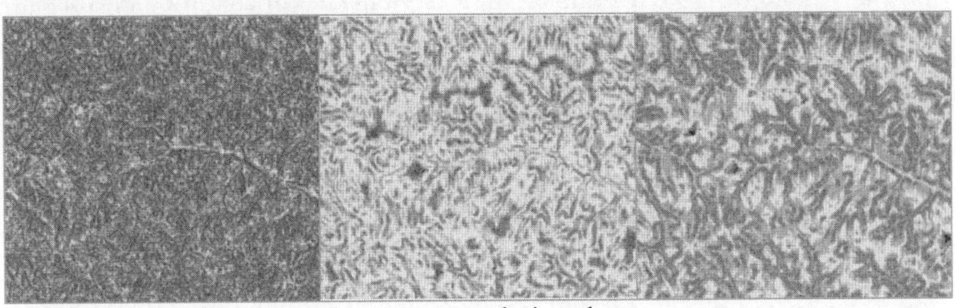

(a) finer resolution slope (SLP_{10}) (b) coarser resolution slope (SLP_{25}) (c) reclassification of SLP_{25}

Slope (deg.)	0-1	1-2.5	2.5-5.0	5.0-10.0	10.0-15.0	15.0-20.0	20.0-25.0	25.0-35.0	>35.0
Reclassified slope (°)	0-0.4	0.4-0.9	0.9-1.7	1.7-3.5	3.5-6.5	6.5-10.5	10..5-16.0	16.0-30.5	30.5-48.0
Legend									

Figure 1. Reduction and reclassification of lower resolution slope (SLP_{25}).

Following effective DEM interpolation, elevation and its derivative (slope) can be treated as properties of spatially continuous surfaces with similar spatial properties to those of the original maps (Hutchinson 2000). So for slope gradient values of s1, s2, …, sn, the frequency distribution (or density function) of slope can be estimated from the slope surface obtained from DEM interpolation of different scale maps. For an area with specific spatial landform properties, the slope characteristics, including the surface and its histogram, exist and are unique. According to the methodology of histogram modification for digital remotely sensed images (Hummel 1975, Harrison and Jupp 1990, Gonzalez and Woods 1992), if one matches the histogram of the coarser slope to the finer one, the slope from coarser resolution DEM, which has had slope value reduction, can be rescaled, with the result that it becomes similar in its statistical and spatial pattern characteristics to the finer slope image. This is the method of rescaling the slope of the coarser slope by matching histograms of slope frequency.

$$F(s) = \int_0^s f(s')ds' \tag{1}$$

In Figure 2, *F (S)* is the cumulative distribution function as a function of *S* as defined in Equation (1), *f (S)* is the density function of a continuous random variable *S* (slope). The transformation function that modifies the coarser resolution slopes so that the histogram of the data following transformation matches that of the finer slopes is sometimes represented as a "look-up" table or LUT. In this form, it is well-known as a method for enhancing digital images for display (Gonzalez and Woods 1992) and matching sensors in image calibration (Horn and Woodham 1979). Here it is shown to be useful for retrieving data scaled in a way that is appropriate for many models.

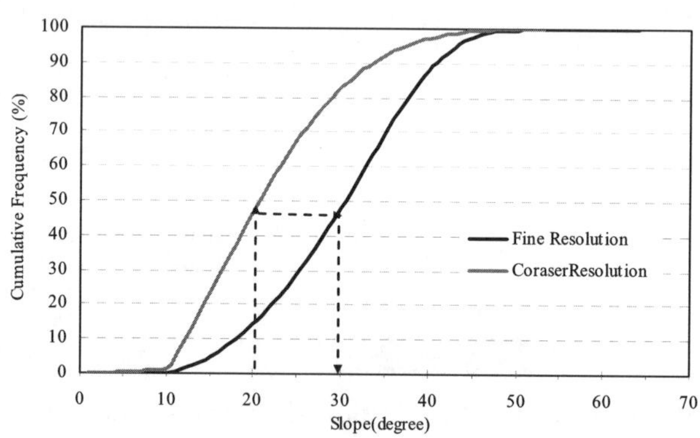

Figure 2. Rescaling coarser resolution slope by histogram matching.

2.2 Datasets and processing

(1) Research site: The research site is located in the loess hilly area of northern Shaanxi province, People's Republic of China. Landforms here are hills and gullies with severe soil erosion. The area with greater than 25 degree slope is more than 50% of the total area (Zhao 1989). The runoff, soil erosion, and land use patterns are all impacted strongly by slope gradient. In the training area (the dark box in Figure 3; 17 by 15km), the slope transform model will be derived and in the test area, the Baota Area (an administrative unit of Yan'an City; 3,549km^2), which is lightly shaded in Figure 3 and includes the training area, the model will be validated.

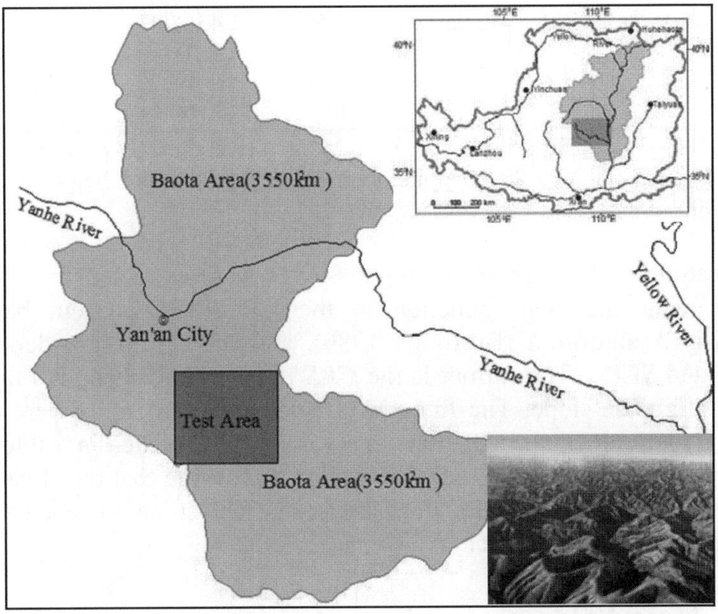

Figure 3. Research Site.

(2) Datasets: This study is based on topographic maps at scales of 1:100,000 and 1:250,000 including contours, spot heights, and stream networks. The basic information for the input datasets is shown in Table 1. The DEMs are interpolated with the maps and ANUDEM software (Hutchinson 1989, Hutchinson 2004). The effective grid cell resolution for the two datasets would be 20 and 50m, respectively, according to a previous study (Yang *et al.* 2005, Yang *et al.* 2006). Considering that the slope reduction is mainly due to the generalization of the topographic map and for ease of matching the frequency distributions and undertaking error analysis of the transformation, a common cell size of 25m is used for the two DEMs. This is valid as long as the interpolation method does not add frequency components in this process that were not in the original map. ANUDEM, by minimizing curvature, will not do this. The DEMs are denoted by DEM_{25} and DEM_{10}. As shown in Figure 1 and Figure 2, although the cell size is the same, the pattern and capacity to depict the higher spatial frequencies in the two DEMs and slope surfaces are clearly different, and the slope in SLP_{25} has generally reduced.

Table 1. Description of input data.

items	riv10	riv25	cntrs10	cntrs25	pnts10	pnts25
No	10	9	1226	285	83	11
Length (m)	29,570	30,059	4,580,968	1,436,907	n/a	n/a
Elevation range	n/a	n/a	1,080-1,400	1,100-1,400	1,255-1,442	1,391-1,434
CI	n/a	n/a	40	100	n/a	n/a

Note: *riv10*—1:100,000 rivers; *riv25*—1:250,000 rivers; *cntrs10*—1:100,000 contours; *cntrs25*—1:250,000 contours; *pnts10*—1:100,000 spot heights; *pnts25*—1:250,000 spot heights; *CI*—contour interval.

(3) Frequency histogram of slope: Slopes were extracted from the two DEMs with the slope function in the ARC/INFO system based on Burrough's algorithm (Burrough 1986), and the results are denoted by SLP_{25} and SLP_{10}. The former is the ***CRS*** to be transformed, and the latter is the reference slope. The frequency tables and histograms were created using Equation (2), and the slope interval scheme is listed in Table 2. The slope range is not constant because we have observed that the slope reduction mainly occurs in gentler slope situations. The column headed SI in the table records the slope interval.

$$f(s) = 100 \times \frac{m}{N} \tag{2}$$

where N is the total number of cells in the slope surface, and m is the cell number with specified slope range.

Table 2. Slope Interval (*SI*) for statistics.

slope range	SI	slope range	SI
0.00 — 1.00	0.05	5.00 — 50.00	1.0
1.00 — 2.00	0.1	50.0 — 70.00	2.5
2.00 — 5.00	0.5	70.0 — 90.00	5.0

2.3 The method of CRS transforming

(1) Co-register slope: Referring to Figure 2, for each ***CRS*** value SLP_i, find the cumulative frequency value in the coarser slope frequency line, move to the finer slope line horizontally (with the same cumulative frequency value), then move down to the x-axes and read a new slope value, SLP_{25}^{c} (Table 3). Define the ratio of SLP_{25}^{c} / SLP_{25} to estimate index (SLP_{25}^{e}) (bold dark in Table 3). Because the lines of SLP_{25}^{e} -- SLP_{25} represent the relative amount of slope correction, they are used to decide how to model and represent the function that has been created by this process. Columns SLP_{25} and SLP_{25}^{c} in Table 2 provide a slope scaling 'look up' table (LUT).

Table 3. Slope, slope frequency, estimated slope and estimated index.

slope (°) (SLP_{25})	slope frequency (%) finer resolution	slope frequency (%) coarser resolution	Estimated Slope (°) (SLP_{25}^{c})	Estimated index (SLP_{25}^{e})
0.10	0.0005	0.0051	0.215	2.150
0.15	0.0022	0.0193	0.342	2.280
0.20	0.0056	0.0399	0.487	2.435
0.25	0.0095	0.0664	0.618	2.472
0.30	0.0146	0.0988	0.775	2.583
0.35	0.0202	0.1424	0.915	2.614
0.40	0.0273	0.1971	1.064	2.660
0.45	0.0339	0.2545	1.185	2.633
…	…	…	…	…
48	0.0925	0.7136	61.5	1.281

(2) Transformation model: To apply the change in slope to any value in the range, a set of regression equations for CRS and SLP_{25}^{e} could be built according to the relationship between SLP_{25} -- SLP_{25}^{e} (Figure 4), and in this study the equations are defined as a piecewise analytic model to operate in sections defined by non-overlapping ranges of slope. The analytic models enable the use of map algebra operations with the independent variable x in the models being the coarser DEM slope surface. For each of these sections, modelled values (SLP_{25}^{m}) of slope can be predicted and errors (absolute and relative) calculated using Equations (3) and (4). Three sections (with ranges of slope ≤ 1.6 degrees, 1.6 < slope ≤ 40 degrees and slope > 40 degrees, respectively) were selected based on error analysis. Table 4 shows that the absolute error is less than 2 degrees, and the relative error less than 10%, and we believe the models are satisfactory for this example.

$$Ea = SLP_{25}^{m} - SLP_{25}^{c} \tag{3}$$

$$Er = 100 \times \frac{SLP_{25}^{m} - SLP_{25}^{c}}{SLP_{25}^{c}} \tag{4}$$

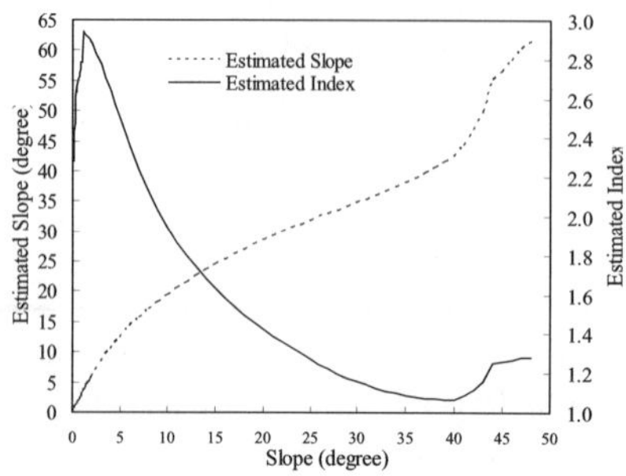

Figure 4a. Slope - estimated slope and slope estimated index.

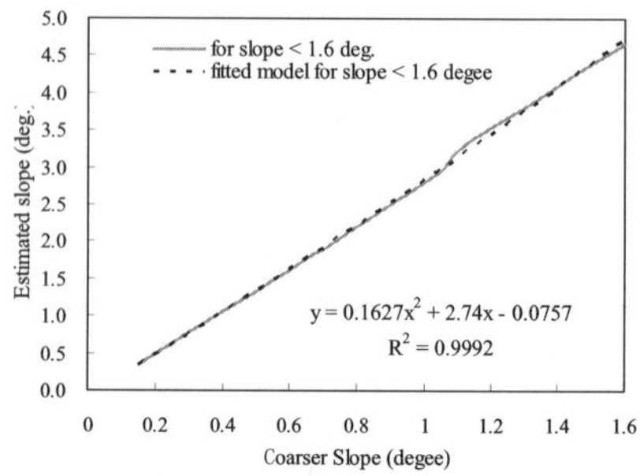

Figure 4b. Fitted model for gentler slope.

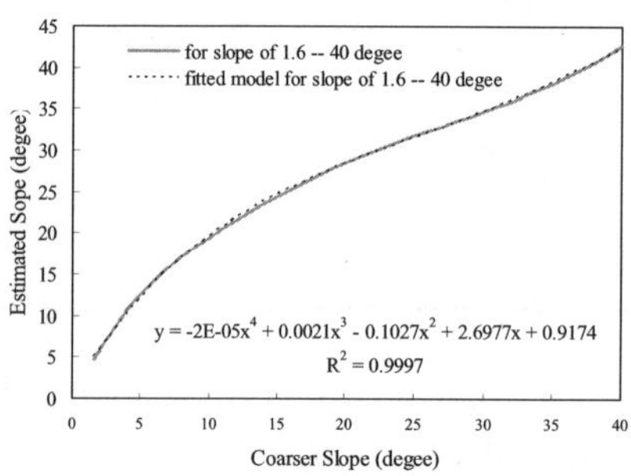

Figure 4c. Fitted model for moderate slope.

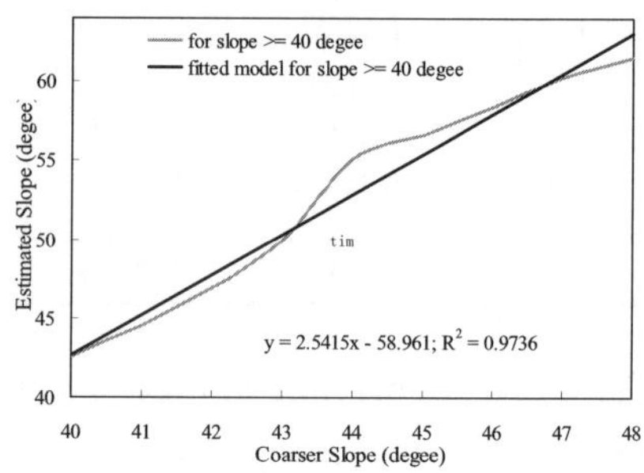

Figure 4d. Fitted model for steep slope.

(3) Slope transformation: Using the model listed in Table 4 and map algebra operations in ARC/INFO, SLP_{25} can be transformed for each of the three slope sections, resulting in intermediate outputs slp_{t_01}, slp_{t_02}, slp_{t_03}, which can then be integrated into one surface SLP_t with the *con* function in an ARC/INFO grid session (see Appendix A where the specific AML code used here is listed). The result is a new surface of slope.

(4) Validation & application: Models listed in Table 4 were used to transform coarser resolution slopes in the Baota area (the light grey shade in Figure 3); the slope is from DEM_{25}, which covers the entire Baota area. The results were compared to the slope map from SLP_{10}, which also covers the entire Baota area.

Table 4. Transform models and the error analysis.

no	models	R^2	suitable slope	absolute errors	relative errors
1	$y = 0.1627x^2 + 2.74x - 0.0757$	0.9992	0-1.6	0.00114	1.53998
2	$y = -2E\text{-}05x^4+0.0021x^3-0.1027x^2+2.6977x+0.9174$	0.9997	1.6-40	0.16029	8.67463
3	$y = 2.5415x - 58.961$	0.9736	>40	1.16828	3.87033

3 Results and Analysis

3.1 Reduction characteristics of SLP_{25}

First looking at the training area, the difference image of SLP_{25} and SLP_{10} (SLP_{25}-SLP_{10}, Figure 5) shows that the values in the surface are mainly negative, although there are some positive ones. This means that overall the slope has been reduced. The mean slope value decreases from 20.636 in SLP_{10} to 12.799 in SLP_{25} (Table 5). The number of cells with a negative value is 314,565, which is 77% of total cells and are mainly at the tops of hills with gentler slope gradients in SLP_{25}, and the cells with positive values are distributed in gullies with steeper slope gradients in SLP_{25}. The maximum decrease and increase is -52.6 and 40.8, respectively, and means that overall the slope surface becomes gentler.

Table 5. Statistics of slope reduction.

Statistics parameter	slope surface		$DLP_{25} - SLP_{10}$
	SLP_{10}	SLP_{25}	
max	0.031	0.015	-52.647
min	60.858	48.000	40.751
mean	20.636	12.799	-7.837
std	8.429	8.109	10.363

Figure 5. Spatial pattern of slope reduction from SLP10 to SLP25.

3.2 Transformed slope surface

Compared to SLP_{25}, the statistical parameters of the transformed slope listed in Table 6, the maximum, the mean slope, and range of slope, all increase, and are close to those of SLP_{10} (Table 6, and Figure 6). The area of gentler slope decreases and steeper slope increases. The range of slope and the scales of slope all increase. The pattern of the slope distribution is the same as in the field, being steeper in the gully areas and gentler on hill tops. Slope of slope (curvature) is calculated by the ARC/INFO slope function (Figure 6). A larger value of the curvature surface means the ground is in gullies, and otherwise on the tops of hills. Slope surfaces before and after transformation (Figure 6b) show that the main patterns of terrain and slope remain, without significant distortion being introduced. The main changes are the range of curvature and cells with higher curvature values are increased, and the mean relief also increases.

3.3 Spatial characteristics of slope changes during transformation

In Figure 7, the difference image (SLP_{10} - SLP_t) shows that the transformation between images is not complete when looked at in the spatial extent. The differences between SLP_{10} and SLP_t come from the generalization of the topographic map. The value of the differences is larger on the hill tops and smaller in the gullies. Differences between the images based on SLP_t and SLP_{25} indicate that the rescaling of slope from SLP_{25} to SLP_t mainly occurred in flatter areas in relative values (Figure 7c), and in the area between hill tops and gullies where the absolute values increase and relative values decrease (Figure 7b). There are only 19 cells in the difference image with negative values, indicating that the general change is to become steeper during slope transformation.

Table 6. Statistics of transformed slope surface.

Statistics parameter	SLP_{25}	SLP_t	SLP_{10}	$SLP_t - SLP_{25}$
min	0.015	0.011	0.031	-0.038
max	48	63.031	60.858	15.031
mean	12.799	19.583	20.636	6.785
std	8.109	8.387	8.429	1.888

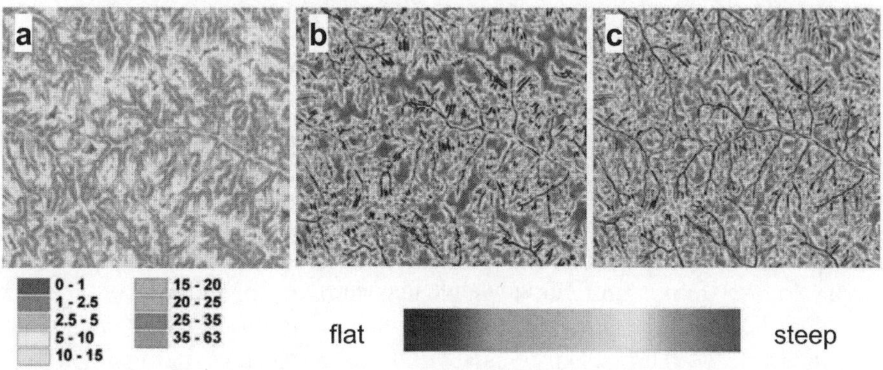

Figure 6. Slope and curvature before and after scaling: (a) rescaled slope; (b) before rescaling (0.003-39.4); (c) after rescaling (0.006-42.02).

3.4 Histogram of slope before and after transforming

Histograms of SLP_{25}, SLP_{10} and SLP_t (Figure 8a and b share the common legend in Figure 8b) indicate that the peak in the frequency moves right, and cells with larger slope values increase. Lines of frequency and the cumulative frequency for SLP_t are all close to that in SLP_{10}. The transformation down-scales the data and the slope properties of the higher resolution slope have been recovered to a greater extent.

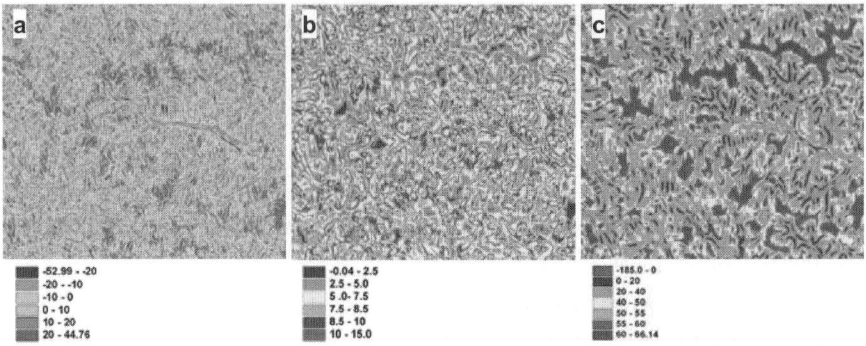

Figure 7. Spatial pattern of re-scaled slope surface: (a) SLP_{10}-SLP_t; (b) SLP_t-SLP_{25}; (c) 100 (SLP_t-SLP_{25}/SLP_t)

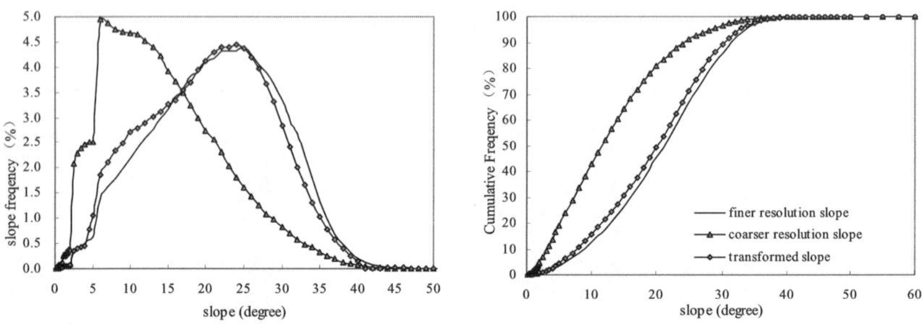

(a) (b)

Figure 8. Frequency distribution of original and scaled slopes: (a) historgram; (b) cumulative frequency.

3.5 Brief assessment of the transformation

The model was then used in the whole of the Baota area to transform ***CRS*** (based on the 1:250,000 topographic map interpolated DEM). The results in Table 7 are similar to the slope from the finer slope DEM (based on 1:100,000 topographic maps). The results are therefore satisfactory when the transformation is extended to a much larger but similar area that contains the training area. We believe that based on this approach, not only can slopes with finer resolution slope accuracy and relief representation power be obtained, but also the requirement for having higher resolution DEMs and associated computations may be decreased greatly as well.

Table 7. Validation of rescaling.

Slope class	Before scaling		After scaling		Finer slope	
	RS	BT	RS	BT	RS	BT
<3	9.08	14.00	1.53	2.87	1.43	2.81
3 – 7	19.69	20.52	6.70	10.04	5.19	7.29
7 – 15	35.79	30.50	22.59	23.51	19.88	20.61
15 – 25	26.50	23.07	40.54	34.25	40.36	35.03
>25	8.93	11.90	28.63	29.33	33.14	34.26

Note: SA: sampling area, BT: Baotao area

In the future, we propose to divide a large area (such as the one illustrated in Figure 9) into a series of transforming areas (TA) and have 2–3 sampling areas in each TA, which can be used to develop a model, and then use the models to predict the higher resolution slope in each of the TAs. Let us say for TA2, which we have already studied, we have S-21 and S-22, and maybe then have 6 models for gentler, moderate and steep slopes, and then use the combined models to scale the slope for the entire TA2 (Figure 9).

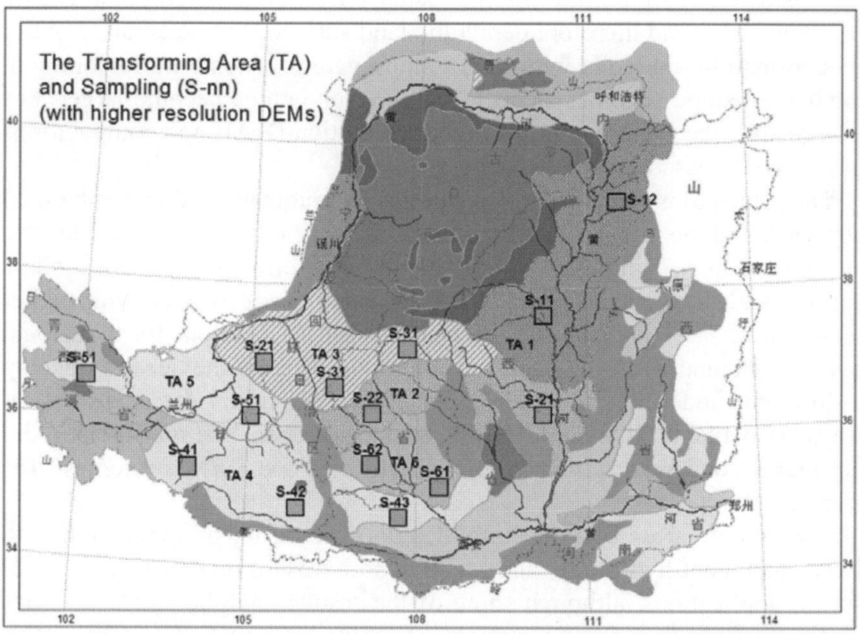

Figure 9. Proposed transforming areas (TA) in the Loess Plateau.

4 Discussion and Conclusions

There are two methods of studying slope: one is slope mapping based on topographic maps, which records the classification code of slope, and the other is to extract slope surfaces based on DEMs and then compute statistics and/or make a map of slope classification or retain the map in raster format and use GIS tools (Tang 1987). These approaches meet the demands for scientific use and policy-making, but have limitations requiring deeper understanding and study of slope issues, especially for slope reduction. In this study, the DEMs and slopes are all regarded as being spatially continuous surfaces that have continuous values and histograms. Based on these ideas, some theories and techniques for digital remote sensed image processing and digital map manipulation, such as histograms, map algebra, etc., can be applied. This is an innovative idea for slope study and also for scaling in geography (Zhang *et al.* 2002, Li and Cai 2005).

In this study, a transformation function has been built by histogram matching. With this approach, the lower resolution slope (***LRS***) can be transformed, and the resultant will have the major terrain characteristics of the histogram and partial surface pattern from finer resolution DEMs (training area), and therefore depict the land surface more accurately. After transformation, the cells with larger slope values become more common, the overall slope becomes steeper, and the slope range become larger. For a large area, the requirement for higher resolution DEMs and computations has been decreased greatly as well.

The principal innovation of this study is its pragmatic and computational approach and the integration of methods of image processing, map algebra, and digital map manipulation, so that the transformation is from a slope surface with lower local relief (***CRS***) to a new slope surface with higher local relief. The resultant surface can be used as an input for a regional scale distributed hydrology and erosion model without the need to develop a different model at a finer scale. Such high resolution data may not always be available. It is also important for researchers and managers in departments concerned with geo-infrastructure to use DEM resources efficiently.

We have to stress that the study has not had sufficient application with large catchment hydrologic and regional soil erosion models to test its result in applications, although some of the results have been used as supporting information for vegetation suitability mapping (Li *et al.* 2005, McVicar *et al.* 2005, Zhang *et al.* 2006).

Although the method is very promising, it is far from perfect. Questions that should be studied further include: (1) the underlying mechanism of slope reduction, based on the theory of geomorphology, cartography, geostatistics, and fractal mathematics; (2) how to modify and extend the models by exploring the histogram matching approach for multiple resolutions and identifying the scale ranges that can be transformed meaningfully; then investigating how to use terrain statistics to recognize areas as similar in the way they scale and the mechanism of scaling using the coarse level data; (3) how to divide the slope sections, or alternative ways to make the transformation function continuous; (4) application and testing of the results in operational hydrology and erosion modelling.

Despite the amount of work to be done, the objective has very practical and useful applications and addresses a current and serious problem for the practical use of existing topographic data in hydrology and erosion applications.

Appendix A

The AML code used here to transform slope in an ARC/INFO GRID session

```
slp01 = 0.1627 * pow (SLP25,2) + 2.74 * SLP25 - 0.0757
slp02 = -0.00002 * pow (SLP25,4) + 0.0023 * pow (SLP25,3) - 0.0998 *
    pow (SLP25,2) + 2.5795 * SLP25 + 0.4602
slp03 = 2.5415 * SLP25 - 58.961
if (SLP25 <= 1.6) then SLPt = slp01
    else if (SLP25 > 1.6 && SLP25 <= 40) then SLPt = slp02
    else SLPt = slp03
endif
kill slp01 all; kill slp02 all; kill slp03 all
```

Note: pow is power function in ARC/INFO GRID session; SLP_{25} refers to coarser resolution slope, so pow (SLP25,2) means SLP_{25}^2.

Acknowledgments

The research was funded by the Project of National Basic Research Program (Water Erosion Processes and Dynamic Changes at the Regional Scale, 2007CB407203), National Key Technologies R&D Program (Key techniques for soil erosion monitoring and assessment, 2006BAD09B05),

CAS Innovation Project (Dynamic Assessment and Scenario Simulation of Soil Erosion in China, KZCX2-YW-401), and Project of The Yellow River Water Conservancy Committee (Research on Regional Soil Erosion Model, 2004SZ01—04). Thanks for help from Dr Tim McVicar, Mr Tom Van Niel, Mr Li Lingtao, Dr John Gallant and Mr Trevor Dowling during the process of generating the hydrologically correct DEMs used in this study. Thanks also to the anonymous reviewers who provided valuable comments, which helped improve the paper.

References

Armstrong, R.N., Martz, L.W., (2003), Topographic parameterization in continental hydrology: a study in scale, *Hydrological Processes*, **(17)** : 3763–3781.

Burrough, P.A., (1986), Principles of Geographical Information Systems for Land Resources Assessment, New York, Oxford University Press.

Chen, Y., Qi, Q.W. and Tang, G.A., (2004), Research on Slope Conversion–atlas in Loess Plateau, *Agricultural Research in the Arid Areas*, (**22**) 3: 180–185 (in Chinese).

Cochrance, T.A. and Flanagan, D.C., (2004), Effect of DEM resolutions in the runoff and soil loss predictions of the WEPP watershed model, *Transactions of the ASAE,* **(47)** 6: 1–12.

David, M.W. and McCabe, G.J., (2000), Differences in topographic characteristics computed from 100- and 1000-m resolution digital elevation model data, *Hydrological Processes,* **14**: 987–1002.

Gao, J. (1997), Resolution and Accuracy of Terrain Representation by Grid DEMs at a Micro-scale, *International Journal of Geographical Information Science,* **(11)** 2: 199–21.

Gonzalez, R.C. and Woods, R.E., (1992), *Digital image processing, 3rd edition,* Reading, Mass: Addison-Wesley.

Hao, Z.C., Chi, C.X., Wang, L., and Wang Y. K. (2005), A preliminary Analysis of DEM Space Data Resolution, *Advances in Earth Science*, **(20)** 5: 499–504 (in Chinese).

Harrison, B.A. and Jupp, D.L.B., (1990), *Introduction to image processing*, Canberra, Australia: CSIRO Division of Water Resources.

Horn, B.K. and Woodham, R.J., (1979), Destriping Landsat MSS images by histogram modification, *Computer Graphics and Image Processing*, **10**: 69–83.

Hummel, R., (1975), Histogram modification techniques, *Computer Graphics and Image Processing,* **4**: 209–224.

Hutchinson, M., (1989), A new procedure for gridding elevation and stream line data with automatic removal of spurious pits, *Journal of Hydrology*, **106**: 211–232.

Hutchinson, M.F., (2000), Digital Elevation Models and Representation of Terrain Shape. In Wilson, J.P. and Gallant, J.C., *Terrain Analysis, Principles and Applications,* New York: John Wiley & Sons, Inc: 29–50.

Hutchinson, M.F., (2004), *ANUDEM Version 5.1,* Canberra: Centre for Resource and Environmental Studies, Australian National University.

Li, L.T., McVicar, T.R. and Van Niel, T.G., (2005), A Bilingual User's Guide for the Decision Support Tool for Managing Re–Vegetation and Its Impact on Hydrology (ReVegIH) in the Coarse Sandy Hilly Catchments of the Loess Plateau, China, Canberra: CSIRO Land and Water.

Li, S.C. and Cai, Y.L., (2005), Some scaling issues of geography, *Geographical Research,* **(24)** 1: 11–18.

Liu, X.H., Yang, Q.K. and Li, R., (2001), Extraction and Application of Relief of China Based on DEM and GIS Method, *Bulletin of Soil and Water Conservation,* **(21)** 1: 57–59 (in Chinese).

Liu, Z.H., (2007), *Assessment of soil and water loss at national scale using RS and GIS techniques*, Doctorial degree thesis of Institute of Soil and Water Conservation, Chinese Academy of Sciences and Ministry of Water Resources, Yangling, China.

Martin, P.L., Lawrence, W.M., Geoff, W.K. and Jurgen, G., (2002), Using digital terrain analysis modelling techniques for the parameterization of a hydrologic model, *Environmental Modelling & Software,* **17**: 127–136.

McVicar, T.R., Wen, Z.M., Li, L.T., Van Niel, T.G. and Jiao, F., (2005), Mapping Perennial Vegetation Suitability and Identifying Targeting Areas for Implementing the Re–Vegetation Program in the Coarse Sandy Hilly Catchments of the Loess Plateau, China, Canberra, Australia: CSIRO Land and Water.

Ren, X.Y., Zhang, X.S., Hao, F.H., Cheng, H.G. and Yang, Z.F., (2004), Effects of DEM Resolution on Simulated Runoff and Sediment Yields, *Research of Soil and Water Conservation* **(1)** 1: 1–4 (in Chinese).

Tang, G.A., (1987), *A Study of Slope Classification and Mapping,* Master degree thesis of Geography Department, Northwest University, Xi'an, China.

Tang, G.A., Yang, Q.K., Zhang, Y., Liu, Y.M. and Liu X.H., (2001), Research on Accuracy of Slope Derived From DEMs of Different Map Scales, *Bulletin of Soil and Water Conservation,* **(21)** 1: 53–56 (in Chinese).

Tang, G.A., Zhao, M.D., Li, T.W., Liu, Y.M, Xie, Y.L., (2003), Modelling Slope Uncertainty Derived from DEMs in Loess Plateau, *Acta Geographica sinica,* **(58)** 6: 824–830 (in Chinese).

Tang, X.M., Li, L., Ji, X.Y, Wang, D.H., Shang, Y.L., Liu, J.J., Song, H.Y., Li, L.M., Liao, A.P., Zhai, Y., Yan, R.H., (2002), Establishment of 1: 10,000 Scale DEM Data Base for Priority Flood Control Areas of China Seven Great River Valleys, *Bulletin of Surveying and Mapping,* **(6)** : 19–22.

Thompson, J.A., Bell, J.C. and Butler, C.A., (2001), Digital elevation model resolution: effects on terrain attribute calculation and quantitative soil–landscape modelling, *Geoderma,* **(100)** : 67–89.

Wang, J.Y., (1993), *Principles of Generalization for Topographic Map,* Beijing: Survey and Mapping Press (in Chinese).

Wu, X.F., Liu, C.M. and Wang, Z.G., (2003), Effect of horizontal resolution of raster DEM on drainage basin characteristics, *Journal of Natural Resources,* **(18)** 2: 168–173 (in Chinese).

Yang, Q.K., Li, R. and Liang, W., (2006), Cartographic Analysis on Terrain Factors for Regional Soil Erosion Modelling, *Research of Soil and Water,* **(13)** 1: 56–58 (in Chinese).

Yang, Q.K., Van Niel, T.G., McVicar, T.R., Hutchinson, M.F. and Li, L.T., (2005), *Developing a digital elevation model using ANUDEM for the Coarse Sandy Hilly Catchments of the Loess Plateau, China,* Canberra, Australia: CSIRO Land and Water Technical Report, 7/05.

Yin, Z.Y. and Wang, X., (1999), A Cross–scale comparison of drainage basin characteristics derived from digital elevation models, *Earth Surface Processes and Landforms*, **24**: 557–562.

Zhang, X., Drake, N. and Wainwright, J., (2002), Scaling land surface parameters for global–scale soil erosion estimation, *Water Resources Research,* **(38)** 9: 1180–1189.

Zhang, X., Drake, N.A., Wainwright, J. and Mulligan, M., (1999), Comparison of slope estimates from low resolution DEMs: scaling issues and a fractal method for their solution, *Earth Surface Processes and Landforms,* **(24)**: 763–779.

Zhang, X.P., Lu, Z., McVicar, T.R., Van Niel, T.G., Li, LT., Li, R., Yang, Q.K., Liang, W., (2006), Modelling the impact of afforestation on water yield in the Loess Plateau, China, *Hydrological Processes* (accepted).

Zhao, C.X., (1989), *Datasets of Slope Gradient of Loess Plateau*, Beijing: Ocean Press (in Chinese).

Zhao, M.D., (2007), *Terrain Factors for Soil Erosion Assessment at the Regional Scale*, Doctorial degree thesis of Institute of Soil and Water Conservation, Chinese Academy of Sciences and Ministry of Water Resources, Yangling, China.

Section 3: Hydrological Terrain Analysis

Water in the Landscape: A Review of Contemporary Flow Routing Algorithms

John P. WILSON, Graeme AGGETT, DENG Yongxin and Christine S. LAM

Abstract

This chapter reviews the various flow routing algorithms that simulate the distribution and flow of water across landscapes. The distinguishing characteristics of nine such algorithms and the experiments that have been conducted to evaluate their performance over the past 15 years are discussed. From there, we consider three sets of enduring challenges: (1) the role of scale and feedback between soil and water, and the need to consider these issues when characterizing the properties of both; (2) the need for dynamic flow routing algorithms and related indices in many landscapes; and (3) some of the as yet unrealized opportunities for treating space and time as continuous variables in the representation of soil water properties. The chapter concludes by noting the current state-of-the-art and where we might go from here.

Keywords: DEMs, flow routing algorithms, soil water relationships.

1 Introduction

A growing body of literature from the 1990s illustrates how flow routing measurements and related topographic attributes can be used in modelling key hydrologic processes controlling the spatial distribution of soil moisture, runoff, and soil erosion in a simplified but realistic manner (e.g. Band 1989, Moore *et al.* 1993, Abbott and Refsgaard 1996, Cluis *et al.* 1996, Maidment 1996, Da Ros and Borga 1997, Beven 1998, Storck *et al.* 1998). The identification of drainage pathways and runoff contributing areas based on DEMs, together with their coupling with hydrological models (e.g. Beven *et al.* 1994, Lee and Chu 1996), provides the means to parameterize spatially distributed, physically-based models, which themselves represent a major approach for incorporating spatial heterogeneity. Digital terrain analysis provides a quantitative and consistent approach to generating inputs for applications of these models as discussed below.

The elevation, slope, and aspect of an area have a strong influence on its microclimate due to insolation and other effects, and topography has a major impact on the hydrological, geomorphological, and biological processes active in landscapes (Moore *et al.* 1993, Dymond *et al.* 1995, Cluis *et al.* 1996, Pickup and Chewings 1996). Geomorphometric parameters derived from DEMs can be used to determine where in a watershed various slope processes, such as landslides and runoff, take place (Montgomery *et al.* 1998). At the local scale, primary geomorphometric parameters can be extracted using standard GIS tools in order to investigate morphometric influences on hydrologic variables such as overland flow depth and velocity. For example, the widely used hydrologic model TOPMODEL is based on the concept of variable source areas contributing to runoff production through saturated overland flow (Beven *et al.* 1994). Formation of the contributing area is related to the topographic index $\ln(A_s/\tan\beta)$, where A_s is the upslope area drained per unit contour length and β is the slope angle. Model inputs are the frequency distribution of $\ln(A_s/\tan\beta)$, daily precipitation and evapotranspiration time series, and several lumped soil and flow routing parameters. Model outputs include the runoff hydrograph, water balances, and contributing areas. Developments in hydrologic models have been greatly facilitated by GIS, which supports the spatial data models that have enabled earth scientists to construct more distributed representations of space than previously possible. Using GIS to parameterize such models has enabled their application across local, watershed, and regional scales, facilitating more realistic model assessment and more accurate process modelling.

The strong influence of elevation and watershed morphology on precipitation, water movement, and slope stability means that DEMs serve as one of the basic building blocks of many environmental model parameterization efforts, and the enhancements made to flow routing measurements have enhanced this capacity. The increasing availability of DEMs, remotely-sensed data, and a dramatic increase in desktop computing power over the past decade have accelerated these developments, enabling researchers to link their chosen process-based model(s) to a spatial database contained within a GIS.

This chapter reviews the most popular flow routing algorithms and what is known about their performance. The choice of algorithm is critical given the key contribution of water distribution and flow in soil development, land cover, soil redistribution, and various forms of mass movement. The remainder of this chapter is divided into three sections. The distinguishing characteristics of nine such algorithms – the D8 (O'Callaghan and Mark 1984), Rho8 (Fairfield and Leymarie 1991), FD8 (Quinn *et al.* 1991), Lea (1992), DEMON (Costa-Cabral and Burgess 1994), ANSWERS (Beasley and Huggins 1978), flux decomposition (Desmet and Govers 1996), D∞ (Tarboton 1997), and MFD-md (Qin *et al.* 2007) algorithms – are first outlined. We then review the continued importance of source data, interpolation algorithms, and the experiments that have been conducted to evaluate their performance over the past 15 years in Section 3. Three groups of studies – those focused on inputs and/or decision rules and those focused on the ability of one or more of the aforementioned algorithms to reproduce the drainage structure or some relevant landscape properties with and without the assistance of field observations – are discussed. The last section concludes by noting the current state-of-the-art and where we might go from here.

2 Basic Characteristics of Flow Routing Algorithms

The automated extraction of surface channel networks from DEMs has grown in popularity during the past 20 years as the availability and resolution of DEMs, as well as the quality of hydrologic modelling tools, has improved. Identifying channel networks and their initiation points is central to hydrology and geomorphology because of the control exerted by climate, topography, soil properties, and other environmental attributes on surface flow paths and erosion potential within a drainage basin (Knighton 1998). The characteristics of a channel network heavily depend on the identification of channel source cells from the digital landscape, and can greatly affect the delineation of catchments or drainage basins (Garbrecht *et al.* 2001). The closer the channels begin to the drainage divide, the greater the number of channels that can occupy a watershed (Montgomery and Dietrich 1988). DEMs generally cannot capture all topographic variations that occur within the landscape, especially if the features are finer than the DEM resolution, and these shortcomings may cause discrepancies between the precise positioning of stream channels in digital landscapes and the real world (Garbrecht *et al.* 2001).

Flow routing algorithms have been used to predict the channel source cells as well as the movement of water, sediment, and nutrients to lower adjacent points or areas in a landscape (e.g. Desmet and Govers 1996). Fundamentally, a flow routing algorithm determines the way in which the outflow from a given cell will be distributed to one or more neighbouring downslope cells. The choice of flow routing algorithm is important as it affects the calculation of the upslope contributing area, the prediction of flow accumulation, and several other topographic and hydrologic attributes. The derivation of these attributes relies on digital elevation source data. Square-grid DEMs are a popular choice for flow routing due to their visual simplicity and ease of computer implementation (Moore *et al.* 1991, Wise 1998a, b, Wilson and Gallant 2000). All nine of the flow routing algorithms discussed below utilize square-grid DEMs as their primary input data, and calculate flow directions and upslope contributing areas using a 3 x 3 moving window.

The D8 (deterministic eight-node) single-flow-direction (SFD) algorithm directs flow from each grid cell to one of eight nearest neighbours based on slope gradient (O'Callaghan and Mark 1984). The aspect Ψ (measured in degrees clockwise from north) marks the direction of steepest descent for each grid cell or point in a catchment, and is the direction in which water would flow from that grid cell or point. Most implementations of D8 utilize the primary flow direction for water moving over the land surface as an approximate replacement for aspect (Moore 1996). The simplest method of calculating primary flow direction is to determine the slope (S_i) to each neighbour and set it to the direction for which S_i is greatest (Gallant and Wilson 2000). The upslope contributing area is the number of cells whose flow reaches the cell of interest multiplied by the cell area, while specific catchment area is the upslope contributing area divided by the contour width, which is assumed to equal the "width" of a grid cell. Some implementations of D8 utilize the grid spacing for both cardinal and diagonal flow assignments, while others, such as TAPES-G (Gallant and Wilson 1996), assume that the grid cell width is a good estimate for flow width in the cardinal directions and that the cell width multiplied by $\sqrt{2}$ is the best estimate of the flow width for flow assignments to diagonal cells. However, there is little theoretical or empirical evidence to support either option (Gallant *et al.* 2000).

The Rho8 (random eight-node) SFD algorithm developed by Fairfield and Leymarie (1991) introduced a degree of randomness to break up the parallel flow paths that D8 tends to produce on planar surfaces (Wilson and Gallant 2000). This algorithm starts by identifying all the neighbouring downslope cells, then calculates the slope gradients in each of these directions, and finally extracts random numbers from a table to direct the flow to one of these candidate cells. The random numbers are allocated on a slope-weighted basis such that the potential flow paths with the steepest gradients have the greatest probability of being selected, and the overall flow pattern more or less matches the one produced with D8. The upslope contributing and specific catchment areas are calculated using the flow width and flow accumulation approaches adopted for D8; however, a different flow network will be produced each time the algorithm is used because of the random assignment of flow among multiple downslope cells (Wilson *et al.* 2000).

The FD8 multiple flow direction (MFD) algorithm developed by Quinn *et al.* (1991) directs water to every adjacent downslope cell on a slope-weighted basis. The slope gradients, slope lengths, and two weights – 0.5 and 0.35 for cardinal and diagonal directions, respectively – are used to direct the flow from the centre cell to each downslope cell in a 3 x 3 moving window. Each cell receives a fraction of the discharge from each upslope cell, and therefore, the upslope contributing area of the receiving cell is typically composed of partial contributions from many different cells. Specific catchment area is calculated as the sum of the contributing areas from upslope cells divided by the cell width for the cardinal flow directions, and by the cell width multiplied by √2 for diagonal flow directions (similar to D8 and Rho8 in TAPES-G). The user can set a maximum cross-grading area threshold in the TAPES-G implementation of FD8 to switch to the D8 algorithm. This approach means that flow dispersion will be terminated whenever the upslope contributing area exceeds this user-specified threshold (Gallant and Wilson 1996, Wilson and Gallant 2000).

Lea's (1992) flow routing algorithm relies on the calculation of the aspect vector and a surface fitting scheme. He argued that flow moves across a planar surface in the direction of the steepest slope, or aspect angle θ, similar to a "rolling ball". The approach has two parts. First, planes are constructed to represent the surface of each cell using estimated elevations at the four corners of each cell. Successively larger windows can be implemented to minimize the occurrence of flat areas (i.e. surface pits). The aspect vector is calculated during the second step in 1° increments (in contrast to the 45° increments used for many implementations of D8) and is utilized to route flow across individual cells. Flow paths are constructed by the repeated application of the algorithm until the catchment outlet is reached or a topographic hollow prevents the continued progress of flow. The contributing area is calculated as the number of flow paths passing through that cell multiplied by the grid cell area, and an arbitrary threshold is utilized to dictate the number of flow paths that need to converge on a pixel for it to be classified as a stream path.

The fifth algorithm called DEMON (Digital Elevation Model Network) was developed by Costa-Cabral and Burgess (1994) and determines flow direction based on the local aspect angle similar to Lea (1992). However, the flow generated over a cell is directed downslope over a two-dimensional flow strip. These flow strips partition catchments into irregularly shaped elements that are defined by pairs of orthogonals and equipotential lines (contour lines). The width of the flow strips increases over divergent topography, decreases over convergent topography, and remains constant over planar surfaces. The flow across each cell is the amount of flow entering that cell plus the flow generated by the cell itself. When flow reaches an edge of a grid cell at a cardinal direction, then all flow is directed to the single neighbour. In other cases, the flow is split amongst the cardinal neighbours. The upslope contributing area for each cell in DEMON is computed by successive addition of the cell areas in each stream tube entering every pixel in the DEM, and the specific catchment area is computed by dividing the upslope contributing area by the flow matrix width. A modified version of DEMON is implemented in TAPES-G in which the nodes of the DEM define the centre of the cells instead of the entire cell area, and the flow direction of a stream tube is defined by the aspect. The upslope contributing and specific catchment areas are calculated in the same way as in the original version of DEMON in TAPES-G (Gallant and Wilson 2000).

The ANSWERS (Areal Nonpoint Source Watershed Environment Response Simulation; Beasley and Huggins 1978) non-point source pollution model describes the runoff, infiltration, subsurface drainage, erosion, and drainage across a watershed during and following single storm events. The watershed is divided into grid cells with parameters provided for each cell, and the continuity equation is used with a stage-discharge curve to calculate the amount of flow that would cross each cell. The cells are split into two parts by a line through one of the cell corners and oriented in the aspect direction of the cell, and the relative proportions used to divide and direct the flow of water into the neighbouring downslope cardinal cells. This algorithm was implemented as a FORTRAN program and coupled with IDRISI (Eastman 1992) by Desmet and Govers (1996) as part of a study comparing the performance of flow routing algorithms in a small catchment near Flanders, Belgium. Two modifications were made to the original ANSWERS algorithm by Desmet and Govers (1996) to solve specific problems such that: (1) flow was assigned to just one of the two cardinal cells when flow was directed to grid points of equal or even higher height; and (2) flow was switched to the D8 steepest descent algorithm (i.e. the diagonal path in a 3 x 3 moving window) when both of the two receiving cells were higher than the central point.

Desmet and Govers (1996) also proposed a new flow routing algorithm based on the decomposition of the flux vector. The flux vector was split into two ordinal components with the magnitude of each component proportional to the sine or cosine of the aspect value. The magnitudes of the two components were normalized by dividing each by the sum of the absolute values of the sine and cosine of the aspect value, and the two modifications noted above for the ANSWERS flow routing algorithm were adopted as a part of this algorithm as well. This algorithm splits the upslope contributing area between two cardinal neighbours and the calculation of specific catchment area is similar to that of D8, FD8 and Rho8 in

TAPES-G, where the contributing area is divided by the effective contour length. The major difference between the ANSWERS and flux decomposition algorithms concerns the routing of flow to the two cardinal neighbours. Beasley and Huggins (1978) chose to divide grid cells based on which cardinal neighbour received flow lines parallel to the aspect direction, whereas Desmet and Govers (1996) relied on the sine and cosine of the aspect vector values to direct flow to these candidate cells.

The D∞ algorithm proposed by Tarboton (1997) incorporates several ideas from DEMON to assign multiple flow directions to selected cells. The flow direction follows the path of steepest descent and is represented as a continuous angle between 0 and 2π radians. Special rules are included to: (1) force flat cells to drain to a neighbour that ultimately drains to a lower elevation; and (2) eliminate loops in the flow direction angles. Grid cells that are flat took flow direction from the D8 method in the original D∞ code, but the latest version uses the method of Garbrecht and Martz (1997) to assign flow directions in flat areas. This algorithm returns NODATA for flow direction in grid cells classified as pits. The upslope area of each cell is taken as its own area plus the fractional areas of upslope neighbours that drain into the cell of interest, similar to FD8 and DEMON. If the angle falls on a cardinal or diagonal direction, then the flow from each cell drains to one neighbour. If the flow direction falls between the direct angles to two adjacent neighbours, the flow is apportioned between the two cells according to how close the flow direction angle is to the direct angle for those cells.

The final MFD-md algorithm proposed by Qin *et al.* (2007) utilizes local topographic conditions to partition the flow between downslope neighbouring cells. This algorithm modifies the flow partition approach of Quinn *et al.* (1991) by utilizing the maximum downslope gradient to model the impact of local terrain on the flow partitioning predicted at each cell. The maximum slope gradient was chosen for inclusion in this algorithm over the minimum and mean downslope gradients because: (1) it is less sensitive to variations in DEM error; and (2) the new algorithm will behave like D8 in steep terrain (Qin *et al.* 2007). The MFD-md flow partitioning scheme uses an exponent that takes values between 1.1 and 10 to model divergent (small flow partition exponent values) and convergent flow (large exponent values) across the landscape (similar to the schemes proposed by Freeman 1991, Holmgren 1994, and Quinn *et al.* 1995).

This proliferation of flow routing algorithms raises an important question; namely, whether one or more of these algorithms performs better than the others in specific landscapes and/or applications. Figure 1 builds on the approach of Qin *et al.* (2007) and shows the routing of flow from the centre cells in three hypothetical DEMs to one or more downslope neighbours for the nine aforementioned flow routing algorithms. These relatively simple examples show how different flow routing algorithms can generate substantially different estimates of upslope contributing area and related attributes (specific catchment area, topographic wetness index, etc.). The major findings from published studies comparing the performance of two or more of these flow routing algorithms are taken up and discussed in more detail below.

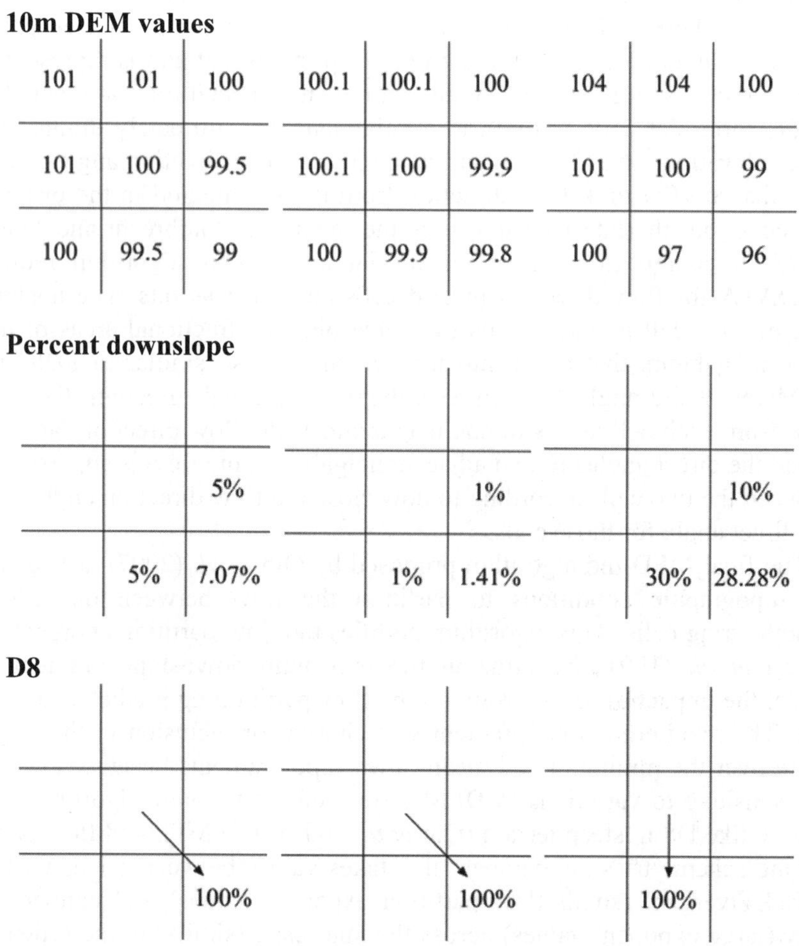

Figure 1. Flow partitioning schemes for nine flow routing algorithms and the three sample DEMs.

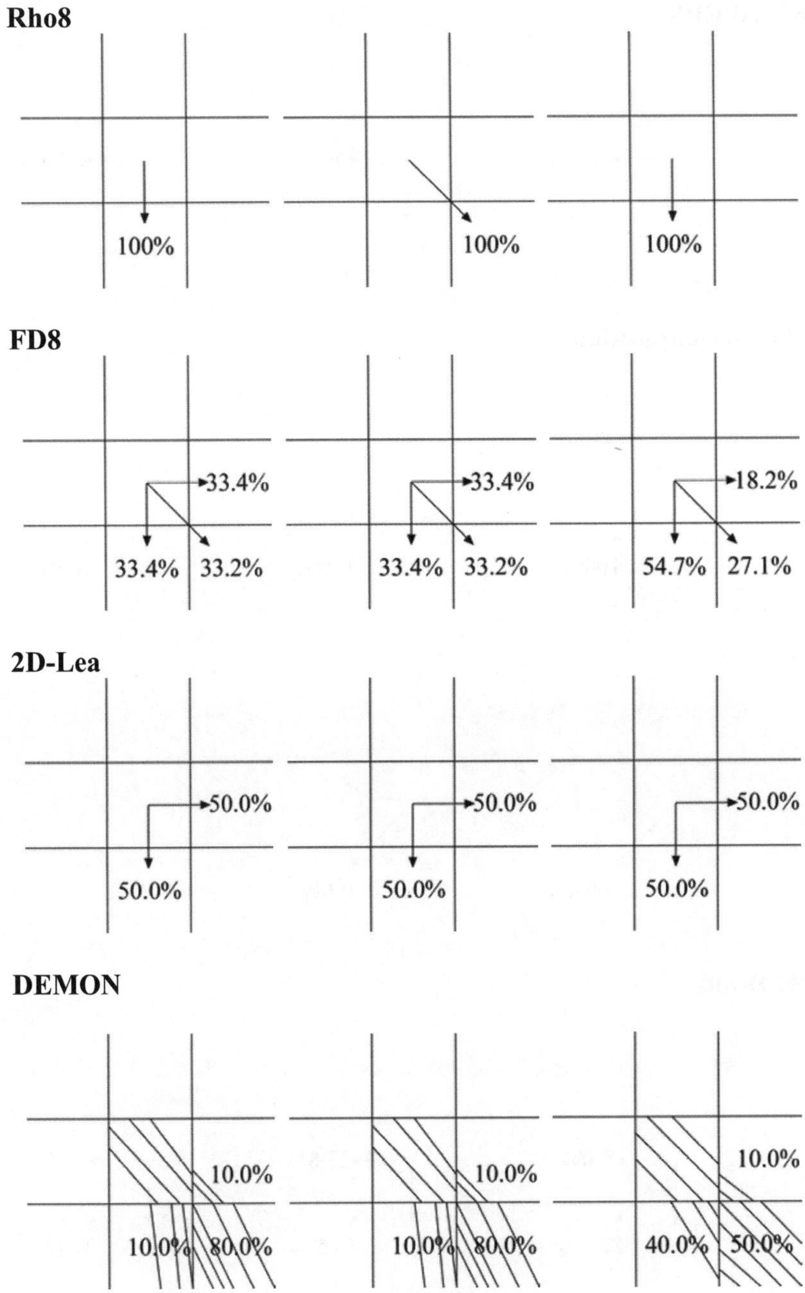

Figure 1. (Continued)

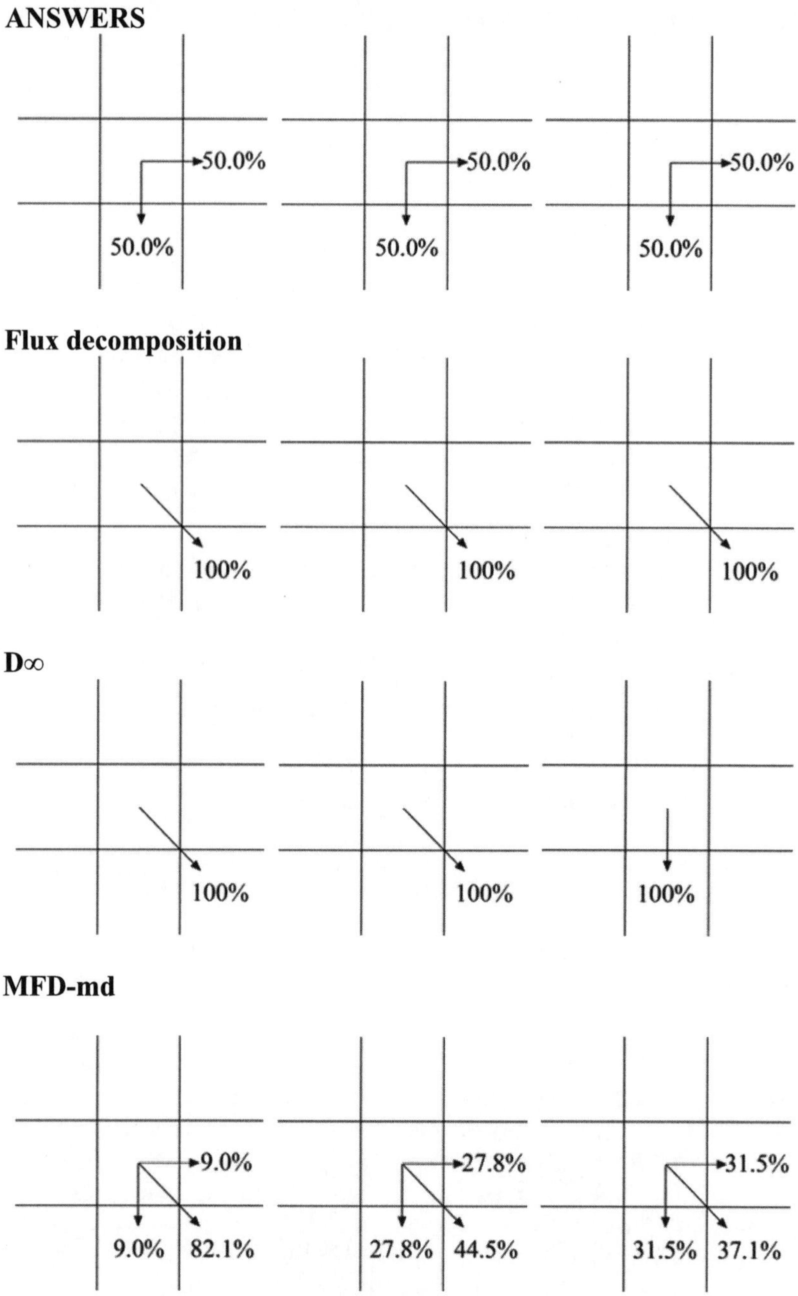

Figure 1. (Continued)

3 Comparisons of Flow Routing Algorithms

The evaluation of flow routing algorithm performance is tricky because of the importance of the underlying source data and the difficulty of separating the impact of the flow routing algorithms from that of the underlying data when reviewing their performance. The elevation data may take one of three forms (square-grid, triangulated irregular, and contour-based networks) although the proliferation of digital elevation sources and preprocessing tools means that the initial choice of data structure is not as critical as it once was (Kemp, 1997a, b). Numerous methods have been proposed to convert digital elevation data from one structure to another, but care must still be exercised with each of these methods to minimize unwanted artefacts (e.g. Krajewski and Gibbs 1994).

Hutchinson (2007) recently documented a 20-year trend of shifting spatial scales – from continental and regional scales (e.g. major drainage divisions), to the mesoscale (e.g. surface climate), and then to the toposcale (e.g. soil properties) – in hydro-ecological applications of topographic data. These scale transitions have mirrored advances in DEM resolutions and improvements in representing local terrain shape and structures. This focus on local details has allowed landforms and hydrological patterns to be differentiated within small watersheds (e.g. 1–100 km^2 in size) and hillslopes (e.g. 100–1,000 m in length). However, larger quantities of data do not necessarily produce better results: Eklundh and Martensson (1995), for example, used ANUDEM (Hutchinson 1989) to derive square grids from contours and demonstrated that point sampling produces faster and more accurate square-grid DEMs than the digitizing of contours. Similarly, Wilson *et al.* (1998) used ANUDEM to derive square grids from irregular point samples and showed that many of the x, y, z data points acquired with a truck-mounted GPS were not required to produce satisfactory square-grid DEMs. ANUDEM calculates ridge and streamlines from points of maximum local curvature on contour lines and incorporates a drainage enforcement algorithm that automatically removes spurious sinks or pits in the fitted elevation surface (Hutchinson 1989). ANUDEM is one of several programs of this type (see Maidment 1996 and Hellweger 1996 for other examples) that modify a DEM to reflect known hydrology, and there are many other methods that have been proposed over the years to automatically extract drainage networks and ridgelines from digital elevation data (e.g. Qian *et al.* 1990, Smith *et al.* 1990).

This proliferation of digital elevation data sources and preprocessing tools is to some extent problematic given the task at hand. Carrara *et al.* (1997), for example, compared several methods for generating DEMs from contour lines and concluded that the range of terrain types, sample structures, and modelling routines is so great that attempts to make generalizations about "best" models is tremendously difficult. Similarly, Callow *et al.* (2007), for example, examined three different algorithms that modify a DEM to reflect known hydrology and showed that these methods permanently altered the source DEM and a variety of computed topographic attributes. Some of the interpolation methods that have been proposed are difficult to use and as a consequence Eklundh and Martensson (1995) recommended that less experienced users focus on the quality of the input data instead of learning sophisticated interpolation methods. Simpler interpolation methods will give satisfactory results so long as the input data are well sampled and sophisticated algorithms are likely to produce unsatisfactory results if applied to poor data (e.g. Wilson *et al.* 1998).

It is perhaps not surprising given this background that many modellers accept the DEMs they work with uncritically despite an ever-increasing literature describing the causes of systematic and random errors in DEMs, and their effects on morphometric and hydrologic parameter estimation (e.g. Lagacherie *et al.* 1996, Lopez 1997, Murillo and Hunter 1997, Wise 1998a, b). If undetected and uncorrected they may propagate into the process models they are used to parameterize, causing considerable uncertainty in the reliability of their simulations. Small errors in elevation or strange behaviour by an interpolator can produce large errors in surface derivatives such as gradient, and topographic surfaces used to define boundary conditions in environmental modelling applications will contain error (Desmet 1997, Liu and Jezek 1999).

Numerous studies have attempted to evaluate the performance of two or more flow routing algorithms notwithstanding the complications introduced by the choice of source data and/or interpolator and the presence of systematic and/or random errors. These studies can be grouped into three sets and their results are discussed in some detail below. The first two studies have examined the sensitivity of flow routing predictions to one or more of the decision rules embedded in the chosen flow routing algorithms.

Wilson *et al.* (2000) examined the effect of DEM source, grid resolution, and choice of flow routing algorithm on three primary and two secondary topographic attributes for a large forested catchment in southwest Montana. The comparisons showed that the D8 and Rho8 SFD algorithms initiated flow from 30–40% of the cells and produced much higher proportions of cells with small upslope contributing areas compared to the FD8 and DEMON MFD algorithms. The results also showed that the choice of cross-grading area threshold, which is utilized in TAPES-G to switch from FD8 to D8, produced very small differences (<5%) in upslope contributing and specific catchment area values. Overall, the results showed that the two MFD algorithms agreed with each other in 71% of the grid cells and that the other algorithms agreed with each other in 49–57% of the pairwise comparisons.

Chirico *et al.* (2005) evaluated several different methods for defining flow width on grids when computing SCA in the second study. Five approaches – D8 with flow width equal to grid size regardless of cell flow direction; modified D8 (D8_v1) with flow width equal to grid size for cardinal flow directions and grid size times $\sqrt{2}$ for diagonal flow directions; modified D8 (D8_v2) with flow width equal to grid size for cardinal flow directions and grid size times $1\sqrt{2}$ for diagonal flow directions; D∞ with flow width equal to grid size regardless of cell flow direction; and modified D∞ (D∞_v1) with flow width varying as a function of flow direction – were tested on sloping planes, inward and outward cones and then compared with theoretical SCA values. Two dimensionless parameters – the global resolution, defined as the ratio of a characteristic length of the study area to the grid size, and the upslope area resolution, defined as the ratio of the local theoretical SCA to the grid size – were used to evaluate the performance of the five approaches. The results, cast in terms of the pattern of errors (i.e. absolute bias, mean absolute error, and local relative error) across different grid sizes indicated that D8 and D∞ performed better than the modified D8 and D∞ algorithms in calculating SCAs.

The second group of studies comparing the performance of two or more of the flow routing algorithms examined their ability to reproduce the drainage structure and/or some other topographic attribute. Desmet and Govers (1996), for example, evaluated six flow routing algorithms in terms of their ability to: (1) reproduce the main structure of the catchment; and (2) predict the location of ephemeral gullies. The D8 and Rho8 SFD routing algorithms produced different spatial and statistical patterns from each other and two pairs of MFD algorithms – the MFD algorithms of Quinn *et al.* (1991) and Freeman (1991), both of which allocate flow to up to eight neighbouring cells, and the ANSWERS (Beasley and Huggins 1978) and flux decomposition (Desmet and Govers 1996) algorithms, which allocate flow to one or two downslope neighbours – for their small study site in Flanders, Belgium. The MFD algorithms produced much smoother images compared to the SFD algorithms (similar to Wolock and McCabe 1995) and taken as a whole, Desmet and Govers (1996) favoured the two algorithms that allowed flow to only one or two downslope neighbours because they (visually) produced a stronger correlation with the main drainage lines. The main structure of the catchment (i.e. the interfluves and main drainage lines) was identified by all six flow direction algorithms and most of the variability occurred in higher elevation areas (based on the maps reproduced in the manuscript).

Desmet and Govers (1996) also examined the effect of the choice of flow direction algorithm on the prediction of ephemeral gullies identified using the methodology of Moore *et al.* (1988). Their results showed that the MFD algorithms were able to identify areas where ephemeral gully erosion is likely, but they could not predict the precise location of the gullies (which never exceeded half the width of the grid cells for this particular study area). The SFD algorithms predicted ephemeral gullies to start higher on the slopes but the correspondence with observed patterns was erratic because these algorithms were very sensitive to small elevation errors.

Zhou and Liu (2002) computed "true" SCAs for ellipsoid, inverse ellipsoid, saddle, and planar simulated surfaces and compared these values to the SCAs derived from the D8, Rho8, Freeman (1991), DEMON, and D∞ flow routing algorithms. The accuracy and spatial distribution of residuals were also analysed by calculating the Root Mean Square Error, mean error, and standard deviation. They found that DEMON generated the lowest randomly distributed error values across all surfaces. Qin *et al.* (2007) used the same simulated surfaces and statistics to compare the performance of their new MFD-md algorithm with D8 and a derivative of the MFD algorithm of Quinn *et al.* (1991). Their results showed that MFD-md produced the lowest error amongst the three algorithms across all four simulated surfaces.

Wilson *et al.* (2007) compared the performance of pairs of SCA grids computed from five flow routing algorithms (D8, Rho8, FD8, DEMON, and D∞) across six user-defined fuzzy landscape classes. Table 1a lists the basic SCA statistics by flow routing algorithm. The minimum values varied because different rules were used to direct flow from each source cell to one or more adjacent downslope cells. The maximum values are similar because they represent watershed outlets at the coast. The mean values varied from a low of 3,429 m^2m^{-1} (DEMON) to a high of 4,356 m^2m^{-1} (FD8), a difference of 27%. Table 1b partitions SCA into a series of classes and indicates the percentage of cells for each flow routing algorithm that was assigned to each class. These results show that D8 and Rho8, and to a lesser extent D∞, have many more "low flow" cells (i.e. SCA $\leq$ 10 m^2 m^{-1}). The same pattern is repeated for the second class although the magnitude of the differences is reduced. The largest number of cells in classes 3 through 6 was generated with different flow routing algorithms – D∞ for class 3, DEMON for class 4, and FD8 for the fifth and sixth classes – although the differences are relatively small.

Table 1. (a) Specific catchment area ($m^2 m^{-1}$) statistics for study area, and (b) percentage of cells per specific catchment area class.

(a)	*Number of cells*	*Minimum*	*Maximum*	*Mean*	*Standard Deviation*
D8	1,263,296	7.07	2237670.25	3715.27	60584.28
Rho8	1,263,296	7.07	2236030.25	3714.18	60469.64
D∞	1,263,296	10.00	2236762.00	3934.18	61469.07
FD8	1,263,296	2.56	2341777.00	4355.83	69911.69
DEMON	1,263,296	7.07	2214353.00	3428.91	55657.18

(b)	*SCA Classes (m^2m^{-1})*						
	1 (≤10.0)	2 (10.1-20)	3 (20.1-40)	4 (40.1-70)	5 (70.1-100)	6 (100.1-1000)	7 (>1000)
D8	12.8	18.5	26.9	16.3	7.2	13.3	5.1
Rho8	13.4	21.6	25.0	14.3	6.7	14.0	5.1
D∞	7.6	12.9	29.9	20.1	7.9	16.0	5.7
FD8	4.5	12.1	24.5	20.7	10.0	23.2	5.2
DEMON	2.7	12.2	29.3	23.6	9.6	17.6	5.0

Table 2 summarizes several noteworthy features about the distribution of low flow cells predicted with the five flow algorithms across the six landscape classes. First, the number of low flow cells predicted with the five flow routing algorithms varied from 169,171 (Rho8) to 33,756 (DEMON). Second, the percentage of low flow cells in the hilltop/ridgeline class varied by a factor of five, from a low of 9% for DEMON to a high of 45% for D8. In general, these percentages indicate the presence of a series of broad hilltops and ridgelines in the study area. Third, Rho8 predicted > 5,000 low flow cells in five of the six landscapes and D8 predicted > 5,000 low flow cells for steep north-facing slopes. Neither of these results is realistic. Overall, the results show that D∞, FD8, and DEMON performed better than D8 and especially Rho8 – the latter algorithm, in particular, had large numbers of low flow cells scattered across most of the fuzzy *k*-means landscape classes – and that the algorithms produced different results in different parts of the catchment.

The final pair of studies that comprise the third group are noteworthy because they compared the performance of the flow routing algorithms to observations of soil wetness and overland flowpaths. Fried *et al.* (2000) estimated the topographic wetness index with four flow routing approaches (static D8 and DEMON, as described in the previous section, and two quasi-dynamic versions of D8, one with dynamic uniform soils and the other with dynamic variable soils) and evaluated the resultant models using field data collected during a post-storm event GPS survey of ponded storm flow accumulations and concentrated storm flow discharge sites for a small first-order catchment in Michigan. The results showed that the quasi-dynamic versions of D8 calculated with DYNWET (Barling *et al.* 1994; see next section for additional discussion of this approach) performed best and that the areas of greatest disagreement were relatively flat, lending credence to the conventional wisdom that flowpath determination by any method is especially challenging in areas of low relief (see Callow *et al.* (2007) for additional insights).

Table 2. Distribution of source cells (SCA $\leq$ 10 m^2m^{-1}) by landscape class.

Topo-climatic Class	Number of Cells	Percentage of Cells with SCA $\leq$ 10 m^2m^{-1}				
		D8	Rho8	D∞	FD8	DEMON
Hilltops/ ridgelines	256,012	114,186	79,789	64,966	39,215	23,583
		(44.6%)	(31.2%)	(25.4%)	(15.3%)	(9.2%)
Steep south-facing slopes	323,989	1,686	25,568	481	107	91
		(0.5%)	(7.9%)	(0.1%)	(0.0%)	(0.0%)
Steep north-facing slopes	231,180	5,630	18,584	331	72	86
		(2.4%)	(8.0%)	(0.1%)	(0.0%)	(0.0%)
Moderately steep lower valley slopes	169,173	37	8,245	175	15	9
		(0.0%)	(4.9%)	(0.1%)	(0.0%)	(0.0%)
Coastal plains /gentle slopes	177,787	39,893	36,526	28,995	16,709	9,960
		(22.4%)	(20.5%)	(16.3%)	(9.4%)	(5.6%)
Stream channels	103,888	35	459	94	62	27
		(0.0%)	(0.4%)	(0.1%)	(0.1%)	(0.0%)
Total Area	1,262,029	161,467	169,171	95,042	56,180	33,756
		(12.8%)	(13.4%)	(7.5%)	(4.5%)	(2.7%)

Endreny and Woods (2003) compared the spatial congruence of observed overland flow paths with those delineated using the D8, FD8, 2D-Lea, 2D-Jensen (Jensen 1996), and D∞ algorithms on agricultural hillslopes in New Jersey. Four new algorithms were created to determine whether the congruence between observed and simulated flow networks improved with changes in the method for allocating flow about the path of steepest descent. D8-buf allowed flow to disperse into all adjacent downslope cells inside a user-specified buffer; D8-2x allowed flow to be split between two downslope cells at the source pixel and then each flow path followed the path of steepest descent; whereas MF(5) and MF(3) constrained the eight possible flow paths available in the FD8 algorithm to the five and three adjacent downslope cells with the steepest gradients, respectively. The results suggest using flow routing algorithms that disperse flow to two or three neighbouring cells when routing runoff across the landscape. The favoured algorithms included D8-buf and MF(3) along with the more sophisticated 2D-Lea, 2D-Jenson, and D∞ algorithms since all five of these algorithms produced the best spatial congruence and kept the commission and omission errors at very low levels.

4 Discussion and Conclusions

It is clear from the aforementioned evaluations that the nine flow routing algorithms produce different results from one another and that the differences can be expected to vary in different parts of the landscape. The nine algorithms take different approaches to fitting a surface to the square-grid DEMs and in terms of the rationale and number of cells to which flow is apportioned. These algorithms all treat flow routing as a function of the topographic surface despite the likelihood that this is only true in a series of relatively rare special cases (i.e. when a land surface is impermeable). The evaluations are noteworthy in that only two studies compared the performance of these algorithms to observations. Both of these studies relied on visual (i.e. qualitative) assessments and recommended using specific algorithms based on "goodness-of-fit" without resort to any theory or knowledge of the soil water relationships that help to direct runoff across the landscape. This is a fundamental shortcoming because the successful deployment of these flow routing algorithms in watershed modelling applications depends ultimately on the amount of spatial variability they are able to reliably measure or account for (Western *et al.* 1999).

Various authors have identified the influence of interpolation errors in DEMs and their propagation through the computation of flowpaths and topographic indices to model output (Desmet 1997, Heuvelink and Goodchild 1998, Holmes *et al.* 2000, McMaster 2002, Van Niel *et al.* 2004). However, while terrain and errors in modelled terrain play an important role in the spatial distribution of surface processes, the spatial patterns of these processes may vary substantially because of the variability of soil and land cover characteristics. Mitas and Mitasova (1998), for example, found that borders between different land cover types (e.g. bare soil and dense grass) caused abrupt changes in flow velocities as well as in transport and detachment capacities, creating effects important for erosion prevention. Management actions may also modify flowpaths since cross-slope furrows tend to channel overland flow directly into concavities, leading to significant flow convergence at points upslope of those that would be identified on the grounds of topography alone (Brown and Quine 1999). There is clearly a need to consider the spatial variability of numerous factors in addition to terrain shape, as illustrated by the example below.

Zhu and Mackay (2001) investigated the effect of using detailed SoLIM spatial soil data in place of traditional soil map data as an input to both lumped and spatially distributed runs of the RHESSys model (Tague and Band 2001). RHESSys is designed to represent surface soil, topographic, and vegetation patterns along with certain hydro-ecological processes at the landscape level, so that the necessary parameters can be realistically estimated to reproduce the dominant patterns of hydro-ecological dynamics (e.g. surface runoff, evapotranspiration, canopy photosynthesis) over the landscape (Band *et al.* 1993). The SoLIM result map described the spatial variation of hydraulic conductivity by identifying contrasts between north and south facing slopes, and between high and low elevations due to the level of soil profile development, thereby providing details that were not visible on traditional soil maps.

The implications of ineffective representation of the spatial details of soil depth and hydraulic conductivity were highlighted when Zhu and Mackay (2001) ran various hydro-ecological models within RHESSys using the detailed (SoLIM) and conventional soil data as input. Underestimation of solum depth using the traditional soil map led to the soils responding quickly and abruptly to precipitation events, producing a rapidly changing and generally unrealistic hydrograph. The soil profile was thus predicted to saturate with less precipitation, while overestimation of hydraulic conductivity simulated water to move too rapidly through the soil column. In the SoLIM scheme, however, soil conditions on side slopes and their deviation from the dominant soil type were considered in model parameterization, and as a result the peaks of the simulated hydrograph were lower and more sensibly characterized the hydrologic response. Interestingly, the simulated streamflow between the two different soil landscape parameterization schemes was small under the distributed approach. Zhu and Mackay (2001) argue this is due to local variation of soil properties (solum depth and hydraulic conductivity) being expressed by the detailed description of other landscape parameters in the distributed approach, particularly the spatial covariation of local topography (elevation and slope gradient) and drainage area on a hillslope.

The spatial variability of soil and land cover characteristics is important because the location and extent of variable source areas is determined by the antecedent soil water content and its spatial distribution within the catchment in many environments (e.g. Walter *et al.* 2007). We therefore need to be able to characterize the spatial variability of soil water content in a simple, yet physically realistic way to generate meaningful hydrologic predictions at the catchment scale (Moore *et al.* 1993). Most models and applications rely on the topographic wetness index ($\ln(A_s/\tan\beta)$) to characterize the soil water distribution, although this form of the wetness index will only serve as a good predictor of soil water content if the drainage flux has reached steady state (i.e. if every point is experiencing drainage from its entire upslope contributing area) (Barling *et al.* 1994).

This last assumption is not true in many watersheds for at least part of the year because the velocity of subsurface flow is so small that most points in a catchment only receive contributions from a small part of their total upslope contributing area and the subsurface flow regime is in a state of dynamic non-equilibrium. Barling *et al.* (1994) proposed a quasi-dynamic wetness index ($\ln(A_e/\tan\beta)$), where A_e is the effective specific catchment area and β is the slope angle, and showed that this approach was a better predictor of soil water content for a small catchment near Wagga Wagga in New South Wales, Australia. This approach, which requires the user to specify drainage times and two soil properties (saturated hydraulic conductivity and drainable porosity), is novel because they considered soil properties in addition to the shape of the topographic surface.

More work along these lines is needed because hydrologic applications utilize flow routing algorithms to connect the precipitation falling on the land surface with the hydrologic response of the catchment. This work will require a greater investment in fieldwork and data modelling than is evident from the flow routing papers published during the past two decades. We need better data models to get the runoff from the land surface to the stream networks (see Kim and Lee 2004 for one such example) and most important of all, we will need field observations in a variety of landscape settings to improve our characterization of the role of topography, soil, and land cover in shaping the hydrologic response of catchments.

References

Abbott, M.B. and Refsgaard, J.C. (eds.), (1996), *Distributed Hydrological Modelling,* Norwell, MA: Kluwer.

Band, L.E., (1989), A terrain-based watershed information system, *Hydrological Processes,* **3**:151–162.

Band, L.E., Patterson, P., Nemani, R. and Running, S.W., (1993), Forest ecosystem processes at the watershed scale: Incorporating hillslope hydrology, *Agriculture, Forestry and Meteorology,* **63:** 93–126.

Barling, R.D., Moore, I.D. and Grayson, R.B., (1994), A quasi-dynamic wetness index for characterizing the spatial distribution of zones of surface saturation and soil water content, *Water Resources Research,* **30:** 1029–1044.

Beasley, D.G. and Huggins, L.F., (1978), ANSWERS: A model for watershed planning, *Proceedings of the Tenth Conference on Water Simulation (Volume 2),* Miami Beach, Florida: 507–515.

Beven, K.J., (ed.), (1998), Distributed Hydrological Modelling: Applications of the TOPMODEL Concept, Chichester: John Wiley and Sons.

Beven, K.J., Lamb, R., Quinn, P., Romanowicz, R. and Freer, J., (1994), TOPMODEL, In Singh, V.P. (ed.): *Computer Models of Watershed Hydrology,* Fort Collins, CO: Water Resources Publications: 627–668.

Brown, A.G. and Quine, T.A., (1999), Fluvial processes and environmental change: An overview. In Brown, A.G. and Quine, T.A. (eds.): *Fluvial Processes and Environmental Change,* Chichester: John Wiley and Sons: 1–28.

Callow J.N., Van Niel, K.P. and Boggs, G.S., (2007), How does modifying a DEM to reflect known hydrology affect subsequent terrain analysis? *Journal of Hydrology,* **332**: 30–39.

Carrara, A., Bitelli, G. and Carla, R., (1997), Comparison of techniques for generating digital terrain models from contour lines, *International Journal of Geographical Information Science,* **11:** 451–473.

Chirico, G.B., Western, A.W., Grayson, R.B. and Günter, B., (2005), On the definition of the flow width for calculating specific catchment area patterns from gridded elevation data, *Hydrological Processes,* **19:** 2539–2556.

Cluis, D., Martz, L., Quentin, E. and Rechatin, C., (1996), Coupling GIS and DEMs to classify the Hortonian pathways of non-point sources to the hydrographic network, In Kovar, K. and Nachtnebel, H.P., (eds.): *HydroGIS '96: Application of Geographic Information Systems in Hydrology and Water Resources Management (Proceedings of the Vienna Conference, (1996),* Wallingford: International Association of Hydrological Sciences: 37–44.

Costa-Cabral, M.C. and Burgess, S.J., (1994), Digital Elevation Model Networks (DEMON): A model of flow over hillslopes for computation of contributing and dispersal areas, *Water Resources Research,* **30:** 1681–1692.

Da Ros, D. and Borga, M., (1997), Use of digital elevation model data for the derivation of the geomorphological instantaneous unit hydrograph, *Hydrological Processes,* **11:** 13–33.

Desmet, P.J.J., (1997), Effects of interpolation errors on the analysis of DEMs, *Earth Surface Processes and Landforms,* **22:** 563–580.

Desmet, P.J.J. and Govers, G., (1996), Comparison of routing algorithms for digital elevation models and their implications for predicting ephemeral gullies, *International Journal of Geographical Information Systems,* **10:** 311–331.

Dymond, J.R., Derose, R.C. and Harmsworth, G.R., (1995), Automated mapping of land components from digital elevation data, *Earth Surface Processes and Landforms,* **20:** 131–137.

Eastman, J.R., (1992), *IDRISI (Version 4.0): Geographic Information System Software and Documentation,* Worcester, MA: Clark University Graduate School of Geography.

Eklundh, L. and Martensson, U., (1995), Rapid generation of digital elevation models from topographic maps, *International Journal of Geographical Information Systems,* **9:** 329–340.

Endreny, T.A. and Woods, E.W., (2003), Maximizing spatial congruence of observed and DEM-delineated overland flow networks, *International Journal of Geographic Information Science,* **17,** 699–713.

Fairfield, J. and Leymarie, P., (1991), Drainage networks from grid digital elevation models, *Water Resources Research,* **27:** 709–717.

Freeman, G.T., (1991), Calculating catchment area with divergent flow based on a regular grid, *Computers and Geosciences,* **17:** 413–422.

Fried, J.S., Brown, D.G., Zweifler, M.O. and Gold, M.A., (2000), Mapping contributing areas for stormwater discharge to streams using terrain analysis, In Wilson, J.P. and Gallant, J.C., (eds.): *Terrain Analysis: Principles and Applications,* New York: John Wiley and Sons: 183–203.

Gallant, J.C. and Wilson, J.P., (1996), TAPES-G: A grid-based terrain analysis program for the environmental sciences, *Computers and Geosciences,* **22:** 713–722.

Gallant, J.C. and Wilson, J.P., (2000), Primary terrain attributes, In Wilson, J.P. and Gallant, J.C., (eds.): *Terrain Analysis: Principles and Applications,* New York: John Wiley and Sons: 51–85.

Gallant, J.C., Hutchinson, M.F. and Wilson, J.P., (2000), Future directions for terrain analysis, In Wilson, J.P. and Gallant, J.C., (eds.): *Terrain Analysis: Principles and Applications,* New York: John Wiley and Sons: 423–427.

Garbrecht, J. and Martz, L.W., (1997), The assignment of drainage directions over flat areas in raster digital elevation models, *Journal of Hydrology,* **193:** 204–213.

Garbrecht, J., Ogden, F.L., DeBarry, P.A. and Maidment, D.R., (2001), GIS and distributed watershed models: I, Data coverages and sources, *Journal of Hydrologic Engineering,* **6:** 506–514.

Hellweger, R., (1996), *Agree.aml,* Austin, TX: Centre for Research in Water Resources, University of Texas.

Heuvelink, G. and Goodchild, M.F., (1998), *Error Propagation in Environmental Modelling with GIS,* London: Taylor and Francis.

Holmes, K.W., Chadwick, O.A. and Kyriakidis, P.C., (2000), Error in a USGS 30-meter digital elevation model and its impact on terrain modelling, *Journal of Hydrology,* **233**: 154–173.

Holmgren, P., (1994), Multiple flow direction algorithms for runoff modelling in grid-based elevation models: An empirical evaluation, *Hydrological Processes,* **8:** 327–334.

Hutchinson, M.F., (1989), A new procedure for gridding elevation and stream line data with automatic removal of pits, *Journal of Hydrology,* **106:** 211–232.

Hutchinson, M.F., (2007), Adding the Z Dimension, In Wilson, J.P. and Fotheringham, A.S., (eds.): *Handbook of Geographic Information Science,* Oxford: Blackwell: 144–68.

Jensen, J.R., (1996), Introductory Digital Image Processing: A Remote Sensing Perspective, Upper Saddle Creek, NJ: Prentice-Hall.

Kemp, K.K., (1997a), Fields as a framework for integrating GIS and environmental process models: I, Representing spatial contiguity, *Transactions in GIS,* **1:** 219–234.

Kemp, K.K., (1997b), Fields as a framework for integrating GIS and environmental process models: II, Specifying field variables, *Transactions in GIS,* **1:** 235–246.

Kim, S. and Lee, H. (2004), A digital elevation analysis: A spatially distributed flow apportioning algorithm, *Hydrological Processes,* **18:** 1777–1794.

Knighton, D., (1998), *Fluvial Forms and Processes,* New York: Oxford University Press.

Krajewski, S.A. and Gibbs, B.L., (1994), Computer contouring generates artefacts, *Geotimes,* **39:** 15–19.

Lagacherie, P., Moussa, R., Cormary, D. and Molenat, J., (1996), Effects of DEM data source and sampling pattern on topographical parameters and on a topography-based hydrological model, In Kovar, K. and Nachtnebel, H.P., (eds.): *HydroGIS '96: Application of Geographic Information Systems in Hydrology and Water Resources Management (Proceedings of the Vienna Conference, (1996),* Wallingford: International Association of Hydrological Sciences): 191–199.

Lea, N.L., (1992), An aspect driven kinematic routing algorithm. In Parsons, A.J. and Abrahams, A.D., (eds.): *Overland Flow: Hydraulics and Erosion Mechanics,* London: UCL Press: 393–407.

Lee, J. and Chu, C.-J., (1996), Spatial structures of digital terrain models and hydrological feature extraction. In Kovar, K. and Nachtnebel, H.P., (eds.): *HydroGIS '96: Application of Geographic Information Systems in Hydrology and Water Resources Management (Proceedings of the Vienna Conference, (1996),* Wallingford: International Association of Hydrological Sciences: 201–6.

Liu, H. and Jezek, K.C., (1999), Investigating DEM error patterns by directional variograms and Fourier analysis, *Geographical Analysis,* **31:** 249–266.

Lopez, C., (1997), Locating some types of random errors in digital terrain models, *International Journal of Geographical Information Science,* **11:** 677–698.

Maidment, D., (1996), Environmental modelling with GIS, In Goodchild, M.F., Steyaert, L.T. and Parks B.O. (eds.): *GIS and Environmental Modelling: Progress and Research Issues,* Fort Collins, CO: GIS World Books: 315–23.

McMaster, K.J., (2002), Effects of digital elevation model resolution on derived stream network positions, *Water Resources Research,* **38**: 13–19.

Mitas, L. and Mitasova, H., (1998), Distributed soil erosion simulation for effective erosion prevention, *Water Resources Research,* **34:** 505–516.

Montgomery, D.R. and Dietrich, W.E., (1988), Where do channels begin? *Nature,* **336:** 232–234.

Montgomery, D.R., Dietrich, W.E. and Sullivan, K., (1998), The role of GIS in watershed analysis, In Lane, S.N., Chandler, J.H. and Richards, K.S., (eds.): *Landform Monitoring, Modelling, and Analysis,* Chichester: John Wiley and Sons: 241–262.

Moore, I.D., (1996), Hydrologic modelling and GIS, In Goodchild, M.F., Steyaert, L.T., Parks, B.O., Johnston, C., Maidment, D., Crane, M. and Glendinning, S., (eds.): *GIS and Environmental Modelling: Progress and Research Issues,* Fort Collins, CO: GIS World Books: 143–148.

Moore, I.D., Burch, G.J. and MacKenzie, D.H., (1988), Topographic effects on the distribution of surface soil water and the location of ephemeral gullies, *Trans. of the American Society of Agricultural Engineers,* **31:** 1098–1107.

Moore, I.D., Grayson, R.B. and Ladson, A.R., (1991), Digital terrain modelling: A review of hydrological, geomorphological, and biological applications, *Hydrological Processes,* **5:** 3–30.

Moore, I.D., Turner, A.K., Wilson, J.P., Jenson, S.K. and Band, L.E., (1993), GIS and land surface-subsurface modelling, In Goodchild, M.F., Parks, B.O. and Steyaert, L.T. (eds.): *Environmental Modelling with GIS,* New York: Oxford University Press: 196–230.

Murillo, M. and Hunter, G.J., (1997), Assessing uncertainty due to elevation error in a landslide susceptibility model, *Transactions in GIS,* **2:** 289–298.

O'Callaghan, J.F. and Mark, D.M., (1984), The extraction of drainage networks from digital elevation data, *Computer Vision, Graphics and Image Processing,* **28:** 328–344.

Pickup, G. and Chewings, V.H., (1996), Correlations between DEM-derived topographic indices and remotely-sensed vegetation cover in rangelands, *Earth Surface Processes and Landforms,* **21,** 517–529.

Qian, J., Ehrich, R.W. and Campbell, J.B., (1990), DNESYS: An expert system for automatic extraction of drainage networks from digital elevation data, *IEEE Transactions on Geoscience and Remote Sensing,* **28:** 29–45.

Qin, C., Zhu, A.-X., Pei, T., Li, B., Zhou, C. and Yang, L., (2007), An adaptive approach to selecting flow partition exponent for multiple flow direction algorithm, *International Journal of Geographical Information Science,* **21:** 443–458.

Quinn, P.F., Beven, K.J., Chevallier, P. and Planchon, O., (1991), The prediction of hillslope flow paths for distributed hydrological modelling using digital terrain models, *Hydrological Processes,* **5,** 59–79.

Quinn, P.F., Beven, K.J. and Lamb, R., (1995), The ln(a/tanβ) index: How to calculate it and how to use it within the TOPMODEL framework, *Hydrological Processes,* **9:** 161–182.

Smith, T.R., Zhun, X. and Gan, P., (1990), A knowledge-based, two-step procedure for extracting channel networks from noisy DEM data, *Computers and Geosciences,* **16:** 777–786.

Storck, P., Bowling, L., Wetherbee, P. and Lettenmaier, D., (1998), Application of a GIS-based distribution hydrology model for prediction of forest harvest effects on peak stream flow in the Pacific Northwest, *Hydrological Processes,* **12:** 889–904.

Tague, C. and Band, L.E., (2001), Evaluating explicit and implicit routing for watershed hydro-ecological models of forest hydrology at the small catchment scale, *Hydrological Processes,* **15:** 1415–1439.

Tarboton, D.G., (1997), A new method for the determination of flow directions and upslope areas in grid digital elevation models, *Water Resources Research,* **33:** 309–319.

Van Niel, K.P., Laffan, S.W. and Lees, B.G., (2004), Effect of error in the DEM on environmental variables for predictive vegetation modelling, *Journal of Vegetation Science,* **15:** 747–756.

Walter, M.T., Walter, M.F., Brooks, E.S., Steenhuis, T.S., Boll, J. and Weiler, K., (2007), Hydrologically sensitive areas: Variable source area hydrology implications for water quality risk assessment, *Journal of Soil and Water Conservation,* **62:** 277–284.

Western, A.W., Grayson, R.B., Bloschl, G., Willgoose, G.R. and McMahon, M.L., (1999), Observed spatial organization of soil moisture and its relation to terrain indices., *Water Resources Research,* **35:** 797–810.

Wilson, J.P. and Gallant, J.C., (2000), Digital terrain analysis, In Wilson, J.P.and Gallant, J.C., (eds.): *Terrain Analysis: Principles and Applications,* New York: John Wiley and Sons: 1–27.

Wilson, J.P., Lam, C.S. and Deng, Y.X., (2007), Comparison of performance of flow-routing algorithms used in Geographic Information Systems, *Hydrological Processes,* **21:** 1026–1044.

Wilson, J.P., Repetto, P.L. and Snyder, R.D., (2000), Effect of data source, grid resolution, and flow-routing method on computed topographic attributes, In Wilson, J.P. and Gallant, J.C., (eds.):*Terrain Analysis: Principles and Applications*, New York: John Wiley and Sons: 133–61.

Wilson, J.P., Spangrud, D.S., Nielsen, G.A., Jacobsen, J.S. and Tyler, D.A., (1998), GPS sampling intensity and pattern effects on computed terrain attributes., *Soil Science Society of America Journal,* **62:** 1410–1417.

Wise, S.M. (1998a), The effect of GIS interpolation errors on the use of digital elevation models in geomorphology, In Lane, S.N., Richards, K.S. and Chandler, J.H., (eds.):*Landform Monitoring, Modelling, and Analysis,* New York: John Wiley and Sons: 139–64.

Wise, S.M., (1998b), Digital terrain models: Traps for the unwary, *Proceedings of the Third International Conference on GeoComputation, University of Bristol, United Kingdom,* Available online at: http://www.geog.port.ac.uk/geocomp/geo98/ (accessed 1 September 2007).

Wolock, D.M. and McCabe, G.J., (1995), Comparison of single and multiple flow direction algorithms for computing topographic parameters in TOPMODEL, *Water Resources Research,* **31:** 1315–1324.

Zhou, Q. and Liu, X., (2002), Error assessment of grid-based flow routing algorithms used in hydrological models, *International Journal of Geographical Information Science,* **16:** 819–842.

Zhu, A-X. and Mackay, D.S., (2001), Effects of spatial detail of soil information on watershed modelling, *Journal of Hydrology,* **248:** 54–77.

An Integrated Raster-TIN Surface Flow Algorithm

Petter PILESJÖ

Abstract

In this chapter, an alternative surface flow algorithm is presented. The basic idea behind the algorithm is the use of the advantages of TIN-based algorithms within a raster based environment. A gridded raster DEM is used to create a 'regular TIN', over which surface flow is estimated. Since each facet in the TIN has a constant slope and slope direction, the estimations of, for example, flow velocity and diversion/convergence are less complicated compared to traditional 'cell based' solutions. The flow is treated as 'water packages', given specific (point) positions on the surface. The number of water packages per cell is initially set to eight, but this number can be increased or decreased. After each time step, the water packages have moved a certain distance (depending on slope and water depth), and new water packages have been created due to precipitation. In order to keep the number of water packages constant (to reduce memory and computer time), all water packages within a TIN facet are merged after each iteration. Parameters in time and space, e.g. precipitation, infiltration, vegetation and elevation, can all be loaded into the software.

Keywords: DEM, TIN, surface flow, topographical modelling, hydrological modelling.

1 Introduction and Background

Catchment topography is critical for models estimating distributed hydrological processes. The key parameter in catchment topography is flow distribution, which tells us how overland flow is distributed over the catchment area. Slope controls flow pathways for surface as well as near surface flow, and influences the sub surface flow pattern substantially.

Flow distribution over a land surface is a crucial parameter in hydrological modelling. The use of Digital Elevation Models (DEM) has made it possible to estimate flow on each location over a surface. Based on the flow distribution estimation on each location represented by a DEM, the drainage pattern over an area, as well as various other hydrological parameters such as catchment area and up-stream flow accumulation, can be modelled.

Generally, surface flow and flow accumulation are estimated by the use of two different types of raster algorithms, either approximating to single, or multiple, flow directions. If working with raster data, multiple flow algorithms assume transport to more than one adjacent cell, while a single flow algorithm only distributes water to one neighbour cell at a time in the raster. In many cases, single directional flow algorithms produce satisfactory results over concave surfaces, while it is often more appropriate to divide the flow into two or more directions if the surface is flat or has a convex form. Combinations of the two types are often preferred when modelling water flow over natural surfaces (Pilesjö *et al.* 1998).

However, the raster algorithms mentioned are often not optimal, and need extensive calibration, mainly due to the problem of how to weight the influence of slope when splitting flow between neighbouring cells (see e.g. Olsson and Pilesjö 2002). Estimation of flow over surfaces with a constant slope and a constant slope direction, like a facet in a TIN, would considerably improve the accuracy of the analysis. On a facet, a one directional flow algorithm is the obvious choice (if the surface is planar the flow does not diverge). However, since a majority of available topographical data is stored in raster format, TIN based algorithms are not widely used. In this chapter, we present an algorithm making it possible to adopt the advantages of TIN modelling in a raster environment. The program produced is able to estimate drainage area/flow accumulation as well as water distribution/Hortonian overland flow at a certain time *t*. Variables and constants like rainfall intensity, infiltration capacity, elevation, and time step can be specified and incorporated by the user.

2 Existing Flow Algorithms

As indicated above, a number of different methods exist for the estimation of flow and flow accumulation. Some of these estimate contributing areas (up-slope) as well as dispersal areas (down-slope). A few of the commonly used methods are presented below.

2.1 The single flow – D8 algorithm

This algorithm was described by O'Callaghan and Mark (1984). It assumes that flow follows only the steepest downhill slope. Using a raster DEM, implementation of this method results in the hydrological flow at a point only following one of the eight possible directions corresponding to the eight neighbouring grid cells (Mark 1984, O'Callaghan and Mark 1984, Band 1986, ESRI 1991). Here we call this approach a 'single flow' algorithm. However, for the quantitative measurement of the flow distribution, this over-simplified assumption must be considered as illogical and would obviously create significant artefacts in the results, as stated by Freeman (1991), Holmgren (1994), Wolock and McCabe (1995), and Pilesjö and Zhou (1996). More complex terrain is supposed to yield more complicated drainage patterns.

The D8 algorithm allows flow from each cell to discharge to only one receiving neighbour. The origin of the flow on the cell is the centre; hence it is treated as a point source. The flow down-slope is described as a one dimensional line. This way of approximating a surface to a point, and a two-dimensional flow tube into a line is, of course, a generalization. The larger the cells and the more undulating the terrain, the larger will be the errors. Also, the fact that only one neighbouring cell can receive water from an out-flow cell is a limiting factor. A flow direction in either a cardinal or diagonal direction does not always reflect reality. For example, parallel flow will only be correctly estimated if the flow direction is a multiple of 45 degrees. The one directional flow problem becomes even more evident in convex terrain, where flow is supposed to diverge. In concave terrain, where flow is supposed to converge, it is less pronounced. This can, for example, be shown by the modelling of water over a circular cone mountain surface and a circular cone crater surface, respectively (see e.g. Costa-Cabral and Burges 1994, Pilesjö and Zhou 1997).

2.2 The Rho8 algorithm

The difference between the D8 algorithm and the Rho8 algorithm, presented by Fairfield and Leymarie (1991), is that the Rho8 also includes a stochastic variable. Water from a cell is discharged to one of its eight neighbours, but the choice of the cell is made stochastically. Depending on slope direction, the probability values of two adjacent cells can be anything between fifty-fifty (0.5 – 0.5, e.g. if the slope direction is 22.5 degrees) and one to zero (e.g. if the slope direction is 0 degrees). The randomness in the method can possibly yield problems like convergence and divergence where flow is supposed to be parallel, as well as "wiggling" patterns over well defined surfaces. The fact that this stochastic method cannot be reproduced with the same result over the same surface is sometimes also considered unfavourable.

2.3 Multiple-direction algorithms

Attempts to solve the problems connected with the 'single flow' algorithms have led to several proposed 'multiple flow direction' algorithms (see e.g. Freeman 1991, Quinn *et al.* 1991, Holmgren 1994, Pilesjö 1994, Pilesjö and Zhou 1996). These algorithms estimate the flow distribution values proportionally to the slope gradient, or risen slope gradient, in each direction. Holmgren (1994) summarizes some of the algorithms as

$$f_i = \frac{(\tan \beta_i)^x}{\sum_{j=1}^{8} (\tan \beta_j)^x}, \text{ for all } \beta > 0 \qquad (1)$$

where i,j = flow directions (1…8), f_i = flow proportion (0…1) in direction i, $\tan \beta_i$ = slope gradient between the centre cell and the cell in direction i, and x = variable exponent.

By changing the exponent (x) in Equation (1), two extreme approaches to estimating flow distribution can be observed. While $x = 1$, flow will be distributed to downhill neighbouring cells proportionally to the slope gradients, as suggested by Quinn *et al.* (1991). The other extreme is when $x \to \infty$, which will approach the 'single flow' drainage distribution mentioned above. Holmgren (1994) suggested an x value between 4 and 6. This gives a result between a very homogeneous flow distribution when $x = 1$, and a distinctive flow, which occurs when x becomes greater than 10.

Pilesjö and Zhou (1997) used mathematical surfaces (a cone, a hemisphere, and an inverse hemisphere) to test different x values. They concluded that an x value of 1 was optimal, especially on convex surfaces. Freeman (1991) proposed an x value of 1.1 after testing for flow over a right circular cone.

One of the problems with the 'standard' multiple-direction algorithm, independent of the value of the exponent x, is the diverging flow over planar surfaces where we expect the flow pattern to be parallel. Depending on aspect, the flow from one cell is almost always (if slope direction is not equal to 0, 90, 180, or 270 degrees) discharged into at least three neighbouring cells down-steam. This results in diverging flow patterns, which is not a correct image of reality. Also, over concave surfaces, where we expect the flow to converge, this problem is evident. The contributing area becomes discontiguous, since cells (too) far away are included. The results are also affected by boundary (to the water divide) proximity.

2.4 A form-based algorithm

This solution, sometimes referred to as the Pilesjö-Zhou algorithm, was presented by Pilesjö *et al.* (1998). Given the limitations and problems of the algorithms presented above, a 'multiple flow direction' approach based on analysis of the form of individual 3 x 3 cell surface facets was proposed. It was assumed that flow diverges over convex surfaces, and converges over concave surfaces. There is no absolute way to determine convexity and concavity of the centre cell in a 3 x 3 cell facet. The possible complexity of the surface often implies approximations. One way to approximate, used in the form-based algorithm, is to employ a trend surface based on the elevation values of all nine cells in the facet. When the topographic form of the centre cell in the facet is judged as concave, the flow is distributed fully to the main drainage direction. If the main drainage direction is not equal to the direction to one of the eight neighbouring cells, the flow distribution has to be split between two cells. This is done by splitting the drainage vector into two diagonal (i.e. 45° apart) vectors. When the topographic form of the centre cell in the facet is judged as convex, the flow is distributed according to Equation (1).

2.5 The DEMON algorithm

The DEMON algorithm was presented by Costa-Cabral and Burges (1994). In order to eliminate the problem with the one-dimensional flow present in the other algorithms, DEMON uses two-dimensional flow tubes in order to trace flow up-stream and down-stream. The direction of flow over each cell is approximated to the aspect value. By connecting (flow) lines, parallel to the aspect values, from the corner points of a cell receiving water, a flow tube can be estimated. In Figure 1, cell 4;3 (row; column) receives water from the cells 4;2, 3;1, 3;2, 3;3, 2;1, 2;2, 2;3, 1;1 and 1;2. The flow is tracked down-stream until it leaves the DEM or enters a sink.

Unlike most other flow algorithms, DEMON does not distribute flow directly to diagonal neighbour cells. This is explained by the fact that the contact with these neighbours is through a point, which has no width. However, this distribution sometimes causes problems. In the example presented in Figure 2 below, we intuitively assume that water from the upper left cell should flow into the lower right cell. Using DEMON, this will not be the case. The reason for this is actually that all cell surfaces are approximated to first order (planar) surfaces. This often results in 'gaps' between individual surfaces, i.e. a discontinuous surface, which is unrealistic. Costa-Cabral and Burges (1994) also write that some problems connected to the method 'can be avoided only if a curved rather than a planar surface is fitted to each pixel'.

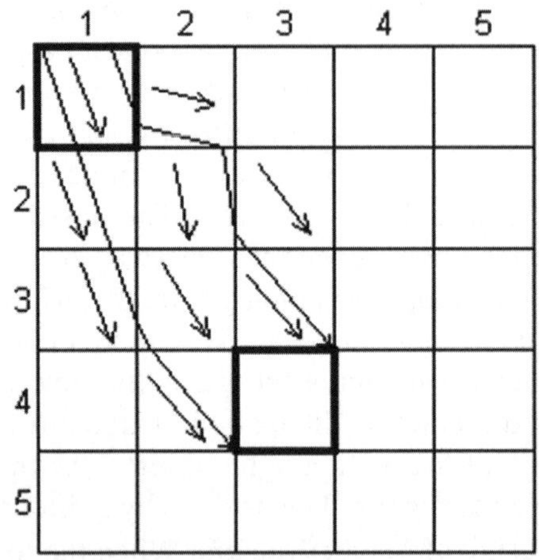

Figure 1. Illustration of how DEMON estimates flow. Modified from Costa-Cabral and Burges (1994).

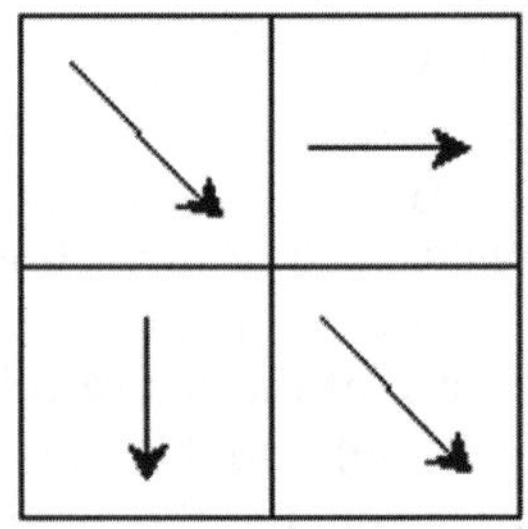

Figure 2. Using the DEMON algorithm, the lower right cell will not be estimated as receiving water from the upper left cell.

2.6 Discussion

The disadvantages of the single flow algorithms are obvious. Directing flow in only one direction from a cell is a strong over-simplification, yielding unreliable results. Also the limitations of the 'standard' multi-directional algorithms are problematic. Estimated contributing drainage areas are discontiguous, and the result is dependent on boundary proximity. Another important source of error, both for single and multiple flow algorithms (including the form-based one), is the point source assumption. Depending on terrain complexity in relation to cell size, this is more or less pronounced. The DEMON algorithm, not based on a point source assumption, may yield significant errors on concave or convex hill-slopes that are large relative to cell size.

Another problem in flow modelling is how to treat sinks and flat surfaces. Sinks are minor depressions in the terrain, creating small drainage areas that sometimes are filled up by water. Flat areas are problematic since all algorithms presented above do not model flow over these, even if most cases seem logical. Water always flows down-slope, and if a flat area borders one or several lower (in elevation) cells we expect water to flow in these direction/directions. The sink problem can be dealt with by eliminating all sinks with an area (or depth) exceeding a defined threshold, or by letting the sinks be filled up with water and then 'spill over'. How to assign flow directions over flat surfaces has been proposed by Pilesjö *et al.* (1998). The method is based on defined flow distribution values in the surroundings. The flow distribution from a cell on a flat area is estimated by vector addition of defined flow directions of the neighbours. If a flat area consists of more than one surface facet, the order of the estimation is related to the number of neighbours with defined flow distribution values. The flat facet with the highest number of defined neighbours is treated first, then the one with the second highest number of neighbours, and so on, until the flow distributions for all the facets on the flat areas have been defined. If the flat area turns out to be a bottom of a sink, the centre cell of the flat area is not allocated any drainage direction. No contradictory drainage directions are ever allowed.

For the proposed algorithm presented in this chapter, we try to keep, and possibly develop, as many as possible of the strengths present in the algorithms presented above, but still keep to an algorithm that is usable in practice. Too complicated models, with a high demand of input data, are often not practical. The main strengths to take charge of, based on the models presented above, are to:

- keep a multiple-flow approach,
- keep a form-based approach,
- model flow over a continuous surface where slope as well as aspects influence the flow,
- try to eliminate or limit the weaknesses connected to point-based solutions, and
- treat flat surfaces in an appropriate way.

Apart from this, we have tried to develop an algorithm that can help us to make estimations of the actual flow (speed as well as amount) at a certain time. It was considered valuable not only to estimate drainage areas, but also the amount of water in time and space. In order to do this, we have to include a number of variables, like precipitation and infiltration. We have also decided to modify but keep the point-based approach by extending the number of point sources in a cell to (at least) eight. By doing this we have the possibility of tracing water packages down-stream, and at any time checking the amount of water in the different cells. The proposed algorithm estimates flow over a gridded DEM, used as a continuous surface by creating triangular facets (i.e. a 'regular TIN') between the cell centres. This 'TIN approach' eliminates gaps between adjacent cells/facets.

3 Method

In this section, the different methodological steps of the algorithm will be presented, from the construction of the TIN to the estimation of the water flow in individual cells.

3.1 TIN construction

In a raster DEM, the cell values represent the elevation in the centre of the cell. Let us imagine that we create triangular facets in a raster DEM by letting the centre point of each cell become a corner in a triangular facet (see Figure 3).

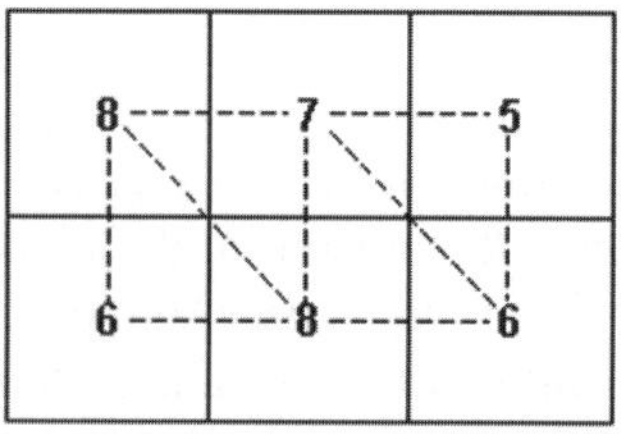

Figure 3. A 'regular TIN' can be created by constructing facets (dashed lines) between the cell centres in the raster DEM. Numbers represents elevation values of the cell centres.

When creating this 'regular TIN', an obvious question is how to perform the triangulation. Since the grid is regular, with equal distances between cells, Delaney triangulation cannot be used. One alternative is to link the two diagonal points (cell centres) that have 'the most equal' elevation values. This means that we assume a ridge between two cells with equally high elevation values (left example in Figure 3), a valley between two cells with equally low elevation values, or 'a more or less horizontal break line' of slope along a slope if the two points with the most equal values are in between the maximum and the minimum value of the cells (right example in Figure 3).

However, since four elevation points only give us limited knowledge about the terrain form, it was decided to expand the window and examine a 4 x 4 cell facet. The idea was to estimate the elevation value in the centre of the facet (where two possible diagonals between the four centre cells in the 4 x 4 cell facet meet) and use this as a help in the triangulation.

One way of approximating the form of the surface (in this case, a 4 x 4 cell facet) is to use a trend surface (TS) based on the elevation values of all cells in the facet. Here a method based on a least squares approximated third-order trend surface is proposed. The first step is to determine the 10 constants in Equation (2) (see below) explaining the surface by the use of the 16 known elevation values in the 4 x 4 cell window. Then the elevation value at the centre of the surface can be computed. To determine the constants, an over-determined equation system needs to be solved. Since the equation system is over-determined, there is no 'true' solution. True in this sense implies that the elevation values for the 16 points lie on the trend surface. The residuals are the discrepancies between the given elevation values from the DEM and the corresponding elevation values computed from the trend surface. The 'optimal' solution minimizes these discrepancies according to a certain criterion. In this study, the least squares method was used to determine the constants of the third-order trend surface.

$$\mathrm{TS}(x_i, y_i) = a_0 + a_1 x_i + a_2 y_i + a_3 x_i^2 + a_4 y_i^2 + a_5 x_i y_i + a_6 x_i^3 + a_7 y_i^3 + a_8 x_i^2 y_i + a_9 x_i y_i^2 \tag{2}$$

where

$i = 1, \ldots, 16$	The index numbers of the 16 cells.
$a_0, \ldots, a_9$	The constants for the third-order trend surface.
x_i, y_i	The cell coordinates in a local system.

When the constants are estimated and the centre point calculated, the elevation value of this point is compared to the four surrounding elevation values. The absolute difference between the calculated value and the average of two diagonal elevation values will decide how to perform the triangulation. The two diagonal elevation points (cell centres) with the least difference to the calculated value will be linked. This approach increases the chances of not interrupting natural valleys and ridges. A screen dump from the developed software, where the cells as well as the facets can be seen, is presented in Figure 4.

When the TIN is defined, the slope and aspect values for each facet can be calculated according to Equations (3) and (4) below. In Figure 5, the slope and slope direction of every facet in the DEM is illustrated.

$$Slope = \arctan\sqrt{\left(a_1^{\ 2} + a_2^{\ 2}\right)} \tag{3}$$

$$Aspect = 180 - \arctan\left(\frac{a_2}{a_1}\right) + 90\left(\frac{a_2}{|a_1|}\right) \tag{4}$$

where a_1 and a_2 are coefficients for the plane.

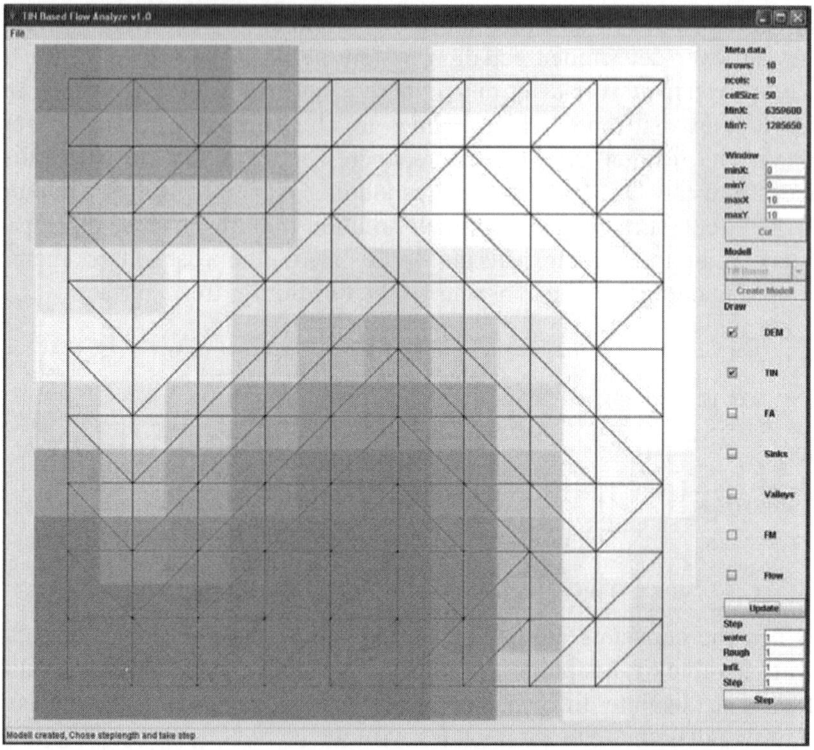

Figure 4. A screen dump from the developed software. In this example, the DEM consists of 10 x 10 cells, coloured differently according to elevation. The TIN structure is superimposed on top. Note that only half (or ¼ in the corners) of the border cells are covered by TIN facets.

3.2 Basic flow estimation

When the TIN is created, the next step is to estimate flow over the surface. The basic idea for doing this is to divide each cell into 8 (or more, or less, depending on the desired accuracy) equally sized areas. The number of areas reflects the level of accuracy in the following flow estimations. The more areas, the more detailed flow estimation, since the area represents the smallest unit over which flow direction and velocity are considered (see below). Each area is represented by a point with a fixed position (at the centre of gravity in each area, see Figure 6). Note that no area can be split by a TIN facet border (compare Figures 4 and 5 above) and consequently no point falls on a border between two facets. The reason for using points in individual areas is that we want to estimate flow by using 'water packages', and we want these water packages to have defined positions (point locations) in space.

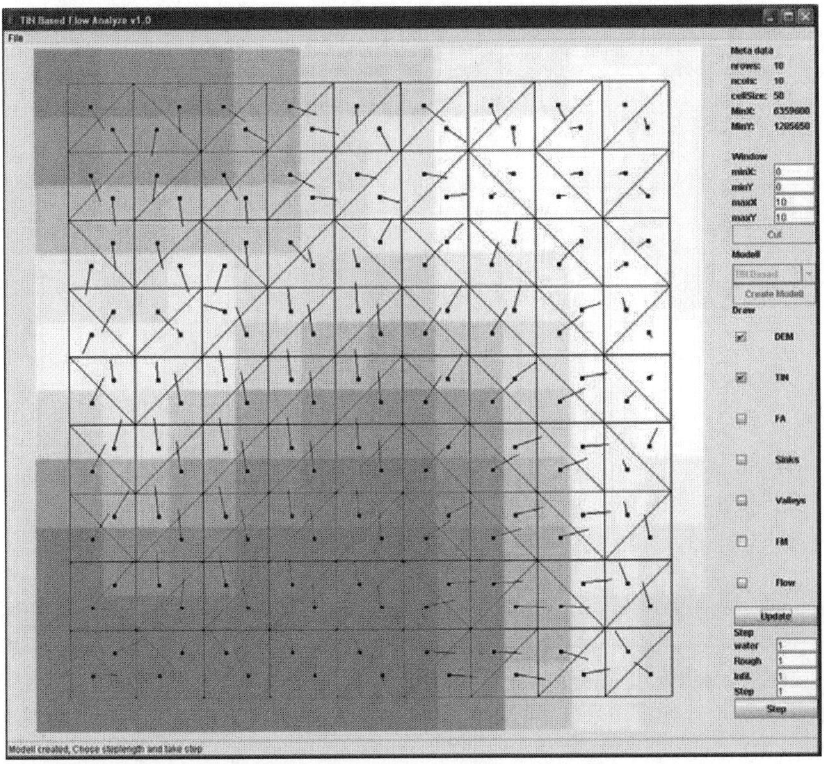

Figure 5. An illustration of a DEM, the TIN structure, slope gradient (proportional to the lengths of the slope vectors), and aspect at every facet.

The flow directions are the same for the four points falling in one facet (the size of a facet is always half the cell size). By drawing all flow directions from all points, we obtain an impression of the flow pattern over the surface (Figure 7).

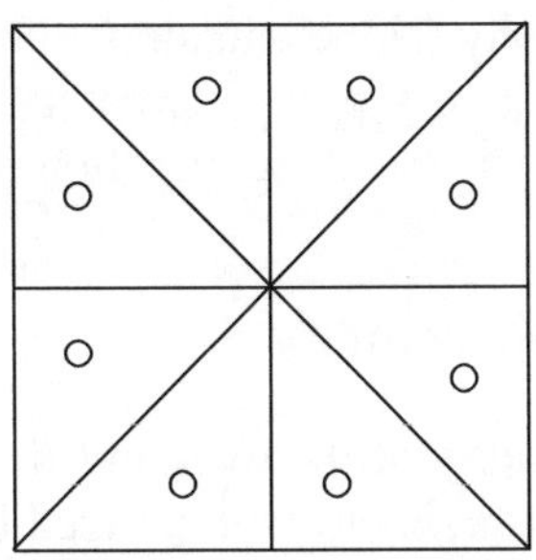

Figure 6. Each cell is divided into eight areas, represented by points with fixed positions at the centre of gravity of each area.

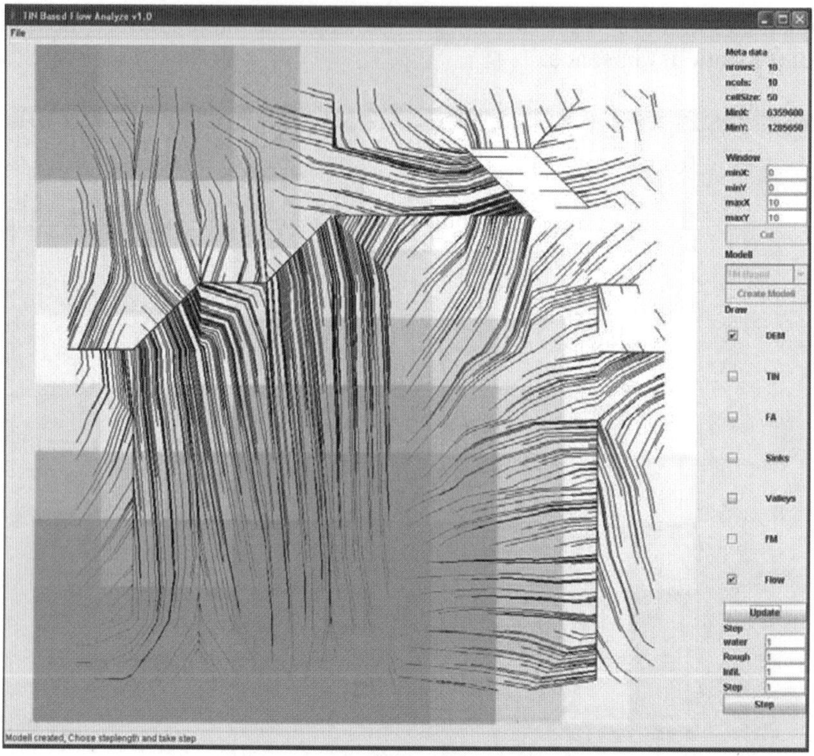

Figure 7. The flow pattern over the surface can be detected by combining flow directions from all initial points (8 in each cell). Note that the flow sometimes follows a valley between two facets.

At time $t = 0$, no flow is present over the surface, and the amount (of water) is equal to an initially given value at every point (and consequently in all areas) in the DEM. At time $t = 1$, the flow has started and mainly the topography (slope and slope direction) has redistributed the water over the surface. Every point is treated as a volume of water, moving in the same direction as the slope direction (aspect) of the facet on which it is situated.

3.3 Flow velocity

The velocity of the overland flow depends on the slope of the surface, the surface roughness (symbolized by a friction factor, *f*), and the water depth (Bengtsson 1997). The relationship between flow velocity, surface roughness, flow depth, and slope can be expressed by Manning's equation:

$$v = M \cdot R^{2/3} \cdot I^{1/2} \tag{5}$$

where v is the water flow velocity, M is Manning's number describing the surface roughness, R is water flow depth, and I is slope.

M varies with different water depth, especially for limited depths on vegetated surfaces (Bengtsson 1997). If the vegetation cover on the surface is known (e.g. by importing a raster file to the program), M can be read from a raster file or a table. R is estimated by summation of water on a facet, and I is known on each facet.

3.4 Infiltration and evaporation

Water loss due to infiltration and evaporation can also be estimated. If we do not estimate flow over long periods of time (e.g. years), loss of flowing surface water due to evaporation is normally negligible. Infiltration, however, normally expressed as infiltration capacity (mm/h), is of high importance for the estimation of Hortonian overland flow. Preferably, infiltration capacity is imported as a raster file to the program. Then the amount of infiltrating water can be subtracted from the water packages, and the Hortonian overland flow can be estimated at every facet.

3.5 Flow over more than one facet

The simplified flow process presented in this chapter can be looked upon as volumes of water (located at points in order to make it simpler) moving over the surface. Depending on slope and vegetation, the velocity, and hence the travel distance of a water volume over a certain time period, of water varies. Whenever we want we can stop or pause the modelling and, by summarizing the points (water volumes) in each cell, obtain information regarding flow as well as flow accumulation.

Depending on flow velocity and time step (i.e. number of seconds, minutes, hours or days) the travel distance of a water package will either not be long enough to 'leave' the facet, or reach the border to another, adjacent, facet. If a water package reaches the facet border, it can either flow into the adjacent facet or follow the facet border. In the latter, case the border line represents a valley.

If a water volume flows into another facet, the speed as well as the flow direction will normally change (if the adjacent facet does not have equal slope, aspect, and roughness as the one the water package has left). Also if it reaches a horizontal facet (i.e. a flat surface) the speed of the water flow will be influenced. Velocity of water will then be more (relatively) influenced by the energy potential between the elevation of the flat surface and the water depth, while the direction will be determined using surrounding facets with defined flow directions (see section 2.6 above).

3.6 Iterative flow estimations

After each time step, the number of water packages (simplified to be located at individual points) in each facet on the surface has either increased, decreased or stayed the same. Note that water added to a facet from above (i.e. precipitation) will be located at the centre of gravitation of each area (covering 1/8 of a cell). If a facet is located in a valley, we can expect the number of water packages to increase as long as the input (precipitation) does not stop. The number of water packages on hill slopes is also expected to increase since eight new water packages will be created in every cell at each time step *t* (if the input is not equal to 0). When the input (precipitation) ends, the number of packages will decrease, starting in the highlands and following the drainage network.

This means that the number of water packages (*n*) over the surface, as long as the input is not equal to 0, equals:

$$n = t * 8 * (ncols - 1) * (nrows - 1) \qquad (6)$$

where *t* is the number of time steps and *ncols* and *nrows* equal the number of columns and rows in the DEM.

The subtraction of one from the number of columns as well as rows is due to the fact that TIN facets cannot be created along the border of the DEM (see Figure 4). If the time step is small (e.g. one minute), the area is large (e.g. 10x10km), the cell size is small (e.g. 10x10m), and we run the program for a long time (e.g. one week), this results in a huge number of water packages (in this case more than 8×10^{10}) that significantly influences the processing time and computing resource. In the present practice, it is impossible to store and track this amount of data.

In order to solve this problem, all water packages on a facet are merged after each iteration. The water volumes are merged into one new position, depending on the amounts of water and positions for the merged volumes. The new position is obtained by weighted average interpolation (see e.g. Burrough and McDonell 1998), where the weighting is proportional to the amount of water in each water volume at each location. The amount of water at the new location is equal to the sum of the merged water volumes. If wanted, water packages on parts of facets (e.g. ¼ instead of a whole facet) can be merged. This increases accuracy in the estimations.

By adding the water volumes in each cell at a certain time *t,* the amount of surface water over the area can be estimated. If the total amount of incoming water is set to 1 unit per cell (i.e. 1/8 at every point), no new incoming water is added, and the amount of water passing (or staying in) each cell is summarized, the result will show the drainage area (in number of cells) to each cell in the DEM.

3.7 Sinks and diverting valleys

One common problem in flow estimation is the presence of sinks, defined as depressions in the terrain without an outflow. At this stage of development of the software, sinks have not been treated individually. This means that a sink will accumulate water until the inflow ends, or until the modelling is stopped. However, in the next version of the program spill-over from sinks will be possible. When the water depth of a sink (consisting of one or more cells) has reach the difference in elevation between the sink and its lowest 'water blocking' neighbour cell, additional water will spill over and flow into that cell.

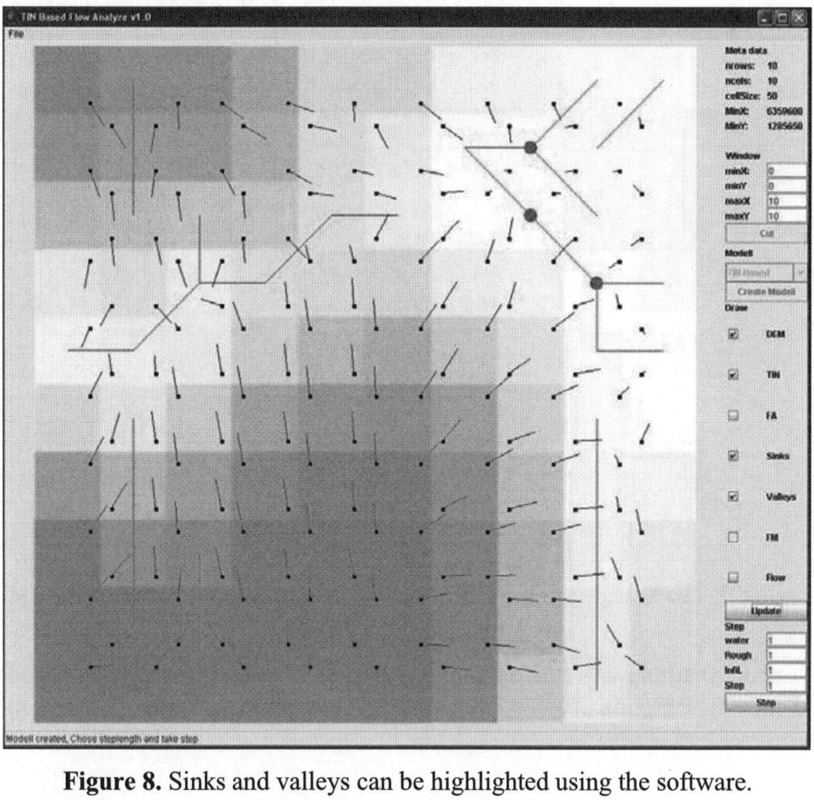

Figure 8. Sinks and valleys can be highlighted using the software.

Apart from sinks, diverting valleys are an important topic to discuss. When flow diverts (i.e. following borders between facets that continue down-slope in more than one direction), it has to be decided how to divide the flow between the alternative paths. This has been done according to Equation (1), with an x value of 1. This means that flow has been judged to be directly proportional to slope gradient. The sinks and valleys can also be highlighted using the software (see Figure 8).

4 Results

In order to make a first test of the algorithm, estimations were performed on a natural surface. In Figure 9, a DEM from an agricultural field in central Sweden, with a resolution of 1 metre, is presented. For a detailed description of the DEM, see Persson (2004). A relatively large number of sinks are present in the DEM, and along the edges there are disturbances probably related to interpolation errors.

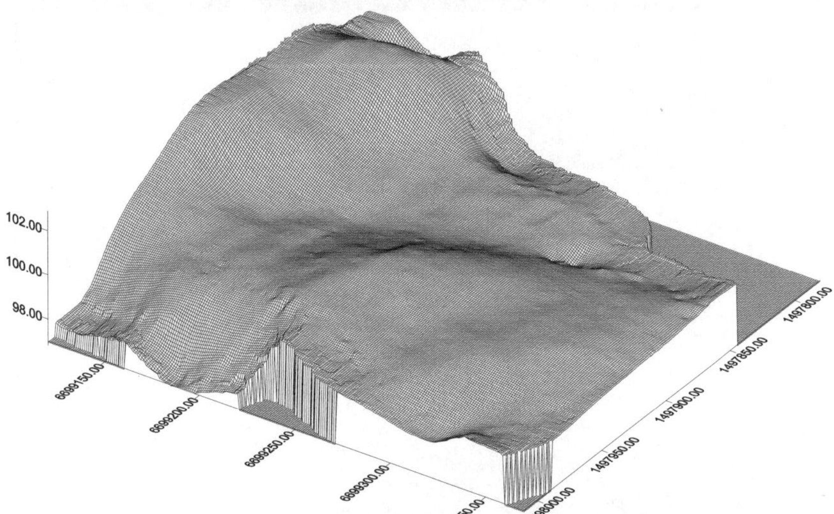

Figure 9. The topography of the test surface. Swedish reference system RT90. Coordinates in metres.

The algorithm was run on the test surface, with constant precipitation and infiltration capacity. The estimated flow accumulation is presented in Figure 10.

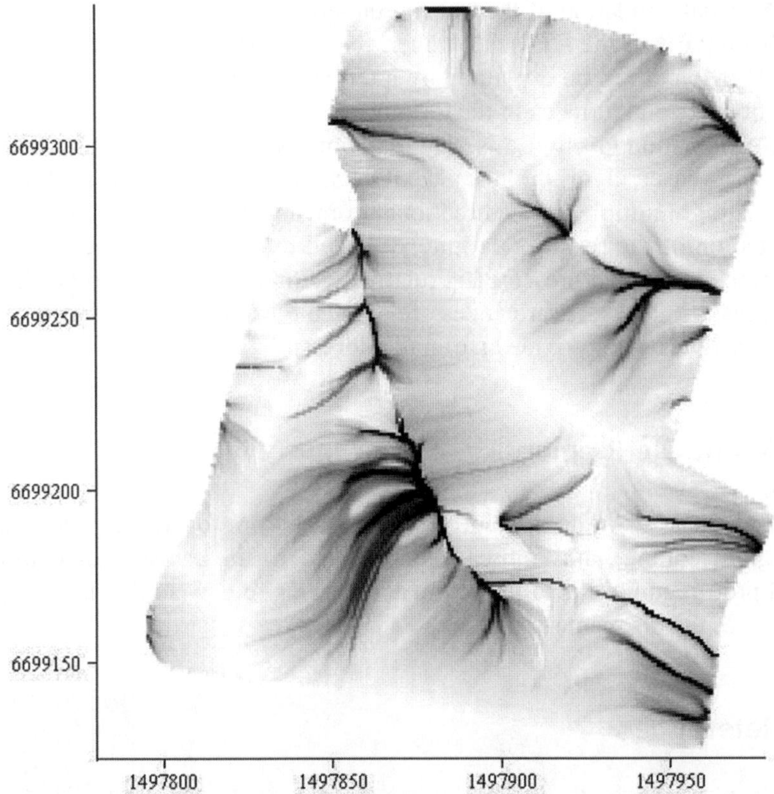

Figure 10. The result of a test run of the algorithm. The figure illustrates flow accumulation (0–1,000 cells). Flow exceeding 1,000 units was set to 1,000. The darker the tone, the higher the flow accumulation. Swedish reference system RT90. Coordinates in metres.

5 Discussion

The algorithm and software presented above give an alternative way of estimating surface flow over digital elevation models. The program enables us to perform estimations in space as well as in time. Input parameters can be both time and space dependent (like precipitation and evaporation), only space dependent (like a static vegetation cover/surface roughness), or only time dependent (like precipitation in a smaller and less documented area).

A visual validation of the test run unveils disturbances in flow in the peripheral parts of the DEM. However, these disturbances are also present in the DEM, and can be explained by interpolation errors and/or a large number of sinks and terraces in the terrain.

It is too soon to say if this way of estimating surface flow surpasses e.g. the form-based algorithm, the DEMON algorithm, and/or other algorithms estimating multiple flow directions from a gridded DEM. A statistical validation of the result is definitely needed to answer this question. However, the algorithm presented is relatively simple and seems to produce robust results. The software codes are available from the author upon request to colleagues who provide an e-mail address.

6 Acknowledgements

I thank Joakim Antonsson and Johan Hasselström for helping me with the program coding, and Andreas Persson and Karin Larsson for comments and proofreading. I also thank the anonymous reviewers for their valuable critiques of this manuscript.

7 References

Band, L.E., (1986), Topographic partition of watersheds with digital elevation models, *Water Resources Research*, **(22)**1: 15–24.

Bengtsson, L., (1997), *Hydrologi – teori och processer,* Svenska Hydrologiska Rådet. Institutet för Teknisk Vattenresurslära, Lund University (in Swedish).

Burrough, P.A. and McDonell, R.A., (1998), *Principles of Geographic Information Systems*, Oxford: Oxford University Press.

Costa-Cabral, M.C. and Burges, S.J., (1994), Digital elevation model networks (DEMON): A model of flow over hillslopes for computation of contributing and dispersal areas, *Water Resources Research*, **(30)**6: 1681–1692.

ESRI, (1991), *Cell-based Modelling with GRID*, Redlands, CA: Environmental System Research Institute.

Fairfield, J., and Leymarie, P., (1991), Drainage network from grid digital elevation models, *Water Resources Research*, **(27)**5: 709–717.

Freeman, T.G., (1991), Calculating catchment area with divergent flow based on a regular grid, *Computers & Geosciences*, **(17)**3: 413–422.

Holmgren, P., (1994), Multiple flow direction algorithms for runoff modelling in grid based elevation models: An empirical evaluation, *Hydrological Processes*, **8**: 327–334.

Mark, D.M., (1984), Automated detection of drainage networks from digital elevation models, *Cartographica*, **(21)**2–3: 168–178.

O'Callaghan, J.F. and Mark, D.M., (1984), The extraction of drainage networks from digital elevation data, *Computer Vision, Graphics, and Image Processing*, **(28)**: 323–344.

Olsson, L. and Pilesjö, P., (2002), Approaches to Spatially Distributed Hydrological Modelling in a GIS Environment, In Skidmore (ed.): *Environmental Modelling with GIS and Remote Sensing*, London: Taylor & Frances.

Persson, D.A., (2004), Hydrological modelling, Topographical Influence and Yield Mapping in Precision Agriculture, PhD Thesis, Lund University, Sweden, ISBN 91–973857–9–4.

Pilesjö, P., (1994), Estimation of drainage directions – A new method, *Proceedings of the 7th Australasian Remote Sensing Conference*, 1-4 March, Melbourne: 328–332.

Pilesjö, P. and Zhou, Q., (1996), A multiple flow direction algorithm and its use for hydrological modelling, *Geoinformatics'96 Proceedings*, April 26–28, West Palm Beach, FL: 366–376.

Pilesjö, P. and Zhou, Q., (1997), Theoretical estimation of flow accumulation from a grid-based digital elevation model, *Proceedings of GIS AM/FM ASIA'97 and Geoinformatics'97 Conference*, 26–29 May, Taipei: 447–456.

Pilesjö, P., Zhou, Q. and Harrie, L., (1998), Estimating flow distribution over Digital Elevation Models using a Form-Based Algorithm, *Geographic Information Science*, **4**: 44–51.

Quinn, P., Beven, K., Chevallier, P. and Planchon, O., (1991), The prediction of hillslope flow paths for distributed hydrological modelling using digital terrain models, *Hydrological Processes*, **5**: 9–79.

Wolock, D.M. and McCabe Jr., G.J., (1995), Comparison of single and multiple flow direction algorithms for computing topographic parameters in TOPMODEL, *Water Resources Research*, **(31)**5: 1315–1324.

DEM-based Modelling and Simulating of Modern Landform Evolution of Loess

TIAN Yuan, WU Lun, GAO Yong, WANG Daming and ZHANG Yi

Abstract

Form and process are the most important contents of geomorphology. For quantitative analysis of form and process, there are respectively methods of statistical analysis and mathematical modelling. At present, the focus and difficulty for modern geomorphology is to establish process-oriented mathematic models based on dynamics and to simulate landscape evolution. This chapter first describes the quantitative analysis method for small catchments, and is divided into three levels, namely form analysis, landform evolution statistical analysis and geomorphological process modelling. And then the Modern Catchment Landform Evolution Model (MCLEM) is presented by which the processes of tectonic elevation, weathering, hillslope and fluvial transport can be described and simulated. Lastly, an experiment is used to demonstrate the advantages of MCLEM.

Keywords: small catchment, quantitative geomorphological analysis, landform evolution, DEM.

1 Introduction

The dynamic changes in the erosion environment of the Loess Plateau, especially in the foothill and ravine regions, are very severe. These changes are both macroscopic and microscopic, just as many other geographic phenomena, and are caused by many different factors, long-term and short-term, natural and artificial. They are extremely difficult to measure by regular statistical methods, which are typically static, quantitative, and simple. By employing geographic information systems (GIS), it is now necessary and possible to make a DEM (digital elevation model) -based quantitative landform analysis and to establish a process-oriented landform evolution model, by which we can simulate the modern loess landform evolution and make a more effective study of its laws.

Table 1. The characteristics of some typical landform evolution models since 1990s.

Name	Processes	DEM Type	Spatial Scale	Time Scale	Flow Direction Algorithm	Transport Rule	Hydrology Model
SIBERIA (Willgoose et al. 1991)	tectonic, slope, stream	grid	N/A	N/A	single flow direction	confined to transit rule	
Precipiton model (Chase 1992)	tectonic, slope, stream	grid	N/A	Geological period	Cascade	confined to transit rule	Random rainfall
Detachment limited model (Howard 1994)	tectonic, slope, stream	grid	N/A	Geological period	single flow direction	confined to both transit and detaching rules	
GOLEM (Tucker and Slingerland 1994)	tectonic, stream	grid	Region (10^3) drainage (50m)	Geological period (10^6 - 10^7 years)	single flow direction	confined to both transit and detaching rules	
CASCADE (Braun and Sambridge 1997)	tectonic, slope, stream	TIN	region (10^3m)	Geological period (10^6 years)	Bucket Passing Algorithm	confined to both transit and detaching rules	
CAESAR (Coulthard et al. 1998)	slope, stream	grid	1-50m	10 - 10^4 years	multi flow direction		driven by rainfall events
ZSCAPE (Densmore et al. 1998)	tectonic elevation, form and transfer of weathered rock, bedrock slide, stream	grid	mountain 10^2 -10^4 m)	10^4 - 10^6 years	single flow direction	confined to both transit and detaching rules	
CHILD (Tucker et al. 1999)	tectonic, slope, stream	TIN	drainage	10^3 - 10^6 years	Cascade	confined to both transit and detaching rules	(exaggeratedly) driven by rainfall events

Many researchers have developed different landform evolution models, with various theoretical structures and goals. The characteristics of some typical landform evolution models are listed in Table 1.

This chapter begins with a description of the quantitative methods of analysis on small catchments in Section 2, which can be divided into three parts: form analysis, landform evolution statistical analysis, and geomorphological process modelling. Based on these studies, a Modern Catchment Landform Evolution Model (MCLEM) is described in Section 3, by which the processes of tectonic elevation, weathering, hillslope, and fluvial transport can be described and simulated. Section 4 gives the details of the outcomes of MCLEM simulation and compares these outcomes with corresponding observation results. We discuss some observations and interpretations arising from the use of the model in Section 5 and draw some conclusions in Section 6.

2 Quantitative Methods of Analysis of Drainage

Quantitative drainage methods use single-phase or multi-phase DEMs to investigate the spatial changes of the form, process of the drainage, and the relations between them, and then predict future changes of the form. A single-phase DEM just represents the landform at the time when the DEM data was collected, but multi-phase DEMs, composed of several successive single-phase DEMs, represent not only the landform at a single time but also the changes in the landform. At present, the development of the technologies of high resolution DEMs and high performance computers offers good opportunities for these quantitative methods.

The changes in the spatial process can be represented by form analysis, which offers information on the basic process. Much attention is focused on the effects of form on geomorphological processes. What are represented by the basic differential equations of the landform changes are the processes determined by the geomorphological forms themselves (Shreve 1972). DEMs can be used to identify places where different hillslope processes occur (Dietrich and Dunne 1993), where hillslope processes change to ravine processes (Tribe 1992), and where bedrock ravines change to depositional ravines (Montgomery and Dietrich 1994), etc. However, the problem is that the distribution of the effects of the process on the form also changes the form, as well as the form controlling process. The form is not only an outcome of the past process, but also a factor which affects the current erosion, so that the current form affects the future form. This feedback is an important factor in landform evolution. Additionally, the effects of the historic geomorphological process and the geomorphological form conditions are both emphasized in current research (Lane and Richards 1997) (see Figure 1). Hence, the contents and the methods of research on quantitative analysis on drainage should be analysing the distribution and speed of the spatial process from the form information and predicting the changes in the landform according to the effects of the spatial process on that landform.

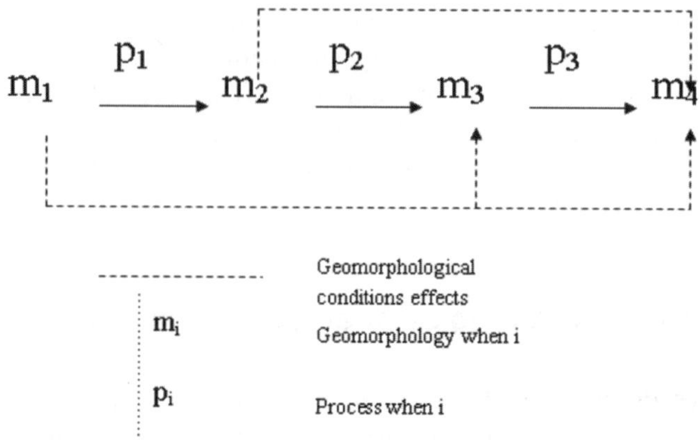

Figure 1. Geomorphology – process feedback and geomorphological condition effects.

The continuous distribution of the landform on the Earth surface offers rich information for quantitative analysis. Regarding current research methods, modern quantitative landform research includes statistical landform analysis and dynamic morphology (Wu 1990). This chapter focuses on the difference between single-phase DEMs and multi-phase DEMs, and divides quantitative methods of landform analysis into landform analysis and landform evolution analysis. Moreover, three levels of quantitative landform analysis on small catchments are defined. They are form analysis, landform evolution statistical analysis, and geomorphological process modelling (see Figure 2).

In form analysis, the factors of the form, such as gradient, aspect, curvature, etc., are first extracted from a single-phase DEM. These factors have significant geomorphological and hydrological meanings. Although these factors do not describe all the features of the form, they are necessary to characterize the important features of the form. Based on form analysis, we can go on to analyse the relations among these basic factors, to reveal the laws of the landform, and to some degree, the geomorphological process of the form.

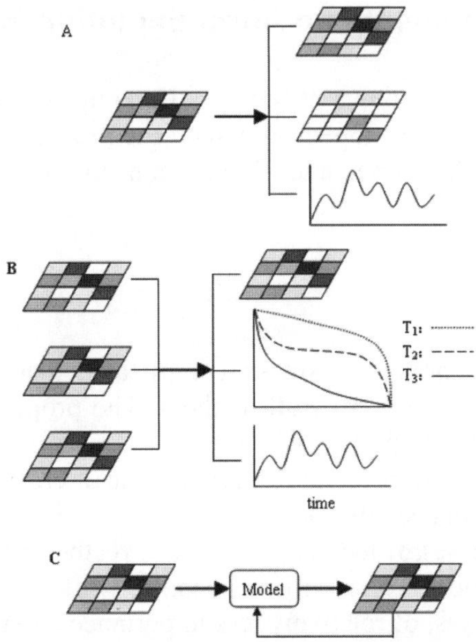

Figure 2. Three levels of the quantitative landform analysis of small catchment. (A) Form analysis; (B) Landform evolution statistical analysis; (C) Geomorphological process modelling.

Landform evolution statistical analysis attempts to quantitatively describe intensity and frequency of geomorphological processes. Successive maps of the geomorphological factors that vary temporally are generated from multi-phase DEMs during landform evolution statistical analysis, such as an Elevation Area Curve Map or a Ravine Density Changes Map. Research on this issue is rarely developed by the geomorphologists due to the lack of multi-phase DEM data of the same drainage. Most geomorphological laws obtained from landform evolution statistical analysis are qualitative. Little information about the future development and feedback is available by using geomorphological methods at this level. Further research methods, which focus on predicting future landforms, should therefore be developed.

Geomorphological process modelling is to predict landform change based on process-oriented mathematical models. The former result is used as a key input to predict the next form. Consequently, an iterative simulating model can be established to predict dynamically the change of the landform. The Modern Catchment Landform Evolution Modelling in the next section gives an example of this kind of model.

3 Modern Catchment Landform Evolution Modelling

Based on the above study, a Modern Catchment Landform Evolution Model (MCLEM) is developed, by which the processes of tectonic elevation, weathering, hillslope, and fluvial transport can be described and simulated.

3.1 Hypothesis

Geomorphological processes are so complicated that no single geomorphological model can describe all of them. The proposed model is based on the following two hypotheses:

1. Physical erosion, rather than chemical erosion, dominates the processes in a small catchment.
2. The processes of tectonic elevation, weathering, hillslope, and fluvial transport are taken into account; while other geomorphological processes, of relatively less importance, are not considered.

3.2 Objectives

According to the problems of existing models and the quantitative landform analysis theory, the objectives of this model are listed below:

1. The model should represent a whole small catchment rather than a cross-section of the river or a section of the river.
2. Fluvial processes are the most important ones and should be represented as accurately as possible.
3. The hillslope processes, such as shallow-focus landslides, should be sufficiently considered.
4. The processes that are limited by transport, detachment, or weathering, should be taken into account synthetically.
5. Modularized programming methods should be employed so as to integrate the other processes, e. g. glacial erosion.
6. The model should be general rather than specific. The different geomorphology of different small catchments and environments should be able to be represented by changing input parameters.

3.3 Modelling

Figure 3 depicts the framework of MCLEM. The small catchment is divided into a series of square cells by finite element and finite difference methods. A group of initial attribute values is assigned to each cell, including elevation, water flow, confluence area, sediment yield, etc. These values are recalculated according to geomorphological processes and relations with the surrounding cells during every iteration step. Five geomorphological processes are considered in the model:

1. The process of hillslope;
2. The process of sediment transferring in the fluvial ravine;
3. The process of erosion of the bedrock;
4. The process of weathering; and
5. The process of tectonic elevation.

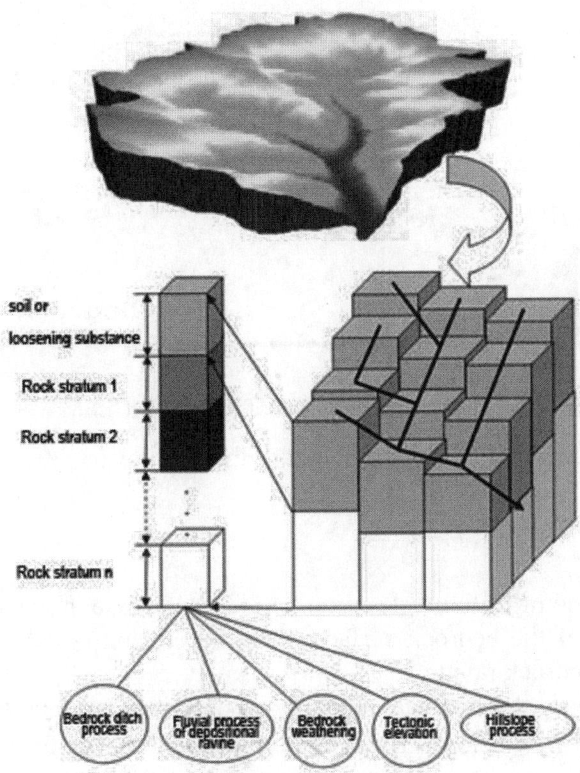

Figure 3. Framework of MCLEM.

Some main and important equations of MCLEM are as follows:

The height of one cell is the sum of the depth of its bedrock and the depth of its weathered rock (see Figure 4):

$$h = R + C \quad (1)$$

h is the height of one cell, ***R*** is the depth of bedrock, ***C*** is the depth of weathered rock.

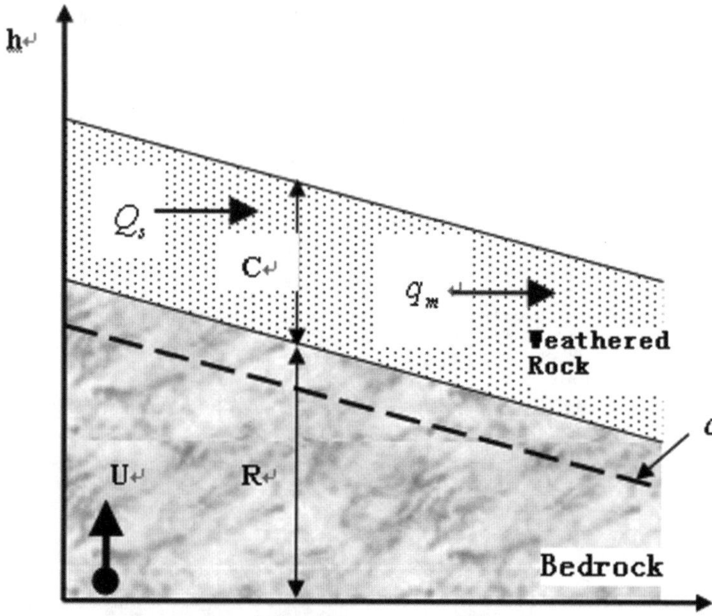

Figure 4. Conservation of mass in MCLEM.

The equation to describe the height of the bedrock is:

$$\frac{\partial R}{\partial t} = U - \omega - E \quad (2)$$

U is the speed of tectonic elevation, ***ω*** is the decrease in height caused by weathering of the bedrock surface, ***E*** is the erosion speed of the fluvial process of bedrock ravine.

In MCLEM, the bedrock lithology is describes by four attributes: ***R***, K_b, K_w, S_t (see below). By assigning different values for these four attributes, different lithologies of bedrock are taken into account.

The equation to describe the depth of the weathered rock is:

$$\frac{\partial C}{\partial t} = \omega - \nabla Q_s - \nabla q_m \quad (3)$$

$\boldsymbol{Q_s}$ is the sediment transport speed of ravine, $\boldsymbol{q_m}$ is sediment transport speed on the slope within a unit width.

Three probability density functions are used to describe the rainfall (Tucker and Bras 2000, Tucker and Whipple 2002).
(1) The probability density function of rainfall intensity is:

$$f(P) = \frac{1}{\overline{P}} \exp(-\frac{P}{\overline{P}}) \tag{4}$$

P is the rainfall intensity.
(2) The probability density function of rainfall duration is:

$$f(T_r) = \frac{1}{\overline{T_r}} \exp(-\frac{T_r}{\overline{T_r}}) \tag{5}$$

T_r is the rainfall duration.
(3) The probability density function of rainfall interval is:

$$f(T_b) = \frac{1}{T_b} \exp(-\frac{T_b}{\overline{T_b}}) \tag{6}$$

***T*b** is the rainfall interval.

MCLEM offers three different methods to calculate the ground runoff rate:
(1) Horton runoff equation(runoff yield under excess infiltration):

$$R = P - I_c,\ P > I_c \tag{7}$$

$$Q = RA = (P - I_c)A \tag{8}$$

R is runoff rate, ***P*** is rainfall intensity, I_c is the loss of rainfall caused by soil infiltration, crown interception, etc., ***Q*** is the amount of runoff, ***A*** is the catchment area.
(2) Dunne runoff equation (runoff yield under excess infiltration).

$$R = \frac{T_r P - D_{sr}}{T_r},\ T_r P > D_{sr} \tag{9}$$

$$Q=RA \tag{10}$$

D_{sr} is the depth of reservoir of crown, soil and land surface.
(3) Saturated surface runoff

$$q_{sub} = \frac{Q_{sub}}{D} \tag{11}$$

$$Q = PA - q_{sub}D, PA > q_{sub}D \tag{12}$$

$\boldsymbol{Q_{sub}}$ is the runoff in the shallow soil, $\boldsymbol{D}$ is the length of the contour line, depth of reservoir of crown, soil and land surface

The equation to describe weathering processes is:

$$\omega = -(K_{w1} \exp(-\alpha_1 C) - K_{w2} \exp(-\alpha_2 C)) \tag{13}$$

and it is simplified to:

$$\omega = \begin{cases} K_w \exp(-\alpha C) & Hillslope \\ 0 & Ravine \end{cases} \tag{14}$$

$\boldsymbol{K_w}$, $\boldsymbol{K_{w1}}$, $\boldsymbol{K_{w2}}$ are the weathering process speeds of the bedrock surface, $\boldsymbol{\alpha}$, $\boldsymbol{\alpha_1}$, $\boldsymbol{\alpha_2}$ are constants dependent on the lithology of weathered rock.

The equation to describe hillslope processes is:

$$\left.\frac{\partial h}{\partial t}\right|_{diffusion} = -\nabla q_m \tag{15}$$

$$q_m = \left[K_d \Re(S) + K_f \left(\frac{1}{1 - \left[|S| / S_t \right]^{\alpha}} - 1 \right) \right] s \tag{16}$$

$$\Re\ (S) = \tan\theta = S$$

$\boldsymbol{q_m}$ is the transport speed of movable weathered rock on the slope, $\boldsymbol{S}$ is the gradient, S_t is the critical gradient, $\boldsymbol{K_d}$ and $\boldsymbol{K_f}$ are constants, $\boldsymbol{s}$ is the unit vector in the direction of the aspect.

The equation to describe fluvial processes of depositional ravines is:

$$\left.\frac{\partial h}{\partial t}\right|_{alluvial} = -\nabla q_s \tag{17}$$

$$Q_s = WK_f (\tau - \tau_c)^p \tag{18}$$

$$W = K_c Q^{\ b} \tag{19}$$

$\boldsymbol{q_s}$ is the sediment transport speed on the slope within a unit width, ***Qs*** is the sum of sediment transport, ***Q*** is the amount of runoff, ***W*** is the width of ravine, τ is the shear stress, τ_c is the critical shear stress, ***p*** and ***b*** and ***Kc*** are constants.

In order to make the mode more flexible, three methods of calculating sediment transport process are offered in MCLEM:

(1) A general model:

$$\tau = K_t Q^m S^n \tag{20}$$

$$Q_s = K_f K_c Q^{1/2} (K_t Q^m S^n - \tau_c)^p \tag{21}$$

$\boldsymbol{K_t}$, ***m***, ***n*** are constants.

(2) Bridge-Dominic equation (Tucker and Slingerland 1997):

$$Q_s = \frac{W a_t}{(\sigma - \rho)\rho^{1/2} g} (\tau - \tau_c)(\tau^{1/2} - \tau_c^{1/2}) \tag{22}$$

g is the acceleration due to gravity, $\boldsymbol{a_t}$ is a constant, σ is the density of mud and sand, ρ is the density of water.

(3) Meyer-Peter and Mueller equation:

$$Q_s = a_m Q^{1/2} (\tau - \tau_c)^{3/2} \tag{23}$$

$\boldsymbol{a_m}$ is a constant.

The equation to describe fluvial processes of bedrock ravines is:

$$\left.\frac{\partial h}{\partial t}\right|_{bedrock} = -K_b (\tau - \tau_c)^u \tag{24}$$

$$\tau = \frac{\rho g^{2/3} f^{1/3}}{2K_c} Q^{1/2} S^{2/3} \tag{25}$$

$\boldsymbol{K_b}$ is the erosion index of the bedrock, ***u*** is a constant.

The equation used to determine the initial origin of the ravine is:

$$A(\tan S)^2 \geq A_{cr} \tag{26}$$

A is the area of the catchment of the current cell. ***S*** is the gradient of the current cell. $\boldsymbol{A_{cr}}$ is a threshold of area. One cell that meets the requirement will be taken as the initial origin of the ravine, when all the cells in the downriver direction will belong to the ravine.

3.4 Advantages of MCLEM

From the above equations, a group of equations was deduced to satisfy the requirements of MCLEM. The MCLEM has several advantages.

1. It differentiates between soil and bed rock. It also differentiates between different lithologies of bedrock by assigning different parameters.
2. It considers and deals with the temporal variability in rainfall and runoff based on the Poisson pulse rainfall model, whereas these were often neglected in the previous models. MCLEM provides three different methods to calculate the ground surface runoff.
3. An explicit differentiation between the hillslope and the channel is made, the transport processes of which are also different.
4. In this model, the growth of the channel network is controlled by a threshold that is nonlinearly dependent on runoff and slope.
5. It can process both the erosional and depositional systems simultaneously and differentiates between the alluvial channel and bedrock channel, which can be converted into each other, whereas most previous models could not.
6. It uses a multiple-flow-direction method to determine the flow directions and designs a multiple-landslide-direction algorithm to simulate landslide processes, whereas most of the previous models used single-flow-direction methods, which do not fit the simulation based on high-resolution DEM data.
7. Most previous models can only give simulation results of the balance state, but applying MCLEM allows researchers to obtain simulation results in any supposed phase.

With these advantages, the MCLEM can be used to predict the short-term and successive form changes, taking the factual form data as initial conditions. The MCLEM can also be implemented as a universal software tool, which provides a quantitative simulation environment to explore the possible consequences of different hypotheses, parameters, and boundary conditions. This will definitely help to deepen our understanding of the effects of different weather, form, and surface conditions on landform evolution.

4 The Simulation Experiment

As part of this research, a simulation experiment on the geomorphology evolution of a small catchment was made in the rainfall simulator hall of the State Key Laboratory of Soil Erosion and Dryland Farming on the

Loess Plateau (Table 2). Some statistical and dynamic-change-mapping analyses were used to extract the features of the small catchment. At the same time, the MCLEM was also used to extract the same features in order to make qualitative and quantitative comparisons and to verify the validity of MCLEM. The results proved a close match between the observation results and the MCLEM-simulated results.

Table 2. Original landform features of the simulated small catchment (Cui 2002).

Projected Area	Catchment Length	Catchment Width	Catch-ment Perimeter	Catch-ment Height Difference	Longitudinal River Slope	Average Gradient	Ravine Classes	Branches Ratio
32.67 m^2	9.1 m	5.8 m	23.3 m	2.57 m	28.24%	150	2	4

4.1 Simulation experiment in the rainfall simulator hall

In the experiment, a fixed-intensity rainfall was simulated as the main driving force of the geomorphology evolution (Table 3). Nine high resolution DEM data sets were gathered by close range photogrammetry in different phases of the experiment. The horizontal resolution of each data set was 1 mm. These DEM data revealed the changes in the ground surface in the experiment very precisely. Based on the DEM data, the law of geomorphology evolution of small catchments was analysed by making statistical and dynamic-change-mapping analysis on hypsometry-catchment area curves (see Figure 5), slope-catchment area curves (see Figure 6), accumulative catchment area distribution (see Figure 7), and width functions (see Figure 8).

4.2 Simulation by MCLEM

Considering the factual status of the Loess Plateau, the MCLEM was customized, where the sediment outflow ratio, hypsometry-catchment area curve, and slope-catchment area curve were elements to be focused on and several parameters were calibrated according to previous research results and observation results. Then, MCLEM was programmed on a GIS platform. Taking the original form data and the rainfall intensity of the rainfall experiment as initial conditions, the MCLEM was run to simulate the geomorphological processes taking place in the rainfall experiment to get landform DEMs of all the 8 phases.

Table 3. Features of simulated rainfall (Cui 2002).

No.	Date	Designed Rainyness (mm/min)	Rating Rainyness (mm/min)	Duration of Rainfall (min)	Quantum of Rainfall (mm)	Sediment Runoff (kg)
1	2001.7.30	0.5	0.54	90.5	48.915	144
2	2001.8.1	0.5	0.52	89.5	46.54	106
3	2001.8.3	0.5	0.49	89.9	44.051	203.5
4	2001.8.5	1.0	1.18	47.52	63.939	277.8
5	2001.8.8	1.0	1.21	45.86	65.05	623.9
6	2001.8.14	2.0	2.41	30.53	73.58	768.5
7	2001.8.17	1.0	1.19	46.17	53.75	624.7
8	2001.8.20	0.5	0.57	90.18	51.528	829.5
9	2001.8.22	0.5	0.59	61.95	36.167	473.2
10	2001.8.28	1.0	1.2	47.92	55.704	819.6
11	2001.8.31	2.0	2.15	31.17	65.36	1,066.2
12	2001.9.3	0.5	0.52	62.94	31.896	311.1
13	2001.9.5	0.5	0.58	61.53	35.96	353.2
14	2001.9.7	0.5	0.56	60.83	34.29	247.7
15	2001.9.11	1.0	1.12	46.82	52.438	500.4
16	2001.9.14	1.0	1.08	45.83	49.896	432.4
17	2001.9.17	1.0	0.98	47.02	46.256	365.5
18	2001.9.20	1.0	1.04	45.37	47.18	282.4
19	2001.9.24	2.0	2.12	30.37	64.384	493.9
20	2001.9.27	2.0	1.98	34.35	67.736	452.2
21	2001.9.30	0.5	0.53	91.27	48.373	265.1
22	2001.10.9	0.5	0.55	90.6	49.83	225.5
23	2001.10.11	0.5	0.60	89.72	53.832	246.6
24	2001.10.15	1.0	1.05	61.35	64.417	313.6
25	2001.10.18	2.0	2.03	31.65	64.249	314.8

4.3 Comparisons of the observation results and MCLEM results

By respectively comparing hypsometrical features, hypsometry-catchment area curves (see Figure 5), slope-catchment area curves (see Figure 6), accumulative catchment area distribution figures (see Figure 7), and width functions (see Figure 8) of each observation DEM and its corresponding simulation result, the MCLEM was tested, both qualitatively and quantitatively. The comparison proved that the MCLEM could simulate the landform evolution of loess quite well. After adjusting a few parameters accordingly, the MCLEM was re-run. The result of the later simulation was more precise than the first one. It proved that MCLEM is an effective and powerful tool for future research, though it does not simulate the geomorphology evolution without any error.

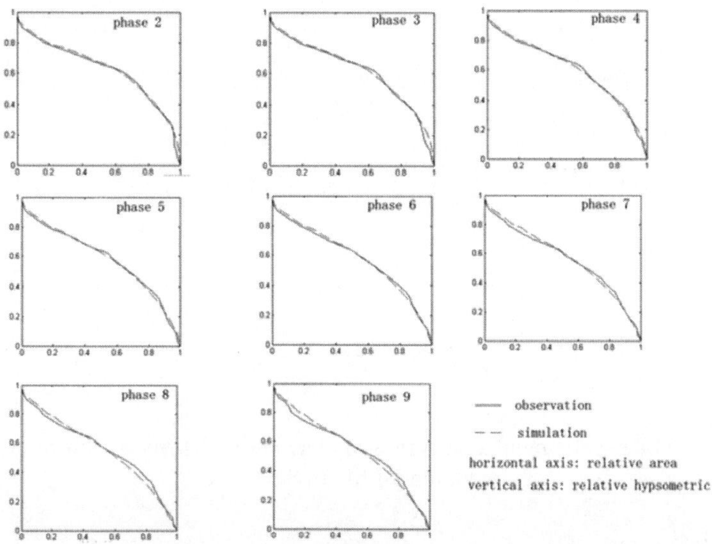

Figure 5. Hypsometry–area curves of observation DEMs and corresponding simulation DEMs of 8 phases.

4.3.1 Qualitative comparison

The comparison of the observation DEM and corresponding MCLEM-simulated DEMs (see Figure 9) demonstrated a quite good qualitative match between MCLEM simulation and the rainfall experiment:

- MCLEM could simulate the formation, position, and development of the ravine quite well.
- The erosion speed calculated by the MCLEM was faster than the observation result and consequently the ravines that were generated by MCLEM were deeper than the ones generated from the rainfall experiment.
- The ravines that were generated from MCLEM were narrower than the ones which were generated from the simulation experiment.

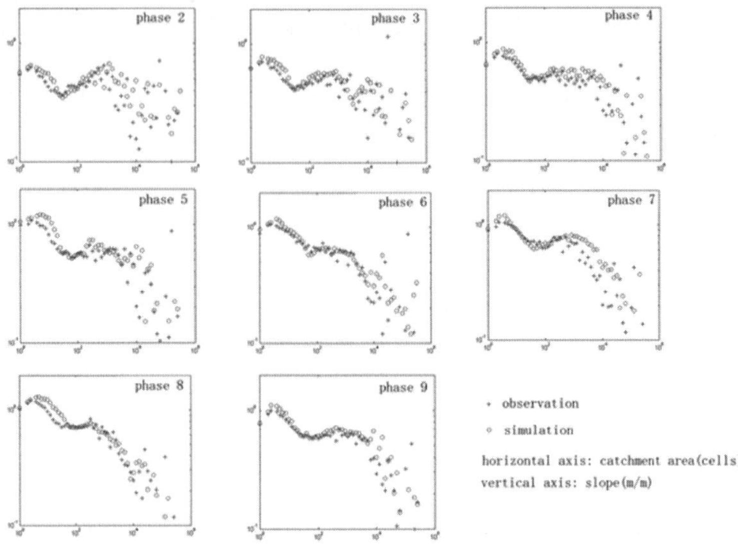

Figure 6. Slope–catchment area curves of observation DEMs and corresponding simulation DEMs of 8 phases.

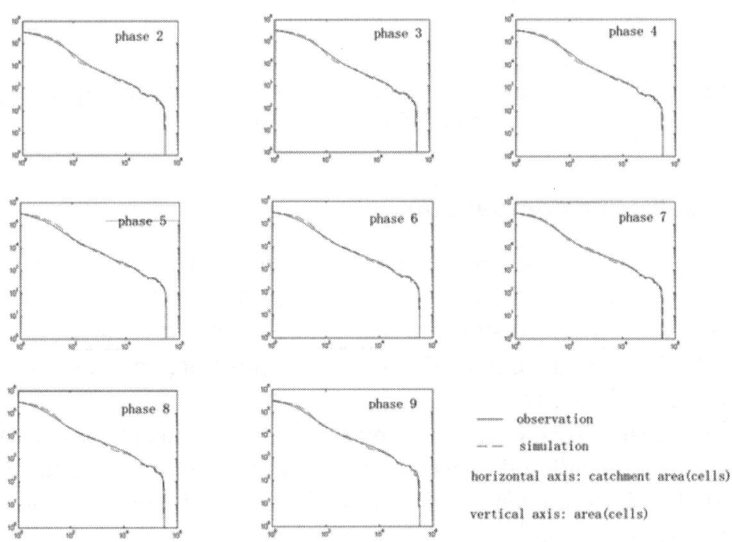

Figure 7. Accumulative catchment area distribution of observation DEMs and corresponding simulation DEMs of 8 phases.

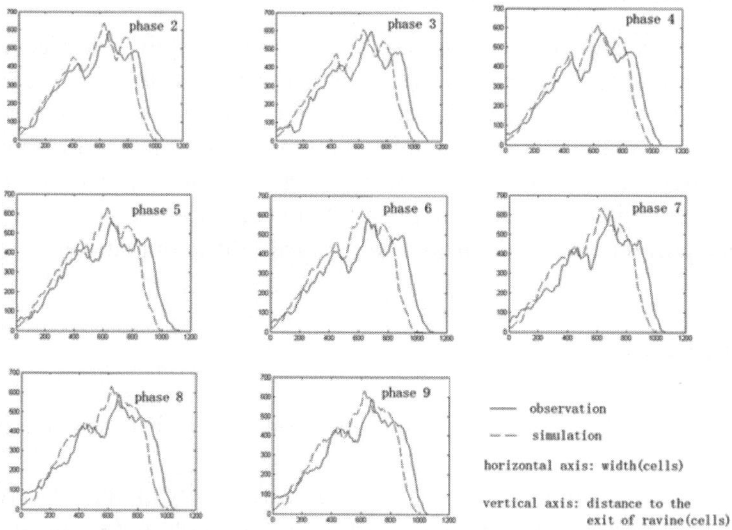

Figure 8. Width function of observation DEMs and corresponding simulation DEMs of 8 phases.

a) 3D display of the observation DEMs b) 3D display of the 8 MCLEM- simulating DEMs

Figure 9. Qualitative comparison between observation DEMs and corresponding simulation DEMs of 8 phases.

4.3.2 Quantitative comparison

Hypsometry-catchment area curves (see Figure 5), slope-catchment area curves (see Figure 6), accumulative catchment area distribution (see Figure 7), and width functions (see Figure 8) were chosen to prove the quantitative match of the results of observation DEMs and corresponding MCLEM-simulating DEM. The comparisons indicated the validity of the MCLEM quantitatively. Moreover, all four comparisons could quantitatively support the three conclusions in the qualitative comparison.

5 Discussion

The following observations and interpretations arose from the use of the model:

- The impact of rainfall on landform varies very much in different phases of a geomorphology process. In the initial phase, the erosion impact caused by rainfall becomes stronger along with the geomorphology process and the ground becomes more and more fragmented. However, it becomes weaker along with the geomorphology process, especially in the last phase.
- The erosion impacts taking place in the upriver, midriver and downriver are very different in different phases of a geomorphology process. In the initial phase, erosion mostly happens in the midriver and downriver; while upriver is the main erosion area in the last phase.
- The change of deposition speed depends most on the geomorphology process itself, when the sediment transport speed and width function are determined by both the phase of the geomorphology process and the intensity of rainfall.
- The width function is also dependent on the phase of the geomorphology process and the rainfall intensity. In the initial phase, the value of the width function increased at a fluctuating rate, but it decreased in the last phase. It is the change of intensity of rainfall that causes the fluctuation.
- Along with the geomorphology process, one ravine may expand gradually at a decreasing speed. The main branch of one ravine always expands prior to the other minor branches. In the initial phase, the branches in downriver areas expand more rapidly than upriver branches. However, in the last phase, upriver branches expand more rapidly.
- MCLEM does have a cell-size dependence, caused by the framework of MCLEM and some of the equations. This could have brought the

errors that can be found easily in Figure 9. Further research should address this problem to make MCLEM more precise.

6 Conclusion

By bringing forward and testing the MCLEM, this chapter deepens the understanding of geomorphological processes, especially the erosion processes, in small catchments and other analogous processes. The research outcome can be used to predict the possible landform evolution in some small catchments where the situation of water and soil loss is severe, which are necessary conditions of making successful plans for water and soil conservation and land use.

References

Braun, J. and Sambridge, M., (1997), Modelling landscape evolution on geological time scales: a new method based on irregular spatial discretization, *Basin Research*, **(9)** 1: 27–52.

Chase, C.G., (1992), Fluvial landsculpting and the fractal dimension of topography, *Geomorphology*, **(5)** 1: 39–57.

Coulthard, T.J., Kirkby, M.J. and Macklin, M.G., (1998), Non-linearity and spatial resolution in a cellular automaton model of a small upland basin, *Hydrological and Earth System Science*, **2**: 257–264.

Cui. L., (2002), The Coupling Relationship Between the Sediment Yield of Rainfall Erosion and the Topographic Feature of Small Watershed on Loess Plateau, Ph.D. dissertation, Institute of Soil and Water Conservation, Chinese Academy of Sciences.

Densmore, A.L., Ellis, M.A. and Anderson, R.S., (1998), Landsliding and the evolution of normal fault bounded ranges, *Journal of Geophysical Research*, **(103)** B7: 15203–15219

Dietrich, W.E. and Dunne, T., (1993), The channel head, In Beven, K. and Kirkby, M.J. (eds) : *Channel Network Hydrology*, New York, NY: Wiley.

Howard, A.D., (1994), A detachment limited model of drainage basin evolution, *Water Resources Research*, **(30)** 7: 2107–2117.

Lane, S.N. and Richards, K.S., (1997), Linking river channel form and process: time, space and causality revisited, *Earth Surface Processes and Landforms*, **(22)** 3: 249–260.

Montgomery, D.R. and Dietrich, W.E., (1994), Landscape dissection and drainage area-slope thresholds, In Kirkby, M.J. (ed.) : *Process Models and Theoretical Geomorphology*, New York: Wiley.

Shreve, R.L., (1972), Movement of water in glaciers, Journal of Glaciology, **(11)** 62: 205–214.

Tribe, A., (1992), Automated recognition of valley lines and drainage networks from grid digital elevation models: a review and a new method, Journal of Hydrology, **139**: 263–293.

Tucker, G.E. and Slingerland, R.L., (1994), Erosional dynamics, flexural isostasy, and longlived escarpments: a numerical modelling study, *Journal of Geophysical Research*, **99**:12,229–12,243.

Tucker, G.E., and Slingerland, R., (1997), Drainage basin responses to climate change, *Water Resources Research*, **(33)** 8: 2031–2047.

Tucker, G.E., Gasparini, N.M, Bras, R.L. and Lancaster, S.L., (1999), *A 3D Computer Simulation Model of Drainage Basin and Floodplain Evolution: Theory and Applications*, Technical report prepared for U.S. Army Corps of Engineers Construction Engineering Research Laboratory.

Tucker, G.E. and Bras, R.L., (2000), A stochastic approach to modelling the role of rainfall variability in drainage basin evolution, *Water Resources Research*, **(36)** 7: 1953–1964.

Tucker, G.E. and Whipple, K.X., (2002), Topographic outcomes predicted by stream erosion models: Sensitivity analysis and intermodel comparison, *Journal of Geophysical Research*, **(107)** B9, 2179, doi:10.1029/2001JB000162.

Willgoose, G., Bras, R.L. and Rodriguez-Iturbe, I., (1991), Results from a new model of river basin evolution, Earth Surface Processes and Landforms, **16**: 237–254.

Wu, L., (1990), Quantitative Research on Regional Morphotectonics and Preliminary Study of Computer-Aided Geomorphological Mapping in Ordos Area, Ph.D. dissertation, Peking University.

Section 4: Uncertainty in Terrain Analysis

Assessing Uncertainties in Derived Slope and Aspect from a Grid DEM

ZHOU Qiming and LIU Xuejun

Abstract

Digital elevation data are the most frequently used for computer-based terrain analysis and they form an integral part of today's GIS data analysis capability. For most GIS-based environmental studies, primary topographic parameters such as slope, aspect, and drainage network are often required for specific environmental models. While derived from digital elevation data, particularly the grid-based Digital Elevation Models (DEM), the parameters often display noticeable uncertainties due to errors (a) in the data, (b) inherent to the data structure, and (c) created by algorithms that derive the parameters from the DEM. Some contradictory results have been reported in evaluating the results of various terrain analysis algorithms, largely because of the variety in the assessment methodologies and the difficulties in separating errors in data from those generated by the algorithms.

This chapter reports on an approach to the assessment of the uncertainties in grid-based Digital Terrain Analysis (DTA) algorithms used for deriving slope and aspect. A quantitative methodology has been developed for objective and data-independent assessment of errors generated from the algorithms that extract morphologic parameters from grid DEMs. The generic approach is to use artificial surfaces that can be described by mathematical models; thus the 'true' output value can be pre-determined to avoid uncertainty caused by uncontrollable data errors. Tests were carried out on a number of algorithms for slope and aspect computation. The actual output values from these algorithms on the mathematical surfaces were compared with the theoretical 'true' values, and the errors were then analysed statistically. The strengths and weaknesses of the selected algorithms are also discussed.

Keywords: digital terrain analysis, uncertainty, error model, DEM derivatives, slope, aspect.

1 Introduction

One critical spatial data set required for digital terrain analysis (DTA) is the surface elevation, which is modelled in GIS as discrete samples for the specific land surface. Although there have been numerous models developed for this purpose, such as Triangulated Irregular Network (TIN), TOPOG model (O'Loughlin 1986) and digital contours, the grid Digital Elevation Model (DEM) has been the most commonly used data source for DTA because of its simple structure, ease of computation, and compatibility with remotely sensed and digital photogrammetry data (Gao 1998, Tang 2000).

For most GIS-based environmental studies, primary topographic parameters such as slope, aspect, catchment area, and drainage network are often required for specific environmental models. When derived from grid-based DEMs, these derivatives often display noticeable uncertainties because of:

1. the data accuracy of the DEM itself including sample errors, interpolation errors, and representation errors (Bolstad and Stowe 1994, Florinsky 1998, Walker and Willgoose 1999, Tang 2000);
2. the spatial data structure of the DEM such as data precision, grid resolution, and orientation (Zhang and Montgomery 1994, Gao 1998, Zhou and Liu 2004b); and
3. the mathematical models and algorithms employed (Holmgren 1994, Desmet and Govers 1996, Zhou and Liu 2002, 2004a).

The focus of this study is on Components 2 and 3 – the uncertainty generated in the implementation of mathematical models and algorithms with the impact of DEM data properties. Although the mathematical definitions of these parameters are quite clear and well reported in the literature (Li *et al.* 2005), their implementation based on grid-based DEMs has varied. Since a grid DEM itself is an approximation to the real-world continuous surface using regularly spaced samples that are restricted by, for example, scale and measurement errors, the implementation of terrain analysis models on it, to derive slope, aspect, curvature, and catchment area, is also represented as an approximation to reality. In practice, numerous assumptions and optimization about the mass transportation and movement on a specific local surface (often represented as a 3 × 3 window) have to be made in order to establish a workable mathematical model (Holmgren 1994), resulting in significantly different approaches and methodologies towards terrain analysis modelling (Wilson and Gallant 2000). The variation in such implementation would only present minor problems in applications such as surface visualization and classification, but its impact on terrain analysis based on quantitative models could be very significant (Zhou and Liu 2002).

This study focuses on a data-independent approach for assessing uncertainties in DEM derivative algorithms such as for slope and aspect computation. The theoretical analyses on error components in the DEM derivatives are discussed first and the results are then tested using the data-independent error assessment method. Analyses are also focused on the impact of data properties and terrain complexity on the selected algorithms. Comparison between algorithms is also made to specify their suitability for various applications and their sensitivity to the properties and quality of data.

2 The Data-Independent Approach

2.1 The need for data-independent experiments

A common approach for error assessment in digital terrain analysis was to apply proposed algorithms on a 'real-world' DEM. The results from the simulation were then visually or statistically examined against 'common knowledge' on what should be expected, or digital data derived from cartographic resources (maps or aerial photographs) (Band 1986, Skidmore 1989, Garbrecht and Martz 1994, Wolock and Price 1994, Zhang and Montgomery 1994, Desmet and Govers 1996, Tang 2000). This approach, however, is challenged by the fact that no real world DEM is perfect so that errors inherent in the DEM largely remain unknown. The conclusion of the assessment therefore can always be questioned because of the uncertainty in the data.

From the point of view of error assessment, there are at least three shortfalls in this evaluation approach.

1. The evaluation results are related to the actual topography. The results derived from one particular landform may be unsuitable for others.
2. The accuracy of the DEM is often difficult to quantify. Thus, the uncertainty in the DEM itself often masks the inherent errors of the models and algorithms, resulting in improper, sometimes controversial, assessment and analytical results (Florinsky 1998).
3. The comparison between different models and algorithms on a real-world DEM can only present their relative accuracy and error distribution (should be termed as 'difference distribution'). Since

the 'true' value is unknown, it is uncertain and subjective to judge the correctness and quality of the models.

Freeman (1991) presented an alternative method that used an artificial surface – a cone – to assess divergent flow simulation algorithms. The results were assessed using the pattern of catchment area that follows a theoretical expectation. Although it is arguable as to whether a cone surface is realistic for real-world cases, this method eliminated the uncertainty in the data and gave more convincing test results.

2.2 The test design

A quantitative, data-independent method for analysing and objectively assessing errors generated from DTA algorithms has been reported by Zhou and Liu (2002, 2004a). The generic approach is to establish a mathematical model to distinguish components of uncertainties and then to conduct tests using artificial surfaces, which can be described by mathematical formulae; thus the 'true' output value can be pre-determined to avoid uncertainty caused by uncontrollable data errors (Jones 1998). Based on this approach, analyses have been reported on the uncertainties caused by data and data processing in derived hydrological parameters (Zhou and Liu 2002), slope and aspect (Zhou and Liu 2004a), the influence of DEM data properties such as precision, resolution, and orientation (Zhou and Liu 2004b), and the impact of terrain complexity (Zhou *et al.* 2006).

The data independent approach is to employ mathematical surfaces in a similar way to Hodgson (1995) and Jones (1998), but with a complexity that represents a closer approximation to the actual land surface. For this purpose, we have selected an inverse ellipsoid (Equation 1) and a Gauss synthetic surface (Equation 2) to define the surfaces and have generated DEMs for a given resolution (Figure 1 and Figure 2).

$$\frac{x^2}{A^2}+\frac{y^2}{B^2}+\frac{z^2}{C^2}=1 \qquad (z<0) \tag{1}$$

$$\begin{aligned} z = {} & A\left[1-\left(\frac{x}{M}\right)^2\right]e^{-\left(\frac{x}{M}\right)^2-\left(\frac{y}{N}+1\right)^2} \\ & -B\left[0.2\left(\frac{x}{M}\right)-\left(\frac{x}{M}\right)^3-\left(\frac{y}{N}\right)^5\right]e^{-\left(\frac{x}{M}\right)^2-\left(\frac{y}{N}\right)^2}-Ce^{-\left(\left(\frac{x}{M}\right)+1\right)^2-\left(\frac{y}{N}\right)^2} \end{aligned} \tag{2}$$

where *A*, *B* and *C* are parameters determining surface relief, and *M, N* in Equation (2) are the parameters controlling the spatial extent of the surface. The 'true values' of slope and aspect can be computed using mathematical inference.

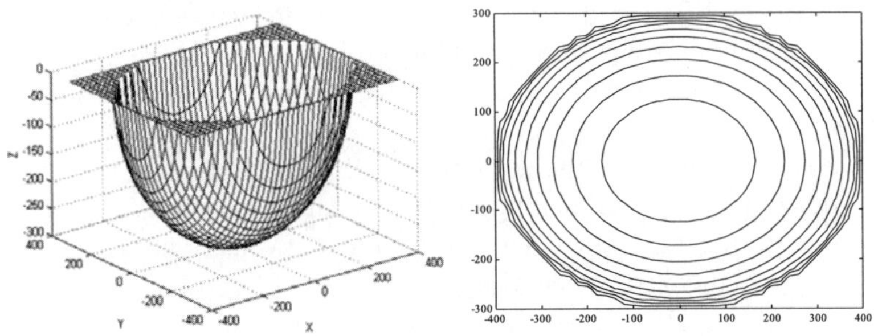

Figure 1. Surface defined by an inverse ellipsoid (A = 400, B = 300 and C = 300).

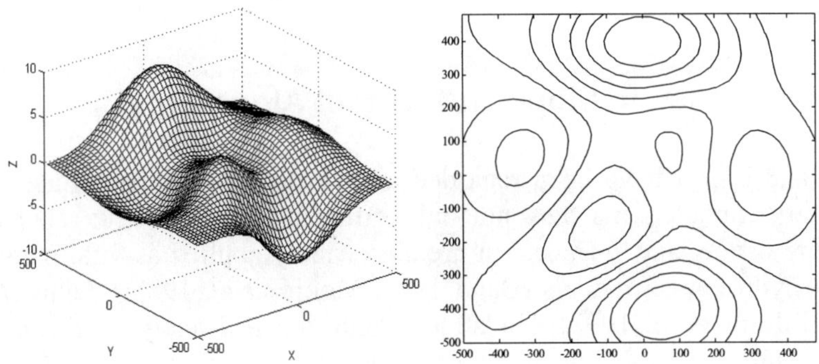

Figure 2. A Gauss synthetic surface (A = 3, B = 10, C = 1/3, $-500 \le X, Y \le 500$).

On the DEM generated from the above surfaces, the selected algorithms have been applied to compute slope and aspect values and statistics were then generated to compare RMSEs between results derived by different algorithms.

Depending on the purpose of the test, alterations on the test surfaces need to be undertaken.

- For analysing the impact of DEM data error, some 'noise' (i.e. random errors) has been added to the generated DEM to simulate DEM data error.
- For testing the sensitivity of the selected algorithms to data precision and resolution, the surfaces have been altered with different grid cell sizes with various precisions at 0.001 m, 0.01 m, 0.1 m, and 1 m levels.

- For assessing the influence of grid orientation, the surfaces have been rotated with an increment of 15° to establish a series of DEMs at directions of 15°, 30°, …, and 345°.
- For evaluating the impact of terrain complexity, the surfaces have been modified with various parameters to create a series of surfaces with various slopes and local relief.

The generic procedure of the data-independent method can be outlined as below:

1. analysing error components on the DEM derivatives;
2. designing and establishing mathematical test surfaces;
3. computing theoretical values for the DEM derivatives using the mathematical formulae of the test surfaces;
4. computing DEM derivatives using the algorithms being evaluated; and
5. comparing the results from mathematical inference and DTA simulation.

3 Uncertainties in Slope and Aspect Algorithms

Slope and aspect have been regarded as two of the most important geomorphic parameters, as they not only efficiently describe the relief and structure of the land surface, but are also widely applied as vital parameters in hydrological models (Band 1986, Moore *et al.* 1988, Quinn *et al.* 1995, Pilesjö *et al.* 1998), landslide monitoring and analysis (Duan and Grant 2000), mass movement and soil erosion studies (Dietrich *et al.* 1993, Mitasova *et al.* 1996, Biesemans *et al.* 2000) and land use planning (Stephen and Irvin 2000).

Numerous studies have been reported that analyse the uncertainties generated by the slope and aspect algorithms with a variety of approaches and methods. One approach emphasizes errors in the DEM while paying little attention to the models themselves (e.g. Skidmore 1989, Hunter and Goodchild 1997), while some other approaches focus on mathematical models while ignoring DEM errors (e.g. Hodgson 1995, Jones 1998). The conclusions from these studies have been quite different, sometimes contradictory. For example, on the spatial distribution of errors, Chang and Tsai (1991) reported that aspect errors are mostly shown in flat areas while slope errors are mainly concentrated in steeper areas. In another study, Carter (1992) stated that both slope and aspect errors occur in flat areas. Davis and Dozier (1990) presented yet another result showing that larger errors in derived slope and aspect would appear in locations where slope and aspect have sudden changes.

Using an analytical approach, Florinsky (1998) developed a test for the precision of four methods for computing partial derivatives of elevations, and produced formulae for root mean square errors (RMSE) of four local topographic variables including slope and aspect. He also used a real-world DEM to map the error distribution for visual analysis, but the contribution from the DEM data error largely remained unknown.

3.1 Slope and aspect algorithms

At a given point on a surface defined by function $z = f(x, y)$, slope (S) and aspect (A) can be represented as its rate of change along the E–W and N–S directions, i.e.

$$S = \arctan\sqrt{p^2 + q^2} \tag{3}$$

$$A = 180° - \arctan\left(\frac{q}{p}\right) + 90°\left(\frac{p}{|p|}\right) \tag{4}$$

where p and q are the gradients at W–E and N–S directions, respectively, and defined as:

$$p = f_x = \frac{\partial f}{\partial x}, \text{ and } q = f_y = \frac{\partial f}{\partial y}.$$

From Equations (3) and (4), it is clear that the key to the computation of slope and aspect at a given point on a surface is the solution of p and q. On a grid DEM, the land surface is represented by discrete sample points and the surface function is usually unknown; thus the computation of p and q often relies on approximations using a moving window (e.g. 3 × 3 window) to derive finite differentials or local surface fitting polynomials for the calculation (Figure 3).

z_7	z_8	z_9
z_4	z_5	z_6
z_1	z_2	z_3

Figure 3. A 3 × 3 window.

There are numerous methods available in the literature to derive the solution of p and q. However, most of them vary in their assignment of the coefficients of the implementing formula, and thus only affect the magnitude of errors in slope and aspect computations (Zhou and Liu 2004a).

Considering the popularity and the use of different algorithms, we have selected six commonly employed algorithms for testing. Letting g denote the grid cell size (i.e. the spatial resolution of the grid), the selected algorithms can be expressed as below:

- Second-order finite difference (2FD – Fleming and Hoffer 1979, Zevenbergen and Thorne 1987, Ritter 1987):

$$
\begin{aligned}
p_g &= f_{x,g} = \frac{1}{2g}(z_6 - z_4) \\
q_g &= f_{y,g} = \frac{1}{2g}(z_8 - z_2)
\end{aligned}
\tag{5}
$$

(where subscript "g" denotes grid spacing of g)

- Third-order finite difference (3FD – Sharpnack and Akin 1969, Horn 1981, Wood 1996):

$$
\begin{aligned}
p_g &= f_{x,g} = \frac{1}{2g}\left[\frac{1}{3}(z_3 + z_6 + z_9) - \frac{1}{3}(z_1 + z_4 + z_7)\right] \\
q_g &= f_{y,g} = \frac{1}{2g}\left[\frac{1}{3}(z_7 + z_8 + z_9) - \frac{1}{3}(z_1 + z_2 + z_3)\right]
\end{aligned}
\tag{6}
$$

- Third-order finite difference weighted by reciprocal of squared distance (3FDWRSD – Horn 1981):

$$
\begin{aligned}
p_g &= f_{x,g} = \frac{1}{2g}\left[\frac{1}{4}(z_3 + 2z_6 + z_9) - \frac{1}{4}(z_1 + 2z_4 + z_7)\right] \\
q_g &= f_{y,g} = \frac{1}{2g}\left[\frac{1}{4}(z_7 + 2z_8 + z_9) - \frac{1}{4}(z_1 + 2z_2 + z_3)\right]
\end{aligned}
\tag{7}
$$

- Third-order finite difference weighted by reciprocal of distance (3FDWRD – Unwin, 1981):

$$\begin{aligned} p_g = f_{x,g} &= \frac{1}{2g}\left[\frac{1}{2+\sqrt{2}}\left(z_3+\sqrt{2}z_6+z_9\right)-\frac{1}{2+\sqrt{2}}\left(z_1+\sqrt{2}z_4+z_7\right)\right] \\ q_g = f_{y,g} &= \frac{1}{2g}\left[\frac{1}{2+\sqrt{2}}\left(z_7+\sqrt{2}z_8+z_9\right)-\frac{1}{2+\sqrt{2}}\left(z_1+\sqrt{2}z_2+z_3\right)\right] \end{aligned} \quad (8)$$

- Frame finite difference (FFD – Chu and Tsai 1995):

$$\begin{aligned} p_g = f_{x,g} &= \frac{1}{2g}\left[\frac{1}{2}\left(z_3+z_9\right)-\frac{1}{2}\left(z_1+z_7\right)\right] \\ q_g = f_{y,g} &= \frac{1}{2g}\left[\frac{1}{2}\left(z_7+z_9\right)-\frac{1}{2}\left(z_1+z_3\right)\right] \end{aligned} \quad (9)$$

- Simple difference (SimpleD – Jones 1998):

$$\begin{aligned} p_g = f_{x,g} &= \frac{1}{g}\left(z_5-z_4\right) \\ q_g = f_{y,g} &= \frac{1}{g}\left(z_5-z_2\right) \end{aligned} \quad (10)$$

3.2 Error components

The accuracy of slope and aspect computations is directly related to the partial derivatives in the *X* and *Y* directions, *p* and *q*, which are estimated by numerous proposed methods. Taking the second-order finite difference (2FD) as an example, let (*x*, *y*) denote the coordinates of the centre cell in a 3 × 3 window, and *g* denote the DEM spatial resolution (i.e. grid cell size), the partial differentials can be expressed as:

$$\begin{aligned} dp &= \frac{g^2}{6}\left(\frac{p''(\xi_x, y)+p''(\gamma_x, y)}{2}\right)+\frac{dz_8-dz_2}{2g} \\ dq &= \frac{g^2}{6}\left(\frac{q''(x, \xi_y)+q''(x, \gamma_y)}{2}\right)+\frac{dz_6-dz_4}{2g} \end{aligned} \quad (11)$$

The first term represents the errors caused by the uncertainty of mathematical model implementation and ξ_x, ξ_y, γ_x, and γ_y are dependences of f''', $\xi_x \in (x, x+g)$, $\xi_y \in (y, y+g)$, $\gamma_x \in (x-g, x)$ and

$\gamma_y \in (y-g, y)$. Since the relationships between these variables and x and y are usually not clear, it is difficult to define their values in applications. Thus, it is common to set the upper limits for f''' instead. Letting M_x and M_y be the upper limits of f''' in terms of x and y, respectively, Equation (11) can be altered to:

$$\begin{aligned} dp &= \frac{g^2}{6}M_x + \frac{dz_8 - dz_2}{2g} \\ dq &= \frac{g^2}{6}M_y + \frac{dz_6 - dz_4}{2g} \end{aligned} \qquad (12)$$

The second term in the equations represents the DEM data error (including data precision).

In Equation (12), M_x and M_y are estimated based on the 'worst scenario', usually representing much larger estimates than the actual case, with a given distribution of probability. Let the root mean square error (RMSE) of M_x and M_y equal M, and m denote the RMSE of the DEM. We then have:

$$m_p^2 = m_q^2 = \left(\frac{g^2}{6}\right)^2 M^2 + \frac{m^2}{2g^2} \qquad (13)$$

Deriving the differentials of slope and aspect equations (Equations (1) and (2)), and considering $S = \arctan\sqrt{p^2+q^2}$ and $\tan^2 S = p^2 + q^2$, we have:

$$dS = \frac{pdp + qdq}{(1+\tan^2 S)\tan S} \text{ and } dA = \frac{qdp - pdq}{\tan^2 S} \qquad (14)$$

Referring to Equation (13), RMSE of slope (m_S) and aspect (m_A) can therefore be expressed as:

$$\begin{aligned} m_S^2 &= \left[\left(\frac{g^2}{6}\right)^2 M^2 + \frac{m^2}{2g^2}\right]\cos^4 S \\ m_A^2 &= \frac{1}{\tan^2 S}\left[\left(\frac{g^2}{6}\right)^2 M^2 + \frac{m^2}{2g^2}\right] \end{aligned} \qquad (15)$$

Note that Equation (15) is for the second-order finite difference method. Letting $a = \frac{1}{6}g^2$ and $b = \frac{1}{\sqrt{2}}g^{-1}$, Equation (15) can then be expressed in a general form as:

$$\begin{aligned} m_S &= \sqrt{a^2M^2 + b^2m^2}\cos^2 S \\ m_A &= \sqrt{a^2M^2 + b^2m^2}\cot S \end{aligned} \tag{16}$$

Similar to the above procedure and referring to Equations (6) to (10), we can derive the RMSE for each selected algorithm as shown in Table 1.

Table 1. The RMSE for selected slope and aspect algorithms.

Algorithm	Coefficient *a* of *M*	Coefficient *b* of *m*
2nd order finite difference	$\frac{1}{6}g^2$	$\frac{1}{\sqrt{2}}g^{-1}$
3rd order finite difference	$\frac{1}{6}g^2$	$\frac{1}{\sqrt{6}}g^{-1}$
3rd order finite difference weighted by reciprocal of squared distance	$\frac{1}{6}g^2$	$\frac{1}{\sqrt{5.33}}g^{-1}$
3rd order finite difference weighted by reciprocal distance	$\frac{1}{6}g^2$	$\frac{1}{\sqrt{5.83}}g^{-1}$
Frame finite difference	$\frac{1}{6}g^2$	$\frac{1}{2}g^{-1}$
Simple difference	$\frac{1}{2}g$	$2g^{-1}$

Equation (16) and Table 1 show that the overall errors of derived slope and aspect come from three sources:

1. Algorithm errors: caused by approximation and sampling of continuous surfaces (variable *M* in Equation (16));
2. DEM data errors: caused by DEM data capture and generation (variable *m* in Equation (16)); and
3. DEM spatial resolution (i.e. the grid cell size – variable *a* and *b* in Equation (16), defined in Table 1).

Note here that the coefficient b of m for 2FD and 3FD confirms the results reported by Florinsky (1998), which represent special cases for the selected algorithms.

By examining Equations (15) and (16), we understand that for a given DEM spatial resolution (g), the accuracy of derived slope and aspect is related to the errors of the algorithm that derives p and q (M), and the DEM error (m). The overall accuracy of slope and aspect depends upon whether M or m dominates the analysis. When analysing the algorithms using a real-world DEM (e.g. Skidmore 1989, Chang and Tsai 1991, Bolstad and Stowe 1994), the DEM error (m) would cause much greater error than that due to the algorithm, so that the method would be more appropriate for analysing the impact of data error on derived slope and aspect values. On the other hand, using a DEM defined by mathematical surfaces (Hodgson 1995, Jones 1998, Carter 1992) would eliminate data error; thus the observed errors would only be caused by algorithms. Since the error sources could not be defined, the results of the studies appear inconclusive.

3.3 Test results and discussion

Tables 2 and 3 summarize RMSE comparison results between the selected six algorithms on two surfaces (inverse ellipsoid and Gauss synthetic surface) with no added error (i.e. DEM data error = 0).

Table 2. Statistics of derived slope with no data error (Unit: degrees).

Algorithm	RMSE		Standard Error		Mean Error		Frequency of positive and negative errors (%)	
	Ellipsoid	Gauss	Ellipsoid	Gauss	Ellipsoid	Gauss	Ellipsoid	Gauss
2FD	0.216	0.003	0.189	0.002	0.104	-0.002	100 / 0	15 / 85
3FD	0.409	0.004	0.003	0.320	0.178	-0.003	100 / 0	9 / 91
3FDWRSD	0.358		0.004	0.345	0.160	-0.003	100 / 0	8 / 92
3FDWRD	0.384		0.004	0.459	0.170	-0.003	100 / 0	9 / 91
FFD	0.507			0.003	0.214	-0.003	100 / 0	10 / 90
SimpleD	1.295	0.046	1.292	0.046	0.078	-0.002	51 / 49	49 / 51

By analysing Tables 2 and 3, it is seen that on a DEM with high accuracy, the error of derived slope and aspect is sourced from the estimates of partial derivatives p and q and sampling errors. In this case, the second-order finite difference method provides a better result than the third-order finite difference methods. For the six algorithms tested, the order from the best to worst is the second-order finite difference (2FD), third-order finite difference weighted by reciprocal of squared distance (3FDWRSD), third-order finite difference weighted by reciprocal of distance (3FDWRD), third-order finite difference (3FD), frame finite difference (FFD), and simple difference (SimpleD).

Table 3. Statistics of derived aspect with no data errors (Unit: degrees).

Algorithm	RMSE		Standard Error		Mean Error		Frequency of positive and negative errors (%)	
	Ellipsoid	Gauss	Ellipsoid	Gauss	Ellipsoid	Gauss	Ellipsoid	Gauss
2FD	0.133	0.117	0.133	0.117	0.000	-0.008	52 / 48	45 / 55
3FD	0.197	0.130	0.197	0.130	-0.001	0.000	52 / 48	51 / 49
3FDWRSD	0.122	0.118	0.122	0.118	-0.001	-0.002	52 / 48	48 / 52
3FDWRD	0.160	0.124	0.160	0.124	-0.001	-0.001	52 / 48	50 / 50
FFD	0.342	0.167	0.342	0.167	-0.002	0.004	52 / 48	52 / 48
SimpleD	20.76	15.13	20.63	15.13	2.347	-0.340	50 / 50	52 / 48

In Equation (16), if we ignore the algorithm error (i.e. the first term), we then derive the Florinsky's (1998) RMSE formulae in general terms:

$$m_S = bm\cos^2 S \text{ and } m_A = bm\cot S \tag{17}$$

Equation (17) suggests that the influence of DEM data error (m) relates to slope (S) and coefficient b, which is determined by grid cell size g. Referring to Table 1, the third-order finite difference methods appear to be less sensitive to data errors for any given g and S.

Algorithm error and DEM data error influence the accuracy of derived slope and aspect differently. In general, all the third-order finite difference methods have applied some smoothing effect on the local data window in order to avoid local relief extremes (Burrough and McDonnell 1998) for better computation results of the surface parameters. On the other hand, the second-order finite difference and simple difference methods only utilize a part of the samples in the local window, so that they are more sensitive to the data error.

The results also confirm the test results reported by Jones (1998). When using an error-free synthetic, trigonometrically defined surface (Morrison's surface III), the 2FD method gave the "best" results for both gradient and aspect. Using a real-world DEM, where data errors were unavoidable, the 3FD performed best.

4 The Impact of Properties of DEM Data

Numerous studies have been reported on the accuracy analysis of slope and aspect algorithms in relation to DEM data errors (Zhou and Liu 2004a), data precision (Theobald 1989), grid resolution, or grid cell size (Chang and Tsai 1991, Garbrecht and Martz 1994, Hodgson 1995, Florinsky and Kuryakova 2000, Tang 2000) and grid orientation (Jones 1998, Zhou and Liu 2004b). Although the errors caused by data precision, grid resolution, and orientation are usually not a concern for the visualization of a 3-dimensional surface, they could create significant impact on the derived surface parameters, such as slope and aspect, which are also largely related to the utilized algorithms.

4.1 The impact of data precision

DEM data precision is indicated by the number of significant digits used for DEM data. In many real-world cases, the DEM precision is defined at the level of one metre, such as the USGS 30 m DEM (Theobald 1989, Carter 1992). In some cases, this precision is required to higher levels, e.g. China's 1:50,000 DEM requires a precision level of 0.1 m (Li *et al.* 2005).

Usually the DEM error caused by the data precision level is quite minimal, except in flat areas where rounding errors could be significant. One test result shows that the RMSE of a DEM is less than one metre at the 1-m precision level. When the precision level is raised to centimetres, the RMSE of the DEM is close to zero (Zhou and Liu 2004b).

The influence of data precision on derived slope and aspect is highly related to the grid resolution. When using a high resolution DEM (e.g. 1 m grid resolution), the influence of data precision becomes quite significant (Figure 4).

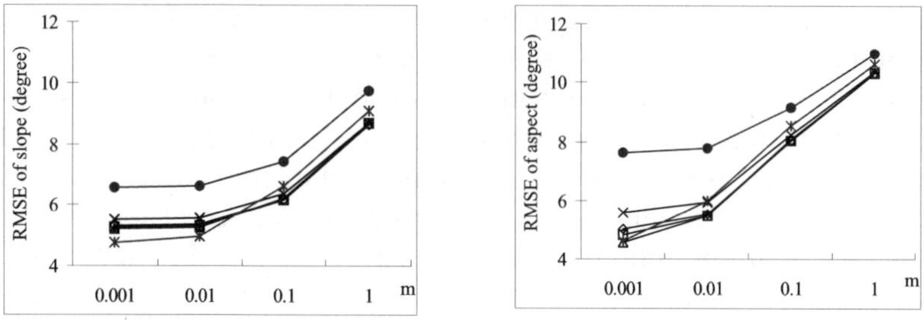

(a) RMSE of derived slope and aspect on an ellipsoid surface
(A = 100, B = C = 60, DEM resolution: 1, Unit: metres)

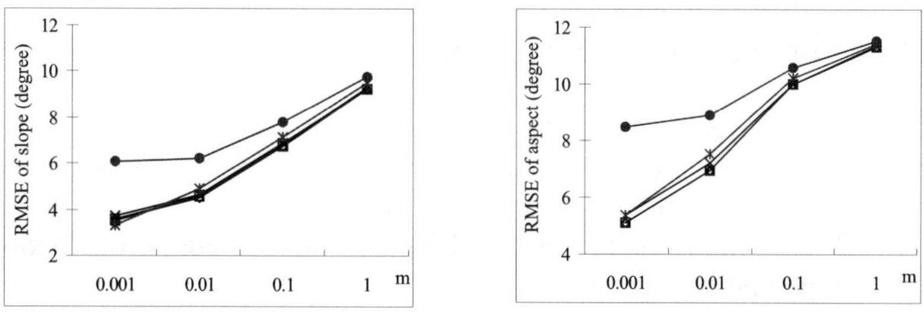

(b) RMSE of derived slope and aspect on a Gauss synthetic surface
(A = 3, B = 10, C = 1/3, DEM resolution: 1, Unit: metres)

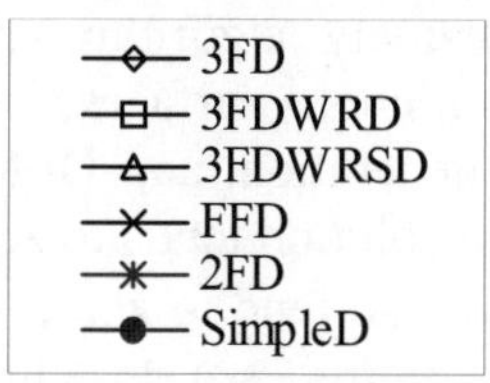

Figure 4. The influence of DEM data precision on derived slope and aspect by selected algorithms. The values of RMSE of slope and aspect have been transformed using the formula: $y = \ln(x \times 1000)$ for illustration.

As illustrated by Figure 4, when the data precision level is reduced from 0.001 to 0.01 m, the change of RMSE of the DEM itself is minimal. Further generalization, however, would cause a much more significant increase in the RMSE. For derived slope and aspect, the SimpleD algorithm tends to create a much greater RMSE than the others with high precision data. On the other hand, the RMSEs of all other algorithms seem to increase constantly with decreasing precision, with 2FD showing the most rapid rate of change. When reaching the precision level of one metre, all algorithms show very similar levels of RMSE, with SimpleD just showing a slightly higher error level than the others.

The results show that data precision may only play a significant role in algorithm performance while the precision level is high. When the precision level is reduced, its influence on different algorithms becomes less important. In reality, errors may occur during different stages of DEM generation, such as data capture, sampling, and interpolation. Compared to these errors, the rounding errors from reducing data precision can be neglected (with an exception in flat areas). In this case, the number of significant digits should not be considered as critical. When DEM data accuracy is higher than the precision, on the other hand, the data precision error must be considered when selecting algorithms.

4.2 The Impact of grid resolution

DEM resolution determines the level of details of the surface being described. It naturally influences the accuracy of the derived surface parameters. In this study, we focus on two questions:

1. Does a high resolution DEM lead to more accurate estimation of slope and aspect?
2. How can we determine an appropriate grid resolution in relation to slope and aspect computation for a given application?

Referring to Equation (16) and Table 1, the overall errors of derived slope and aspect come from three sources, namely, algorithm errors (*M*) caused by approximation and sampling of continuous surfaces, DEM data error (*m*) caused by DEM data capture and generation, and DEM spatial resolution (*g*, i.e. grid cell size). When DEM resolution tends to zero (i.e. $g \to 0$), $a \to 0$, so that the algorithm error will also tend to zero, while the influence of the DEM error (*m*) will tend to infinity ($+\infty$). In other words, for slope and aspect computation, the impact of the algorithm error is positively proportional to DEM resolution, while the influence of the DEM error is negatively proportional to DEM resolution. With a higher DEM resolution, the level of detail is increasing (i.e. the surface is better represented), but the influence of data error is also increased at the same time. With a lower DEM resolution, the impact of any data error is decreased, but algorithm errors will cause more significant error on the derived results.

When ignoring algorithm error *M*, we can determine an appropriate DEM resolution according to the known DEM error (*m*) and the average slope (*S*) of the region:

$$g = \frac{bm}{m_S} \times \frac{180^\circ}{\pi} \cos^2 S \tag{18}$$

Table 4 illustrates the computed DEM resolution for DEMs with various error levels (specified by m – RMSE of DEM data). To consider the influence of DEM resolution on selected algorithms, we have conducted tests to analyse the relationships of slope and aspect errors and DEM resolution, with and without DEM data error. The test results show that when DEM data error is minimal, the RMSE of slope and aspect increases with lower resolution (i.e. larger grid cell size), regardless of which algorithm is used. The RMSE of the derived slope and aspect is positively proportional to grid resolution. When DEM data error is significant, the RMSE of the derived slope and aspect decreases with lower resolution, showing a negative proportional relationship to the DEM resolution.

Table 4. Computed DEM resolution using RMSE of DEM data and average slope.

m (m)	Test surface	RMSE of slope (°)	Mean slope (°)	Computed DEM resolution (m)
0.6	Ellipsoid	1.0	37.0	9.0
	Gauss	1.4	1.1	9.9
2.1	Ellipsoid	3.2	37.0	9.8
	Gauss	6.1	1.1	8.2
6.6	Ellipsoid	9.2	37.0	10.7
	Gauss	19.4	1.1	8.0
9.5	Ellipsoid	12.4	37.0	11.4
	Gauss	26.4	1.1	8.5
15.1	Ellipsoid	18.0	37.0	12.5
	Gauss	37.4	1.1	9.5
19.5	Ellipsoid	21.3	37.0	13.6
	Gauss	44.0	1.1	10.4

The test results confirm the relationship described by Equation (16). We therefore conclude that a high-resolution DEM does not assure higher slope and aspect accuracy. Better results may only be possible with high DEM data accuracy. In reality, DEM data often contain significant levels of error. It is therefore argued that a higher resolution DEM does not lead to higher accuracy of estimated slope and aspect. Rather, the accuracy of the derived slope and aspect is increasing with lower DEM resolution.

4.3 The impact of grid orientation

At any given point on a surface, slope and aspect are constant parameters, which do not change with grid orientation. However, as a DEM organizes ground elevations using regularly distributed sample points, different grid orientations may result in errors in computing partial derivatives for slope and aspect computation.

The selected algorithms have been applied to the rotated surfaces as described in Section 2 to derive a graphical representation of the test results, from which the following findings are revealed (Zhou and Liu 2004b):

1. DEM orientation has greater influence on third-order finite difference algorithms (including 3FD, 3FDWRD and 3FDWRSD) than other algorithms (2FD, FFD and SimpleD). 2FD, FFD and SimpleD have shown almost constant error levels in all directions, while the 3FD series has shown great changes in association with directions.
2. All algorithms show extreme values at 45° × k (k = 0, 1, …, 7), i.e. demonstrating the octant pattern in directional distribution.
3. Errors in slope and aspect synchronize with each other and reach the extreme values with the same directional pattern. This is because the errors in slope and aspect are related to slope itself, as described by Equation (16).

5 The Impact of Terrain Complexity

From Equation (16), we understand that the uncertainties of derived slope and aspect may also be significantly influenced by the slope (S), which we refer to here as an indicator of terrain complexity. Terrain complexity is defined using Gauss synthetic surfaces defined by Equation (4) with varied coefficients to simulate land surfaces with various mean slopes and curvatures (Zhou *et al.* 2006). The third-order finite difference algorithm (3FD) is used to generate slope and aspect from the test surfaces and its results are then compared to the theoretical 'true values' inferred from the surface formulae.

Referring to Equation (16) and Table 1, the RMSE of slope (m_S) and aspect (m_A) of the 3FD method can therefore be expressed as:

$$m_S = \pm\frac{m}{\sqrt{6}g}\cos^2 S \text{ and } m_A = \pm\frac{m}{\sqrt{6}g}\cot S \tag{19}$$

From Equation (19), it is clear that for a given grid resolution (g) and RMSE of DEM data (m), the RMSE of derived slope (m_S) and aspect (m_A) is positively correlated with slope (S). As slope values range within

$0 - 90°$, $\cos^2 S$ and $\cot S$ will always be greater than 0. Taking the ratio of m_S and m_A, we have:

$$\frac{m_A}{m_S} = \frac{2}{\sin 2S} \qquad (0 < S \leq 45°) \tag{20}$$

Referring to Figure 5 and considering Equations (19) and (20), it is known that:

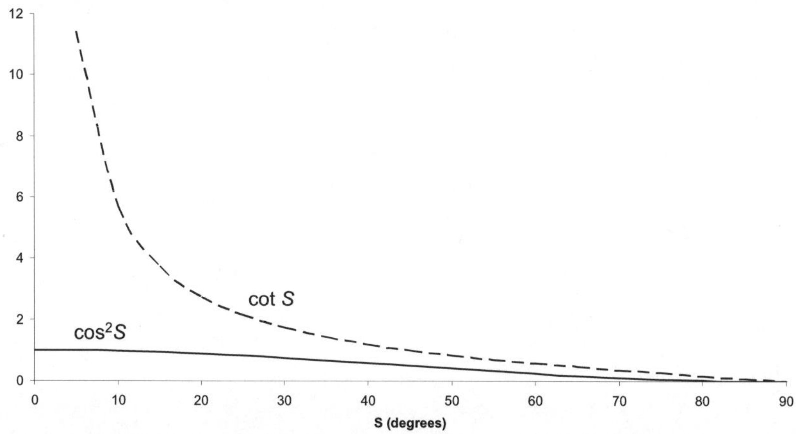

Figure 5. Relationships between slope (*S*), $\cos^2 S$ and $\cot S$.

- The larger is *S*, the smaller are the m_S and m_A values and vice versa; and
- m_A is much more sensitive than m_S, but its impact is much greater when the *S* value is small.

The above characteristics explain that the errors of the derived slope and aspect occur mainly in flatter areas. Moreover, since $\cot S$ has a faster decreasing rate with increasing slope than that of $\cos^2 S$, in a flatter area even a small change in slope will lead to a great change of $\cot S$. We therefore conclude that the error in the derived aspect is more sensitive to relief change than that of slope.

To investigate the impact of changing slope on errors in the derived morphological parameters, a flattening rate of the ellipsoid (*E*), which is defined as the ratio of the vertical axis (*C*) over the long horizontal axis (*A*), is used. Referring to Figure 6, by varying *E* from 0.1 to 4.0, on a DEM with 10 m resolution, the average slope increased from 8° to 74°, while the

RMSE of derived slope and aspect decreased from 10.4° to 1.9°, and from 130.8° to 19.3°, respectively. This suggests that on a simple surface, the derived slope and aspect tend to contain greater uncertainty with decreasing slope.

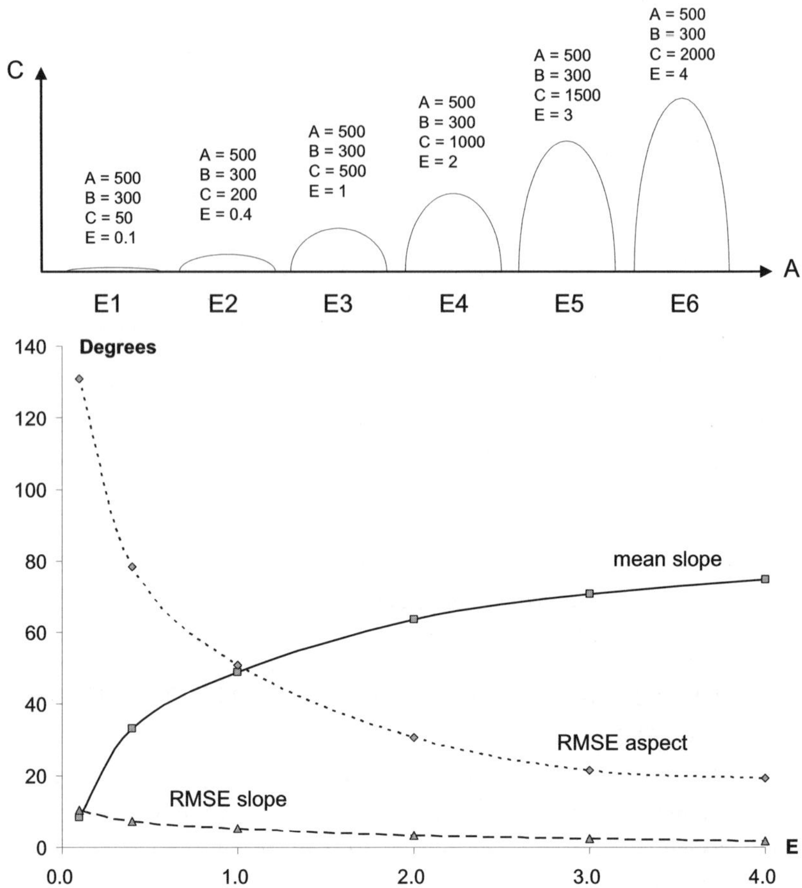

Figure 6. RMSE of derived slope and aspect corresponding to the flattening rate of an inverse ellipsoid (DEM resolution 10 m, RMSE of DEM data 4.56 m).

Using the above flattening rate is adequate to describe the impact of surface complexity on a simple surface. However, on a more complex surface, as we usually find in the natural world, more parameters need to be used for better simulation. Gao (1998) proposed the use of local relief (i.e. the difference between local highest and lowest points), standard deviation of elevation, and contour density of unit area (i.e. total length of contour lines over area) to describe surface complexity. This method, however, presented some limitations, as acquisition of total contour length is not only related to computation algorithms, but also influenced by the map scale and DEM resolution. Besides, the use of relief and standard deviation alone may be too simplistic and would produce too much bias for a satisfactory description of surface complexity.

In this study, we propose to use relief, standard deviation, and slope frequency for the description of surface complexity. By changing the parameters of the Gauss Synthetic surface (Equation (2)), a series of simulated surfaces were created for the test as described by Zhou *et al.* (2006). Based on these surfaces, parameters for describing surface complexity were computed and listed in Table 5. Corresponding slope frequency was also derived and presented in Figure 7.

Table 5. The complexity parameters of Gauss Synthetic surfaces (DEM resolution 10 m).

Surface	Average slope (°)	Relief (m)	Standard deviation (m)
G1	6.3	20.4	3.4
G2	12.1	152.1	26.3
G3	20.5	298.5	52.4
G4	28.0	444.9	78.5
G5	34.2	591.3	104.7
G6	39.5	737.5	131.2

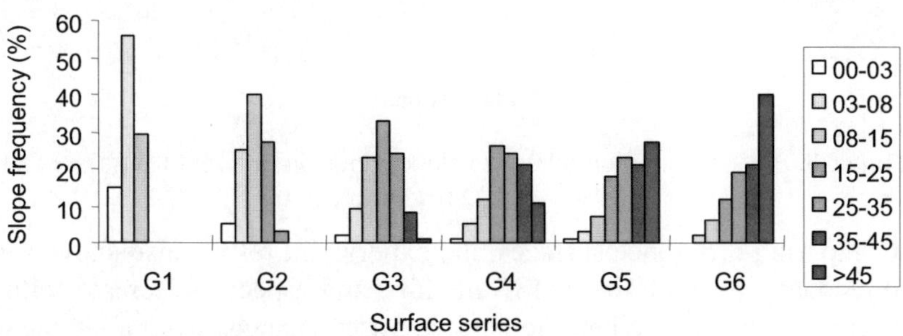

Figure 7. Slope frequency on different Gauss Synthetic surfaces (DEM resolution 10 m).

Referring to Figure 7 and Table 5, it is known that surface G1 has the simplest and flatter form with the smallest relief (20.4 m) and standard deviation (3.36 m), with slopes mainly falling in the range of 0 – 8° (70%). Surfaces G2 and G3 have a moderate complexity with slopes of 3 – 25°. In contrast, surfaces G4, G5 and G6 represent rather rugged terrain with slopes greater than 25°. For the extreme case, G6 has a relief of 737.5 m and standard deviation of 131.2 m. It can therefore be considered that the surface series G1 to G6 represent an increasing complexity of surface morphology.

Based on the surface series, statistical analysis of error distributions related to slope has been conducted and the result is shown in Figure 8, where surface slopes are divided in slope bands (classes) and the RMSE of derived slope and aspect for each slope band has been computed for surfaces with different complexities. Within each surface, Figure 8 shows a general tendency for both the RMSE of slope and aspect to decrease with increasing slope, and the RMSE of aspect to be more sensitive (i.e. decreases more rapidly) than that of slope.

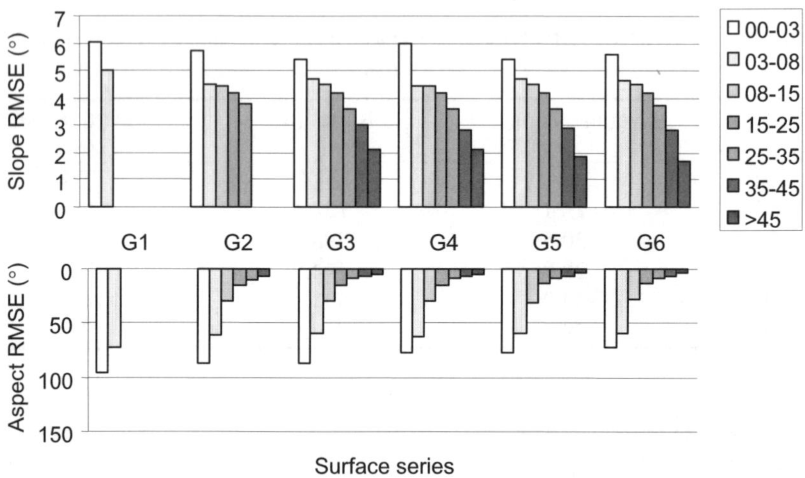

Figure 8. Error distribution related to slope bands on different Gauss Synthetic surfaces (DEM resolution 10 m).

Comparing difference surfaces, the experiment results also show a general tendency for both the RMSE of slope and aspect to decrease with increasing complexity. When the surface series changes from a simple surface (G1) to the most complex surface (G6), we observed the RMSE of slope and aspect decreasing from 6.1 → 4.5 → 4.1 → 3.7 → 3.4 → 3.0° and 95.6 → 46.7 → 29.4 → 21.9 → 17.5 → 14.5°, respectively. This means that the more complex the land surface, the more accurate is the derived slope and aspect from a grid-based DEM. The result also shows that when the surface becomes more complex, the RMSE of derived parameters tends to be stable. For example, when surface G1 to G6 with increasing complexity defined by a 25% increase of relief and standard deviation

(Table 5), the corresponding increments of RMSE of slope and aspect are -25% → -9% → -9% → -9% → -9% and -50% → -37% → -26% → -20% → -17%, respectively. From this we can further confirm the relationships shown by Equation 6, namely, that both RMSE of slope and aspect are mainly distributed in flatter areas, and aspect is more sensitive than slope on the impact of surface complexity.

The results from the experiment can be further illustrated by a test using a real-world DEM. Figure 9 shows the test results. On the DEM represented by Figure 9a, a slope map has been derived (Figure 9b). An analysis of the results of the RMSE of surface derivatives including slope and aspect are shown in Figure 9c (RMSE of slope) and Figure 9d (RMSE of aspect). The results also confirm those from the experiment based on mathematical surfaces, as higher RMSE values are observed in the flatter areas for both slope and aspect and the magnitude of increase for the aspect RMSE is much greater than that of the slope RMSE.

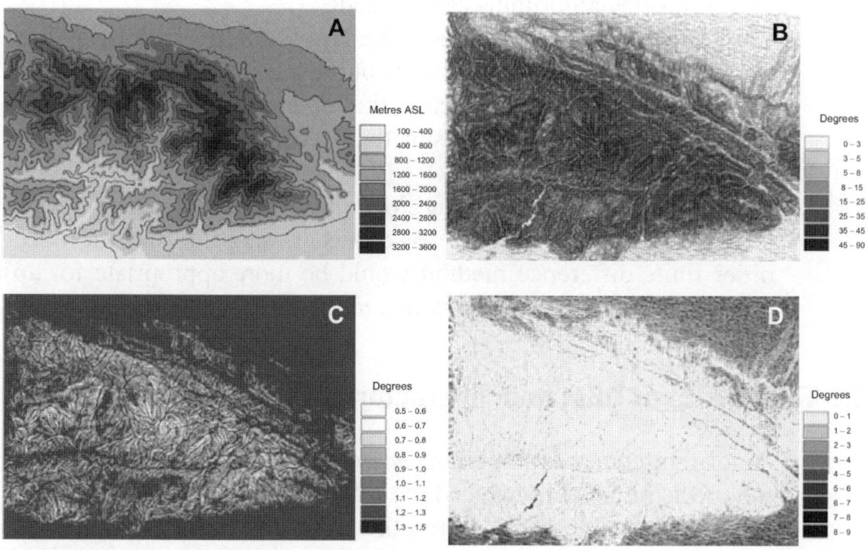

Figure 9. Error distribution on a real-world DEM. (a) The original DEM (RMSE = 20 m, grid resolution = 60 m); (b) derived slope; (c) RMSE of derived slope; (d) RMSE of derived aspect.

6 Conclusions

Slope and aspect are two of the most frequently utilized variables in GIS-based terrain analysis and geographical modelling. There have been numerous analyses on the accuracy of these variables derived from grid-based DEMs. The reported findings, however, did not always agree and sometimes they were controversial. This study attempts to evaluate the issues and establish a 'fair' quantitative measure for assessing various slope and aspect algorithms, the impact of data properties including precision, resolution and orientation, as well as the impact of terrain complexity. From the findings of this study, we can draw the following conclusions.

6.1 The general nature of uncertainties in DEM derivatives

1. It is important to identify the sources and nature of errors of derived slope and aspect in order to evaluate algorithms and mathematical models.
2. Evaluation of algorithms and models must be based on an objective, data-independent methodology so that a 'fair' comparison between selected algorithms can be made.
3. On a DEM with high accuracy, the error of derived slope and aspect is sourced from the estimates of the partial derivatives p and q and sampling errors. In this case, the second-order finite difference method provides a better result than the third-order finite difference methods.
4. In reality, the influence of DEM data error is, in general, much larger than the algorithm errors; thus it suggests that the third-order finite difference method would be more appropriate for applications since it is least sensitive to the DEM data error.

6.2 The impact of DEM data characteristics

1. Algorithm choice is important when data precision is high. When the precision level is reduced, its influence on different algorithms becomes less important. When the error level in a DEM is high, the round-up errors due to reduced data precision can be neglected.
2. A high-resolution DEM does not assure higher slope and aspect accuracy. Better results may only be possible with high DEM data accuracy. In reality, where DEM data often contains errors, the accuracy of derived slope and aspect increases with lower DEM resolution.
3. Grid orientation does cause directional bias on derived slope and aspect, and the third-order finite difference algorithm series has shown the most significant errors due to the change of grid orientation.

6.3 The impact of terrain complexity

1. The RMSE of derived slope and aspect is negatively correlated with slope steepness.
2. The RMSE of derived aspect is more sensitive to terrain complexity than that of derived slope.
3. The errors in derived slope and aspect tend to be found in flatter areas, and decrease with increasing terrain complexity.

In general, when the terrain is rugged with steep slopes, the error of derived parameters is quite minimal. In a flatter area, the derived parameters, particularly the aspect, may contain considerable errors, causing significant limitations in applying the analytical results.

6.4 The evaluation of the approach

In this study, we have argued that although primary surface parameters can be well defined mathematically, the implementation of those mathematical models in a GIS environment may generate considerable errors related to DEM data quality, algorithm selection, DEM data properties such as precision, resolution and orientation, and terrain complexity.

This study has shown that using a selection of mathematical surfaces with controlled parameters and data errors, digital terrain analysis algorithms can be objectively compared and evaluated, independently from the data and the bias of the human analyst. It is also shown that the impact of individual factors can be independently examined by this approach so that appropriate justification can be made according to the application requirements and data characteristics.

It should be noted that the study attempts to assess the uncertainty of slope and aspect from DEM via a simple simulation defined by mathematical surfaces. The real-world slope may present much more complicated features such as micro-relief, which closely relates to the scale of topography being studied. When applying the test results to real-world problems, variation between the artificial and real-world slopes must be considered. Further studies also aim to analyse the uncertainties in other DEM derivatives, such as slope profile and curvature, catchment areas, drainage networks, and other derived geomorphic parameters. Real-world tests are also needed to compare with the findings of theoretical analysis. Based on these analyses, the ultimate goal is to set a conclusive guideline for deriving geomorphic parameters from DEM for a given application project.

Acknowledgement

This study is supported by the National Natural Science Foundation of China (No.40571120), Hong Kong Research Grants Council Earmarked Research Grant (HKBU 2029/07P) and Hong Kong Baptist University Faculty Research Grant (FRG/06-07/II-76).

References

Band, L.E., (1986), Topographic partition of watershed with digital elevation models, *Water Resources Research*, **22**(1): 15–24.

Biesemans, J., Meirvenne, M.V. and Gabriels, D., (2000), Extending the RUSLE with the Monte Carlo error propagation technique to predict long-term average off-site sediment accumulation, *Journal of Soil and Water Conservation*, **55**(1): 35–42.

Bolstad, P.V. and Stowe, T.J., (1994). An evaluation of DEM accuracy: elevation, slope and aspect, *Photogrammetric Engineering and Remote Sensing,* **(60)**11: 1327–1332.

Burrough, P.A. and McDonnell, R.A., (1998), *Principles of Geographical Information System*, Oxford, UK: Oxford University Press.

Carter, J., (1992), The effect of data precision on the calculation of slope and aspect using gridded DEMs, *Cartographica*, (29)1: 22–34.

Chang, K. and Tsai, B., (1991), The effect of DEM resolution on slope and aspect mapping, *Cartography and Geographic Information Systems*, **18**: 69–77.

Chu, T.H. and Tsai, T.H., (1995), Comparison of accuracy and algorithms of slope and aspect measures from DEM, *Proceedings of GIS AM/FM ASIA'95*, 21–24 August, Bangkok: I–1–1 – I–1–11.

Davis, F.W. and Dozier, J., (1990), Information analysis of spatial database for ecological land classification, *Photogrammetric Engineering and Remote Sensing*, **56**: 605–613.

Desmet, P.J.J. and Govers, G., (1996), Comparison of routing algorithms for digital elevation models and their implication for predicting ephemeral gullies, *International Journal of Geographical Information science*, **(10)**10: 311–331.

Dietrich, W.E., Wilson, C.J., Montgomery, D.R. and McKean, J., (1993), Analysis of erosion thresholds, channel networks, and landscape morphology using a digital terrain model, *The Journal of Geology*, **10**: 259–278.

Duan, J. and Grant, G.E., (2000), Shallow landslide delineation for steep forest watersheds based on topographic attributes and probability analysis, In Wilson, J.P. and Gallant, J.C. (eds.): *Terrain Analysis: Principles and Application*, New York: John Wiley & Sons: 311–330.

Fleming, M.D. and Hoffer, R.M., (1979), *Machine processing of Landsat MSS data and LARS Technical Report 062879*, West Lafayette, Indiana: Laboratory for Applications of Remote Sensing, Purdue University.

Florinsky I.V., (1998), Accuracy of local topographic variables derived from digital elevation models, *International Journal of Geographical Information Science*, **(12)**1: 47–61.

Florinsky, I.V. and Kuryakova, G.A., (2000), Determination of grid size for digital terrain modeling in landscape investigations—exemplified by soil moisture distribution at a mirco-scale, *International Journal of Geographical Information Science*, **(14)**8: 815–832.

Freeman, T.G., (1991), Calculating catchment area with divergent flow based on a regular grid, *Computer and Geosciences*, **(17)**3: 413–422.

Gao, J., (1998), Impact of sampling intervals on the reliability of topographic variables mapped from grid DEMs at a micro-scale, *International Journal of Geographical Information Science*, **(12)**8: 875–890.

Garbrecht, J. and Martz, L.W., (1994), Grid size dependency of parameters extracted from digital elevation models, *Computer and Geosciences*, **(20)**1: 85–87.

Hodgson, M.E., (1995), What cell size does the computed slope/aspect angle represent? *Photogrammetric Engineering and Remote Sensing,* **61**: 513–517.

Holmgren, P., (1994), Multiple flow direction algorithm for runoff modelling in grid based elevation models: an empirical evaluation, *Hydrology Processes*, **8**: 327–334.

Horn, B.K.P., (1981), Hill shading and the reflectance map, *Proceedings of IEEE*, **(69)**1: 14–47.

Hunter, G.J. and Goodchild, M.F., (1997), Modeling the uncertainty of slope and aspect estimates derived from spatial databases, *Geographical Analysis*, **(29)**1: 35–49.

Jones, K.H., (1998), A comparison of algorithms used to compute hill slope as a property of the DEM, *Computer and Geosciences*, **(24)**4: 315–323.

Li, Z., Zhu, Q. and Gold, C., (2005), *Digital Terrain Modelling: Principles and Methodology*, Boca Raton: CRC Press.

Mitasova, H., Hofierka, J., Zlocha, M. and Iverson, L.R., (1996), Modelling topographic potential for erosion deposition using GIS, *International Journal of Geographical Information Systems*, **(10)**5: 629–641.

Moore, I.D., O'Loughlin, E. M. and Burch, G.J., (1988), A contour-based topographic model for huydrological and ecological applications, *Earth Surface Processes and Landforms*, **13**: 305–320.

O'Loughlin, E.M., (1986), Prediction of surface saturation zones in natural catchments by topographic analysis, *Water Resources Research*, **22**: 794–804.

Pilesjö, P., Zhou, Q. and Harrie, L., (1998), Estimating flow distribution over digital elevation models using a form-based algorithm, *Geographical Information Science*, **(4)**1: 44–51.

Quinn , P.F., Beven, K.J. and Lamb, R., (1995), The ln(a/tanβ) index: how to calculate it and how to use it within the TOPMODEL framwork, *Hydrological Processes*, **9**: 161–182.

Ritter, D., (1987), A vector-based slope and aspect generation algorithm. *Photogrammetric Engineering and Remote Sensing,* **(53)**8: 1109–1111.

Sharpnack, D.A. and Akin, G., (1969), An Algorithm for computing slope and aspect from elevations, *Photogrammetric Survey*, **35**: 247–248.

Skidmore, A.K., (1989), A comparison of techniques for the calculation of gradient and aspect from a griided digital evevation model, *International Journal of Geographical Information Systems*, **(3)**4: 323–334.

Stephen, J.V. and Irvin, B.J., (2000), Automated landform classification methods for soil-landscape studies, In Wilson, J.P. and Gallant, J.C. (eds.): *Terrain Analysis: Principles and Application*, New York: John Wiley & Sons: 267–294.

Tang, G., (2000), A Research on the Accuracy of Digital Elevation Models, Beijing: Science Press.

Theobald, D.M., (1989), Accuracy and bias issues in surface representation, In Goodchild, M.F. and Gopal, S. (eds.): *The Accuracy of Spatial Database*, New York: Taylor and Francis: 99–106.

Unwin, D., (1981), *Introductory Spatial Analysis*, London and New York: Methuen.

Walker, J.P. and Willgoose, G.R., (1999), On the effect of digital elevation model accuracy on hydrology and geomorphology, *Water Resources Research*, **(35)**7: 2259–2268.

Wilson, J.P. and Gallant, J.C. (eds.), (2000), *Terrain Analysis: Principles and Applications*, New York: John Wiley & Sons.

Wolock, D.M. and Price, C.V., (1994), Effects of digital elevation model and map scale and data resolution on a topography-based watershed model, *Water Resources Research*, **30**: 3041–3052.

Wood, J.D., (1996), The Geomorphological Characterisation of Digital Elevation Model, PhD Thesis, University of Leicester.

Zevenbergen, L.W. and Thorne, C.R., (1987), Quantitative analysis of land surface topography, *Earth Surface processes and Landforms*, **12**: 47–56.

Zhang, W.H. and Montgomery, D.R., (1994), Digital elevation models grid size, landscape representation, and hydrologic simulations, *Water Resources Research*, **(30)**4: 1019–1028.

Zhou, Q. and Liu, X., (2002), Error assessment of grid-based flow routing algorithms used in hydrological models, *International Journal of Geographical Information Science*, **(16)**8: 819–842.

Zhou, Q. and Liu, X., (2004a), Error analysis on grid-based slope and aspect algorithms, *Photogrammetric Engineering and Remote Sensing*, **(70)**8: 957–962.

Zhou, Q. and Liu, X., (2004b), Analysis of errors of derived slope and aspect related to DEM data properties, *Computers and Geosciences*, **30**: 369–378.

Zhou, Q., Liu, X. and Sun, Y., (2006), Terrain complexity and uncertainties in grid-based digital terrain analysis, *International Journal of Geographical Information Science*, **20**(10): 1137–1147.

Accuracy Assessment of DEM Slope Algorithms Related to Spatial Autocorrelation of DEM Errors

LIU Xuejun and BIAN Lu

Abstract

Digital elevation models (DEM) are representations of topography but with unavoidable inherent errors, which result in DEM uncertainty. Entities are spatially dependent and with autocorrelation related to distance. The spatial autocorrelation of the DEM error will systematically affect the result of terrain analysis. It should be taken into account when calculating slope from a grid-based DEM. The effect of the accuracy of slope calculation models on terrain analysis based on quantitative models could be very significant. DEM uncertainty and error propagation is hard to describe with a single, fixed and analytical error based on spatial databases. The Monte Carlo simulation technique is an effective method to simulate error fields and can be used to research the DEM error's spatial autocorrelation and DEM uncertainty. The accuracy of slopes is studied and four commonly used slope algorithms, namely, the 2nd Finite Difference, 3rd order Finite Difference, 3rd order Finite Difference Weighted by Reciprocal of Squared Distance, and 3rd order Finite Difference Weighted by Reciprocal of Distance, are compared theoretically in this paper based on the error being spatially dependent and autocorrelated. The theoretical results are confirmed by Monte Carlo simulation experiments using three digital elevation model data sets.

Keywords: DEM, slope algorithm, accuracy, spatial autocorrelation.

1 Introduction

Digital Elevation Models (DEM) are representations of topography with unavoidable inherent errors, which result in DEM uncertainty. The use of DEMs as the principal database for deriving geomorphic parameters such as slope, aspect, curvature, and hydrological features has been well documented by several authors (Moore *et al.* 1991, Fisher 1994).

Slope, which has been regarded as the most important geomorphic parameter, does not only efficiently describe the relief and structure of the land surface, but is also widely applied as a vital parameter in hydrological models, landslide monitoring and analysis, mass movement and soil erosion studies, and land use planning. Today in GIS software and most applications, slope is often derived from a grid-based DEM with some numerical algorithms such as difference, surface fitting, and vector techniques. It is well known that unavoidable DEM errors have serious effects on slope and many studies have been conducted to analyse these. In effect, the slope algorithms can also influence the slope itself for its implementation based on grid-based DEM can be varied, since some assumptions must be made on how the continuous surface is approximated by discrete sample points (i.e. grid cells). The variation in the implementation would only present minor problems in applications such as surface visualization and classification, but its impact on terrain analysis based on quantitative models could be very significant (Zhou and Liu 2002). It was pointed out that the selection of algorithms could be a critical factor that might create a major impact on the results of the analysis (Moore 1996, Burrough and McDonnell 1998).

Some studies have been made to analyse the errors created by the slope algorithms with a variety of approaches and methodologies. Liu (2002), through a data-independent error analysis of DEM interpretation, stated that from the existing slope algorithms the third-order finite difference (3FD) method (Horn 1981) can give more accurate slope estimations on real DEMs. Unfortunately, Liu (2002), as most current studies related to DEM error, neglected the spatial autocorrelation of the DEM error.

The first law of geography points out that "Everything is related to everything else, but near things are more related than distant things". Entities in space are spatially dependent and autocorrelated. A familiar example of spatial autocorrelation is that of house price. The location of the house has a major effect on the price of the house. This is true for elevation data, as well as for errors in elevation data. Therefore, the DEM error is also spatially dependent and autocorrelated in the interpolation method used in DEM creation. In effect, Hunter and Goodchild (1997) pointed out that neglecting the error's spatial autocorrelation leads to a "worst-case scenario" in spatial analysis. They also derived analytical equations describing the slope variance with RMSE and the correlation coefficient of the DEM error. Later work is contributed by Florinsky (1998a), Ehlschlaeger (1998), Wechsler (1999) and Oksanen and Sarjakoshi (2005). Florinsky (1998a) described the analytical propagation analysis of DEM, in which he derived a number of variance equations for different calculations of slope and curvature. The weakness of that work was that the DEM error was assumed to be uncorrelated. Ehlschlaeger (1998) pointed out near entities are more related to each other than those further away and this is spatial autocorrelation in nature. Wechsler (1999) presented a methodology for simulating the DEM error correlation though the Monte Carlo simulation method, and evaluated the effects of DEM error correlation on elevation and three derived parameters that are often used in hydrological analyses. Recently, Oksanen and Sarjakoshi (2005) reported their study on error propagation of DEM based surface derivatives with Monte Carlo simulation and analytical methods. It should be made clear that the above studies have emphasized the DEM error's spatial autocorrelation model and its effect on derivatives, such as slope, aspect, and curvature derived from the DEM.

DEM uncertainty refers to a lack of knowledge about specific factors, parameters, or models, and includes the parameter uncertainty, model uncertainty, and scenario uncertainty. It is hard to describe DEM uncertainty, error spatial autocorrelation and the law of error propagation with single, fixed or analytical errors during the handling of spatial databases. Simulation error technique is one effective method to study error distribution and the law of error propagation. The Monte Carlo simulation analytical technique is a commonly used simulation method. Stochastic simulation is a generalized and flexible technique for modelling uncertainty in the output of any spatial analysis (Openshaw *et al.* 1991). Many scholars (Fisher 1991, Lee *et al.* 1992, Lee 1996, Liu 1994, Ehlschlaeger and Shortridge 1996, Heuvelink 1998) have researched DEM uncertainty with the Monte Carlo simulation method. Fisher (1991) estimated the effect of DEM error on viewshed analysis with the Monte Carlo simulation method. Lee (1992) and Lee *et al.* (1996) simulated errors in a grid DEM and determined that small errors introduced into the database significantly affect the quality of extracted hydrological features. Liu (1994) utilized the Monte Carlo method to simulate errors in DEM to evaluate uncertainty in a forest harvesting model. Ehlschlaeger and Shortridge (1996) stochastically simulated error in a DEM to evaluate the impact of DEM uncertainty on a least-cost-path application. Heuvelink (1998) analysed accuracy of two slope algorithms with Monte Carlo simulation.

With DEM error being independent, Liu (2002) discussed the rank of the DEM slope algorithms. The ranking of slope algorithms is third-order finite difference (3FD), (Sharpnack and Akin 1969, Horn 1981, Wood 1996), third-order finite difference weighted by reciprocal of squared distance (3FDWRSD), (Horn 1981), third-order finite difference weighted by reciprocal of distance (3FDWRD), (Unwin 1981) and second-order finite difference (2FD), (Fleming and Hoffer 1979, Zevenbergen and Thorne 1987, Ritter 1987). The remaining question is whether this ranking stays unchanged if the DEM error is spatially autocorrelated? This paper estimates the accuracy of these four slope algorithms when the error is spatially autocorrelated. To prove the veracity of the ranking of slope algorithms, this paper designed a Monte Carlo simulation experiment and validated the results with three sets of DEM data.

2 Error analysis

At a given point on a surface $z = f(x, y)$, the slope (S) is defined as a function of gradients at X and Y (i.e. N–S and W–E) directions, i.e.:

$$S = arctg\sqrt{p^2 + q^2} \quad (1)$$

where p and q are the gradients at N–S and W–E directions, respectively.

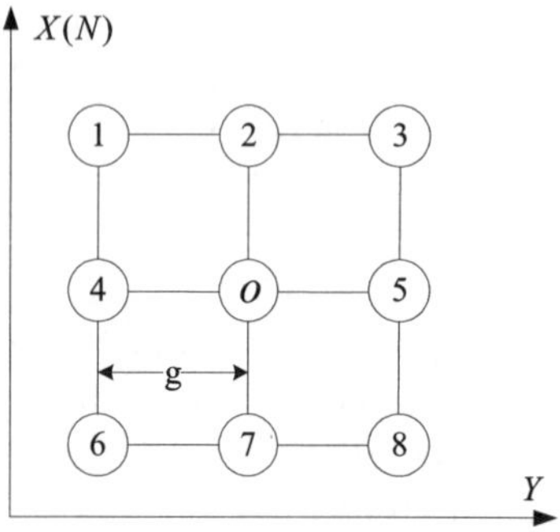

Figure 1. 3 × 3 local moving window.

From the above equation, it is clear that the key step for slope computation is to calculate p and q. Using a grid-based DEM, the common approach is to use a moving 3 × 3 window (Figure 1, point O is the given point, g is the resolution of the grid) to derive finite differentials or local surface fitting polynomials for the calculation (Skidmore 1989, Florinsky 1998a, b). Table 1 shows the four selected slope methods commonly utilized and found in the literature and the GIS software that is used to calculate slope and a comparison of the accuracy in this investigation.

From Table 1, we know that p and q are functions of neighbouring points (z_i, i=1,…,8) at the given point (O). Although the slope algorithms have different expressions with p and q, they can be written uniformly as follows:

$$\left.\begin{aligned} p &= LK_p Z \\ q &= LK_q Z \end{aligned}\right\} \quad (2)$$

where $Z=(z_1, z_2, z_3, z_4, z_5, z_6, z_7, z_8)^T$, K_p and K_q are coefficient matrixes, L is a constant that lies on slope models. For example, in the 2FD slope algorithm, the coefficient matrix (K_p) of p is (0,1,0,0,0,0,-1,0), the coefficient matrix (K_q) of q is (0,0,0,0,-1,1,0,0) and L =1/(2g). Table 2 shows the coefficient matrix Z of p and q and constant L of the four slope algorithms.

Table 1. Slope Algorithms (z_i denotes elevation value of grid point).

Slope algorithms	p	q
2nd Finite Difference (2FD: Fleming and Hoffer 1979, Zevenbergen and Thorne 1987, Ritter 1987)	$\frac{z_2 - z_7}{2g}$	$\frac{z_5 - z_4}{2g}$
3rd order Finite Difference (3FD: Sharpnack and Akin 1969, Horn 1981, Wood 1996)	$\frac{z_1 - z_6 + z_2 - z_7 + z_3 - z_8}{6g}$	$\frac{z_3 - z_1 + z_5 - z_4 + z_8 - z_6}{6g}$
3rd order Finite Difference Weighted by Reciprocal of Squared Distance (3FDWRSD: Horn 1981)	$\frac{z_1 - z_6 + 2(z_2 - z_7) + z_3 - z_8}{8g}$	$\frac{z_3 - z_1 + 2(z_5 - z_4) + z_8 - z_6}{8g}$
3rd order Finite Difference Weighted by Reciprocal of Distance (3FDWRD: Unwin 1981)	$\frac{z_1 - z_6 + \sqrt{2}(z_2 - z_7) + z_3 - z_8}{(4 + 2\sqrt{2})g}$	$\frac{z_3 - z_1 + \sqrt{2}(z_5 - z_4) + z_8 - z_6}{(4 + 2\sqrt{2})g}$

Table 2. Coefficient matrix Z of p and q & constant L.

Slope algorithms		Coefficient matrix Z								Constant L
		z_1	z_2	z_3	z_4	z_5	z_6	z_7	z_8	
2FD	p	0	1	0	0	0	0	-1	0	$\frac{1}{2g}$
	q	0	0	0	-1	1	0	0	0	
3FD	p	1	1	1	0	0	-1	-1	-1	$\frac{1}{6g}$
	q	-1	0	1	-1	1	-1	0	1	
3FDWRSD	p	1	2	1	0	0	-1	-2	-1	$\frac{1}{8g}$
	q	-1	0	1	-2	2	-1	0	1	
3FDWRD	p	1	$\sqrt{2}$	1	0	0	-1	$-\sqrt{2}$	-1	$\frac{1}{(4+2\sqrt{2})g}$

Considering that p and q are independent of each other, $S = arctg\sqrt{p^2 + q^2}$ and $tg^2 S = p^2 + q^2$, the variance of slope (m_s^2), according to the general law of propagation of variance, can be derived as:

$$m_S^2 = \frac{p^2 m_p^2 + q^2 m_q^2}{tg^2 S} \tag{3}$$

where $m_p^{\ 2}$ is the variance of p and $m_q^{\ 2}$ is the variance of q.

Supposing the Root Mean Square Error (RMSE) of DEM is m_i and the covariance of two points (i and j) is ρ_{ij}, $m_p^{\ 2}$ and $m_q^{\ 2}$ can be expressed as follows:

$$\left.\begin{aligned} m_p^2 &= K_p D_Z K_P^T \\ m_q^2 &= K_q D_Z K_q^T \end{aligned}\right\} \tag{4}$$

where D_Z is the variance-covariance matrix of Z in the calculation window, which here is the neighbourhood of 3 × 3 elevation points (see Figure1). The variance-covariance matrix of Z can be expressed as follows:

$$D_z = \begin{bmatrix} m_1^2 & \rho_{12} & \rho_{13} & \rho_{14} & \rho_{15} & \rho_{16} & \rho_{17} & \rho_{18} \\ \rho_{12} & m_2^2 & \rho_{23} & \rho_{24} & \rho_{25} & \rho_{26} & \rho_{27} & \rho_{28} \\ \rho_{13} & \rho_{23} & m_3^2 & \rho_{34} & \rho_{35} & \rho_{36} & \rho_{37} & \rho_{38} \\ \rho_{14} & \rho_{24} & \rho_{34} & m_4^2 & \rho_{45} & \rho_{46} & \rho_{47} & \rho_{48} \\ \rho_{15} & \rho_{25} & \rho_{35} & \rho_{45} & m_5^2 & \rho_{56} & \rho_{57} & \rho_{58} \\ \rho_{16} & \rho_{26} & \rho_{36} & \rho_{46} & \rho_{56} & m_6^2 & \rho_{67} & \rho_{68} \\ \rho_{17} & \rho_{27} & \rho_{37} & \rho_{47} & \rho_{57} & \rho_{67} & m_7^2 & \rho_{78} \\ \rho_{18} & \rho_{28} & \rho_{38} & \rho_{48} & \rho_{58} & \rho_{68} & \rho_{78} & m_8^2 \end{bmatrix} \tag{5}$$

Generally, the DEM error is often assumed as a second-order stationary Gaussian random field and the covariance is only dependent on the distance between two locations. The covariance of two locations decreases as the distance increases. Therefore, the Exponential, Gaussian and linear spatial autocorrelation models are often selected to represent the correlation structure of the DEM error. Considering that the aim of this paper is to investigate the accuracy of slope algorithms under different spatial autocorrelation circumstances, the linear and Gaussian spatial autocorrelation models are used to test the slope algorithms. The expressions for two spatial autocorrelation models are:

$$\rho_{ij} = \begin{cases} C - kd_{ij} & \textit{Linear Function model} \\ Ce^{-k^2 d_{ij}^2} & \textit{Gaussian Funtcion model} \end{cases} \tag{6}$$

where C and k are constant, d_{ij} is the horizontal distance between the two locations that can be regarded as the distance between two grid points (i and j) in the DEM.

Since the DEM error was assumed to be isotropic and the variance of each grid point is m^2, the variance-covariance D_Z is also a symmetrical matrix, the equation of the variance for both p and q are equal, and thus the equations can be expressed by only m_p^2 or m_q^2:

$$\left.\begin{aligned} m_p &= m_q \\ m_S^2 &= m_p^2 \cos^4 S \end{aligned}\right\} \tag{7}$$

So Equation (7) is the accuracy model of slope derived from grid DEM.

From Equation (7) and Table 1, we know that different slope algorithms have different variances of p. The variance of p, m_p^2, can be regarded as an indicator for assessing the slope algorithms. The key for calculating m_p^2 is Equations (5) and (6). From Figure 1, there exist five different distances between two grid points (i and j) in a 3 × 3 local moving window, and they can be summarized as shown in Table 3.

Table 3. Distances of two grid points in 3 × 3 local moving window.

Distance	*Corresponding grid points (see Figure 1)*
g	(1,2) (1,4) (2,3) (3,5) (4,6) (5,8) (6,7) (7,8)
$\sqrt{2}g$	(2,4) (2,5) (4,7) (5,7)
$2g$	(1,3) (1,6) (2,7) (3,8) (4,5) (6,8)
$\sqrt{5}g$	(1,3) (1,5) (1,7) (2,6) (2,8) (3,4) (4,8) (5,6)
$\sqrt{8}g$	(1,8) (3,6)

According to the linear and Gaussian spatial autocorrelation model shown in Equation (6) and the distance between two points (i and j) listed in Table 3, the variance-covariance matrices of Z of linear and Gaussian spatial autocorrelation can be expressed by Equations (8) and (9), respectively, as follows:

$$D_Z = \begin{pmatrix} m^2 & C-kg & C-2kg & C-kg & C-\sqrt{5}kg & C-2kg & C-\sqrt{5}kg & C-\sqrt{8}kg \\ C-kg & m^2 & C-kg & C-\sqrt{2}kg & C-\sqrt{2}kg & C-\sqrt{5}kg & C-2kg & C-\sqrt{5}kg \\ C-2kg & C-kg & m^2 & C-\sqrt{5}kg & C-kg & C-\sqrt{8}kg & C-\sqrt{5}kg & C-2kg \\ C-kg & C-\sqrt{2}kg & C-\sqrt{5}kg & m^2 & C-2kg & C-kg & C-\sqrt{2}kg & C-\sqrt{5}kg \\ C-\sqrt{5}kg & C-\sqrt{2}kg & C-kg & C-2kg & m^2 & C-\sqrt{5}kg & C-\sqrt{2}kg & C-kg \\ C-2kg & C-\sqrt{5}kg & C-\sqrt{8}kg & C-kg & C-\sqrt{5}kg & m^2 & C-kg & C-2kg \\ C-\sqrt{5}kg & C-2kg & C-\sqrt{5}kg & C-\sqrt{2}kg & C-\sqrt{2}kg & C-kg & m^2 & C-kg \\ C-\sqrt{8}kg & C-\sqrt{5}kg & C-2kg & C-\sqrt{5}kg & C-kg & C-2kg & C-kg & m^2 \end{pmatrix} \tag{8}$$

$$D_Z=\begin{pmatrix} m^2 & Ce^{-k^2g^2} & Ce^{-4k^2g^2} & Ce^{-k^2g^2} & Ce^{-5k^2g^2} & Ce^{-4k^2g^2} & Ce^{-5k^2g^2} & Ce^{-8k^2g^2} \\ Ce^{-k^2g^2} & m^2 & Ce^{-k^2g^2} & Ce^{-2k^2g^2} & Ce^{-2k^2g^2} & Ce^{-5k^2g^2} & Ce^{-4k^2g^2} & Ce^{-5k^2g^2} \\ Ce^{-4k^2g^2} & Ce^{-k^2g^2} & m^2 & Ce^{-5k^2g^2} & Ce^{-k^2g^2} & Ce^{-8k^2g^2} & Ce^{-5k^2g^2} & Ce^{-4k^2g^2} \\ Ce^{-k^2g^2} & Ce^{-2k^2g^2} & Ce^{-5k^2g^2} & m^2 & Ce^{-4k^2g^2} & Ce^{-k^2g^2} & Ce^{-2k^2g^2} & Ce^{-5k^2g^2} \\ Ce^{-5k^2g^2} & Ce^{-2k^2g^2} & Ce^{-k^2g^2} & Ce^{-4k^2g^2} & m^2 & Ce^{-5k^2g^2} & Ce^{-2k^2g^2} & Ce^{-k^2g^2} \\ Ce^{-4k^2g^2} & Ce^{-5k^2g^2} & Ce^{-8k^2g^2} & Ce^{-k^2g^2} & Ce^{-5k^2g^2} & m^2 & Ce^{-k^2g^2} & Ce^{-4k^2g^2} \\ Ce^{-5k^2g^2} & Ce^{-4k^2g^2} & Ce^{-5k^2g^2} & Ce^{-2k^2g^2} & Ce^{-2k^2g^2} & Ce^{-k^2g^2} & m^2 & Ce^{-k^2g^2} \\ Ce^{-8k^2g^2} & Ce^{-5k^2g^2} & Ce^{-4k^2g^2} & Ce^{-5k^2g^2} & Ce^{-k^2g^2} & Ce^{-4k^2g^2} & Ce^{-k^2g^2} & m^2 \end{pmatrix} \quad (9)$$

According to the coefficient matrix Z of p and q and the constant L shown in Table 2, the distances between the two locations shown in Table 3 and the Equation (7), (8) and (9), the accuracy models of the selected slope algorithms can be described as shown in Table 4.

Table 4. Accuracy model of slope algorithms.

Slope algorithms	Spatial autocorrelation models	$m_p^2=m_q^2$
2FD	Linear	$0.5m^2-0.5C+gk$
	Gaussian	$0.5m^2-0.5ce^{-4k^2g^2}$
3FD	Linear	$0.167m^2-0.167C+0.7gk$
	Gaussian	$0.167m^2+0.222ce^{-k^2g^2}-0.056ce^{-4k^2g^2}-0.222ce^{-5k^2g^2}-0.111ce^{-8k^2g^2}$
3FDWRSD	Linear	$0.188m^2-0.188C+0.738gk$
	Gaussian	$0.188m^2+0.25ce^{-k^2g^2}-0.125ce^{-4k^2g^2}-0.25ce^{-5k^2g^2}-0.063ce^{-8k^2g^2}$
3FDWRD	Linear	$0.172m^2-0.172C+0.728gk$
	Gaussian	$0.172m^2+0.234ce^{-k^2g^2}-0.086ce^{-4k^2g^2}-0.234ce^{-5k^2g^2}-0.086ce^{-8k^2g^2}$

Since C, k and g are constant, the accuracy of the selected slope algorithms can be assessed by comparing the coefficient of equations in Table 4. Some results are:

- The ranking of the DEM slope algorithms with the DEM error's spatial autocorrelation is the same as that with the DEM errors uncorrelated (Liu 2002); that is the 3rd order Finite Difference (3FD) slope algorithm can get more accurate slope estimations than any other slope algorithm. Then follow the 3rd order Finite Difference Weighted by Reciprocal of Distance (3FDWRD) and the 3rd order Finite Difference Weighted by Reciprocal of Squared Distance (3FDWRSD). The 2nd Finite Difference (2FD) gives the least accurate estimation.

- The structure of the spatial autocorrelation models of the DEM error, whether exponential, Gaussian, or linear, has no effect on the ranking of the slope algorithms.
- The number of grid points neighbouring the centre point has an effect on the accuracy of slope estimation. For example, in the 3 × 3 local moving window, the 3FD, 3FDWRSD, 3FDWRD used the eight points of the centres to calculate the centre slope, so those algorithms can estimate slope more accurately than that obtained by 2FD, which only uses four points.
- With the same number of points neighbouring the centre, such as 3FD, 3FDWRSD, and 3FDWRD slope algorithms, the accuracy of the slope depends on the weight used in p and q (see Table 1). For terrain surfaces assumed with the characteristics of self-similarity and isotropy, equal-weight in p and q estimation is more appropriate than unequal weight. The 3FD obtains the best accuracy estimation, followed by 3FDWRD and 3FDWSRD.

3 Experiments and results

Theoretical conclusions will be confirmed by experiments utilizing Monte Carlo stochastic simulation techniques. Wechsler (1999) introduced four methods, which are neighbourhood autocorrelation, mean spatial dependence, weighted spatial dependence, and interpolated spatial dependence to generate spatial autocorrelation error fields and this chapter adopts the neighbourhood autocorrelation method, which can generate spatial autocorrelation errors reliably. The neighbourhood autocorrelation method increases the spatial autocorrelation of the random error field by passing a mean low pass filter in a 3 × 3 (or 5 × 5, 7 × 7, … , n × n) neighbourhood over the error field surface (See Figure 2). Each cell in the random field is replaced by the mean of the random values in the surrounding nine cells. This method increases the spatial autocorrelation of the random surface and decreases the standard deviation of the values.

Figure 2 shows the procedure of the Monte Carlo simulation experiment. Firstly, a random error field is generated by Monte Carlo simulation according to relief type of the selected study areas Secondly, spatial autocorrelation error fields are created by the neighbourhood autocorrelation method. The output DEMs are generated by adding spatial autocorrelation error fields to the original DEMs. Thirdly, four selected slope algorithms are used to calculate slope mapping on the output DEM. All the above approaches are executed N times. Measurement of vertical accuracy in DEMs is usually represented statistically in the form of the Root Mean Square Error (RMSE) (Equation 10).

$$RMSE = \sqrt{\frac{\sum_{i=1}^{N}(z_i - z_t)}{N-1}} \tag{10}$$

where z_i is the observed elevation, z_t is the true elevation, and N is the number of sample points. This paper regards the Root Mean Square Error (RMSE) as the estimation standard of the accuracy of the four slope algorithms. Therefore, the last step is the calculation of the RMSE values.

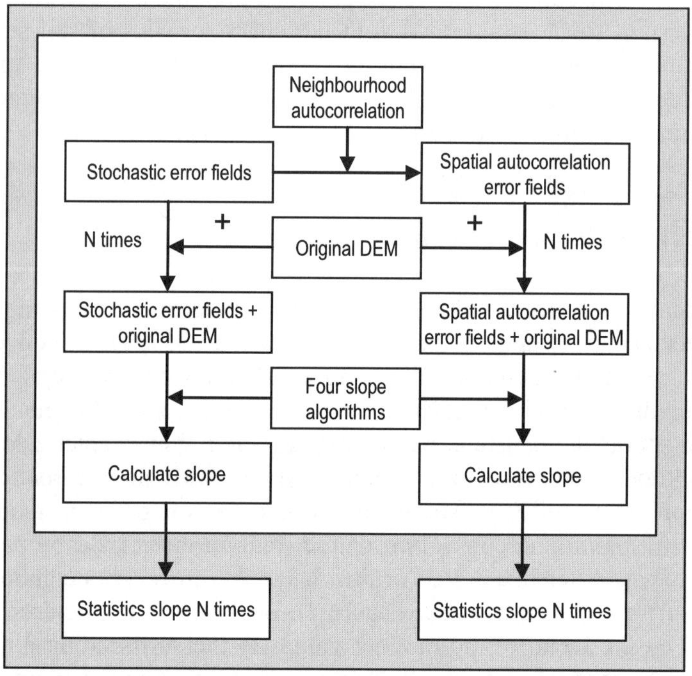

Figure 2. Procedures of Monte Carlo simulation experiment.

The process in the Monte Carlo simulation experiment shown on the left in Figure 2 is used to derive slopes from DEMs with stochastic error fields.

Three DEMs of study areas were selected to analyse the accuracy of the selected slope computation models. To validate that the theoretical conclusion has high reliability, these DEMs have different scales, different grid resolutions, different hypsography and belong to different terrain type (Table 5 gives their particulars). Figure 3 shows the DEMs of the study areas.

Table 5. Description of study areas.

Study area	*Scale*	*Grid resolution of DEM*	*Hypsography extent*	*Terrain type*
Study area 1	1:50,000	30 m	410 m–1,990 m	Mixed
Study area 2	1:10,000	5 m	894 m–1,185 m	Hilly country
Study area 3	1:50,000	25 m	973 m–2,725 m	High mountain

Study area 1 Study area 2 Study area 3

Figure 3. DEMs of study areas.

Statistics of stochastic error fields and spatial autocorrelation error fields of these study areas are shown in Table 6. The standard deviation (SD) of stochastic simulation error fields is given when generating them. According to the DEM grid resolutions and terrain types of the selected study areas, the SD of study area 1 is 7 metres, SD of study area 2 is 3.5 metres and SD of study area 3 is 11 metres. These stochastic error fields are non-spatially autocorrelated. To obtain the spatial autocorrelation error fields, an n × n local moving window is used when dealing with stochastic error fields and the spatial autocorrelation error fields are generated using neighbourhood autocorrelation introduced by Wechsler (1999). In this chapter, a 3 × 3 local moving window is adopted to filter study areas 1 and 3 and a 5 × 5 local moving window is adopted to filter for study area 2.

Here Moran's I, which is an index that assesses the intensity of spatial autocorrelation, is used to assess the level of spatial autocorrelation generated by neighbourhood autocorrelation at different local moving windows.

Table 6 clearly shows Moran's I derived from spatial autocorrelation error fields to be higher than the Moran's I derived from stochastic error fields. This is because the former take the effect of neighbourhood points on the centre point into account and the latter is generated stochastically. The local moving window sizes used in generating neighbourhood autocorrelation will affect the intensity of spatial autocorrelation of the study areas. Study area 2, which used a 5 × 5 local moving window, got more spatial autocorrelation than the other two study areas using a 3 × 3 local moving window. The neighbourhood autocorrelation increased the spatial autocorrelation and decreased the standard deviation of the values. This method can generate spatial autocorrelation fields steadily and these spatial autocorrelation error fields can be used reliably in the experiment.

Table 6. Statistics of error fields of study areas.

Error fields		*Study area 1*	*Study area 2*	*Study area 3*
Stochastic error fields	MIN	-30.2699	-15.4060	-47.9298
	MAX	30.5959	15.1985	48.2075
	MEAN	0.0048	0.0020	-0.0009
	SD	7.0257	3.5128	11.0474
	Moran's I	0.0022	-0.0006	0.0001
Spatial autocorrelation error fields	MIN	-11.4059	-3.7466	-17.5982
	MAX	11.6661	3.5866	17.8861
	MEAN	0.0052	0.0021	-0.0017
	SD	2.3505	0.7050	3.6928
	Moran's I	0.6669	0.7995	0.6673

The output DEMs are of two kinds; one is with stochastic error fields and the other is with spatial autocorrelation error fields. The method of generating these DEMs is by adding the stochastic error fields and spatial autocorrelation error fields to the original DEMs, respectively. And then the slope derived from either DEMs can be calculated using the four selected calculation models. Figure 4 provides a visual comparison of RMSE of slope. From Figure 4 some results are:

1. The ranking of the RMSE of slope is the same as the result shown in Table 4 whether the DEM error field is stochastic or is spatial autocorrelation. The 3FD can produce the most accurate estimation, followed by 3FDWRD and 3FDWRSD. The 2FD gives the lowest accuracy.
2. The RMSE of slope derived from a DEM with the error spatially autocorrelated is better than that from a DEM with the error uncorrelated. The Standard Deviation of the error field is obviously decreased by neighbourhood autocorrelation. The gap between the

third-order finite difference and the second-order finite difference is smaller after neighbourhood autocorrelation.

3. For the same DEM, the difference of RMSE derived from the four slope algorithms when the error field is spatially autocorrelated is smaller than when it is not spatially autocorrelated.

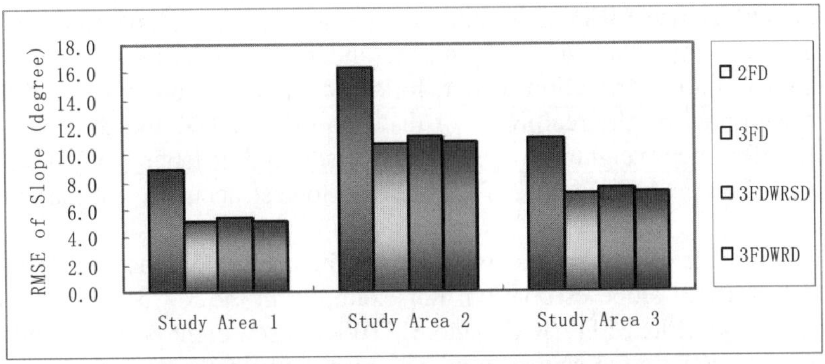

(a) DEM with stochastic error fields.

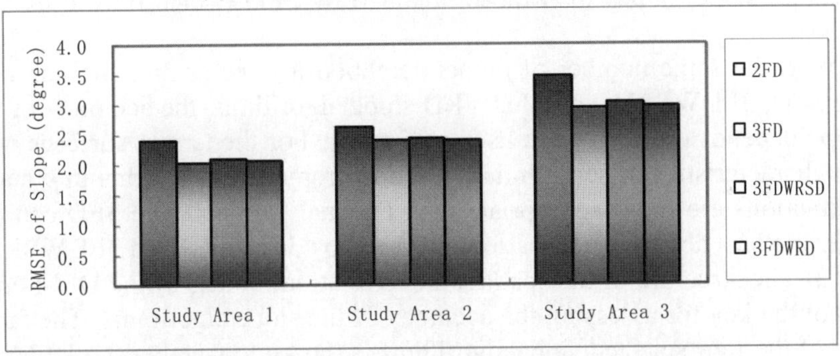

(b) DEM with spatial autocorrelation error fields.

Figure 4. RMSE of slope derived from four slope algorithms.

4 Conclusions

Slope is the most frequently utilized derivation in GIS-based terrain analysis and geographical modelling. The slope algorithm can influence slope in different ways by its implementation based on grid-based DEM and its impact on terrain analysis based on quantitative models can be very significant. DEM error has serious effects on slope; thus choosing an appropriate and accurate slope algorithm could be a critical factor that might create significant impact on the analysis results.

This paper analysed the effect of error spatial autocorrelation on the accuracy of the four selected slope algorithms and assessed the accuracy of slope algorithms when the error is spatially autocorrelated. From the findings of this study, we can draw the following conclusions:

(1) The conclusion derived from theoretical deduction accords with the result gained from the Monte Carlo simulation experiment. The 3rd order Finite Differencc (3FD) slopc algorithm can estimate slope more accurately than any other slope algorithm under any circumstance, whether with spatial autocorrelation or not, followed by the 3rd order finite difference weighted by the reciprocal of distance (3FDWRD) and the 3rd order finite difference weighted by reciprocal of squared distance (3FDWRSD). The 2nd Finite Difference (2FD) gives the lowest accuracy estimation of slope.

The number of grid points neighbouring the centre point has an effect on the accuracy of slope estimation. For example, in the 3 × 3 local moving window, the 3FD, 3FDWRSD, and 3FDWRD used eight points round the centre to calculate the centre slope, so those algorithms can estimate more accurate slope values than those obtained by 2FD, which only uses four points.

With the same number of points neighbouring the centre, such as with the 3FD, 3FDWRSD, and 3FDWRD slope algorithms, the accuracy of the slope depends on the weight used in p and q. For the terrain surfaces with the characteristics of self-similarity and isotropy, equal weights in p and q estimations are more appropriate than unequal weights. The 3FD can deliver the best accuracy estimation, followed by 3FDWRD and 3FDWSRD.

(2) The structure of the spatial autocorrelation models of the DEM error is not the key for assessing the accuracy of the slope algorithms. The ranking of the four selected slope algorithms is the same whether the DEM error is exponential, Gaussian, or a linear function.

(3) Neighbourhood autocorrelation can generate reliable spatial autocorrelation error fields and can decrease the standard deviation of the spatial autocorrelation error fields. By this method, the intensity of spatial autocorrelation is related to the window size passing a mean low pass filter in a 3 × 3 (or 5 × 5, 7 × 7, … , n × n) neighbourhood over the error field surface. The larger the window size, the higher the intensity of the spatial autocorrelation. The accuracy of slope estimation from DEMs whose errors are spatially autocorrelated is better than that from DEMs with errors uncorrelated.

Further study will focus on the character of spatial autocorrelation DEM error and how it affects slope.

Acknowledgement

This study is supported by the National Natural Science Foundation of China (No. 40571120)

References

Burrough, P.A. and McDonnell, R.A., (1998), *Principles of Geographical Information System*, Oxford, UK: Oxford University Press.

Ehlschlaeger, C.R. and Shortridge, A., (1996), Modeling Elevation Uncertainty in Geographical Analyses, *Proceedings of the International Symposium on Spatial Data Handling*, Delft, Netherland, 9B: 15–25.

Ehlschlaeger, C.R., (1998), *The Stochastic Simulation Approach: Tools for Representing Spatial Application Uncertainty*, Ph.D. Dissertation, University of California, Santa Barbara.

Fisher, P.F., (1991), First experiments in viewshed uncertainty: The accuracy of the viewshed area, *Photogrammetric Engineering and Remote Sensing*, **(57)**10: 1321–1327.

Fisher, P. F., (1994), Probable and fuzzy models of the viewshed operation. In Worboys, W. F. (ed.): *Innovations in GIS* 1: 161–176.

Fleming, M.D. and Hoffer, R.M., (1979), *Machine processing of Landsat MSS data and DMA topographic data for forest cover type mapping: LARS Technical Report 062879,* West Lafayette, IN: Purdue, University, Laboratory for Applications of Remote Sensing.

Florinsky, I.V., (1998a), Accuracy of local topographic variables derived from digital elevation models, *INT. J. Geographical Information Science*, **(12)**1: 47–61.

Florinsky, I.V., (1998b) Combined analysis of digital terrain models and remotely sensed data in landscape investigations, *Progress in Physical Geography*, **(22)**1: 33–60.

Heuvelink, B.M., (1998), Error Propagation in Environmental Modeling with GIS, Taylor and Francis.

Horn, B.K.P. (1981), Hill shading and the reflectance map, *Proceedings of IEEE*, **(69)**1: 14–47.

Hunter, G. J. and Goodchild, M.F., (1997) Modeling the uncertainty of slope and aspect estimates derived from spatial databases, *Geographical Analysis*, **(29)**1: 35–49.

Lee, J., Snyder, P.K. and Fisher, P.F., (1992) Modeling the effect of data errors on feature extraction from digital elevation models, *Photogrammetric Engineering and Remote Sensing*, **(58)**10: 1461–1467.

Lee, J., (1996) Digital Elevation Models: Issues of Data Accuracy and Applications, ESRI User Conference, www.esri.com/resources/userconf/proc96.

Liu, R., (1994), *The Effects of Spatial Data Errors on the Grid-Based Forest management Decisions*, Ph.D. Dissertation, State University Of New York College Of Environmental Science and Forestry, Syracuse, NY.

Liu Xuejun, (2002), On the Accuracy of the Algorithms for Interpreting Grid-based Digital Terrain Model, PhD Thesis, Wuhan Univeisity.

Moore, I.D., Grayson, R.B. and Ladson, A.R., (1991), Digital terrain modelling: A review of hydrological, *geomorphological and biological applications*, *Hydrological Processes*, **5**, 3–30.

Moore, I.D., (1996), Hydrologic Modeling and GIS, In Gooldchild, M.F., Steyaert, L.T., Parks, B.O., Crane, M.P., Maidment, D.R. and Glendinning, S. (eds.): *GIS and Environmental modeling: Progress and Research Issuse*, Fort Collins, CO: GIS World Books: 143–148.

Oksanen, J. and Sarjakoshi, T., (2005) Error propagation analysis of DEM-based surface derivatives, *Computers & Geosciences*, **31**: 1015–1027.

Openshaw, S., Charlton, M. and Carver, S., (1991) Error Propagation: a Monte Carlo simulation, In Masser, I. and Blakemore, M., (eds.): *Handling Geographic Information*, New York, NY: John Wiley & Sons, Inc.

Ritter, D., (1987), A vector-based slope and aspect generation algorithm, *Photogrammetric Engineering and Remote Sensing,* **(53)**8: 1109–1111.

Sharpnack, D.A. and Akin, G., (1969), An Algorithm for computing slope and aspect from elevations, *Photogrammetric Engineering and Remote Sensing,* **5**: 247–248.

Skidmore, A.K., (1989), A comparison of techniques for the calculation of gradient and aspect from a gridded digital elevation model, *International Journal of Geographical Information Systems*, 3: 323-334.

Unwin, D., (1981), *Introductory Spatial Analysis*, London and New York: Methuen.

Wechsler, S.W., (1999), A methodology for digital elevation model (DEM) uncertainty evaluation: the effect DEM uncertainty on topographic parameters, *URISA 99 Annual Conference Proceeding.*

Wood, J.D., (1996), The geomorphological characterisation of digital elevation model, PhD Thesis, University of Leicester.

Zevenbergen, L.W. and Thorne, C.R., (1987), Quantitative analysis of land surface topography, *Earth Surface Processes and Landforms,* **12**: 47–56.

Zhou, Q. and Liu, X., (2002), Error assessment of grid-based flow routing algorithms used in hydrological models, *International Journal of Geographical Information Science*, **(16)**8: 819–842.

Modelling Slope Field Uncertainty Derived From DEM in the Loess Plateau

DENG Fengdong, WANG Lili, ZHUO Jing and LIU Anlin

Abstract

This chapter explains how land use in the Loess Plateau area was classified using TM (Thematic Mapper) remote sensing data. Slope fields were extracted from DEMs (Digital Elevation Models) at different scales and the effects of different scales on slope field precision were also analysed. Results show that a DEM with a scale of 1:50,000 (with a resolution of 25 m) was basically sufficient to extract slope fields with the required precision. Although the accuracy of slopes extracted from a scale of 1:10,000 is higher than DEMs of other scales, the detailed land appearance did not show obvious improvement on the slope field precision because of the special land use in the studied area, especially the considerable effect of land use type on plantation layout (plantations focused on Loess Mao of small slopes and large areas). Furthermore, with the constraints of using the resolution of 30 m for the land use, DEMs with the scale of 1:10,000 could not show its advantage of high precision and increased the redundancy and computing load; thus it was not appropriate for obtaining slope fields of large areas. The DEM with a scale of 1:250,000 could not be used to extract slope field graphs effectively either, because of its large generalizations over the study area.

Keywords: Loess Plateau, slope field, DEM, uncertainty, remote sensing.

1 Introduction

A DEM (Digital Elevation Model) is convenient for representing the continuously varying topographic surface of the Earth, together with common terrain attributes that are readily computed from a DEM including slope gradient, slope aspect, slope curvature, upslope length, specific catchment areas (upslope contributing area divided by the grid cell size), and the compound topographic index (CTI – a hydrological based index that is related to zones of surface saturation; Moore *et al.* 1993). In recent years,

DEMs have been widely accepted and applied in ecology environmental investigations (Bolstad and Stowe 1994). As the development of research into DEMs, especially the theory concerning uncertainty of DEM and DEM-derived quantities, has greatly improved, the accuracy of DEMs became increasingly important as the application of DEM were extended for slope farmland extraction and ecosystem environmental investigations. The accuracy of the DEM and the DEM-derived products depend on (i) the source of the elevation data, including the techniques for measuring elevation, either on the ground or remotely, the locations of samples, and the density of samples; (ii) the methods used to create the DEM from this elevation data; (iii) the data model, or structure of the elevation data grid, contour, triangular irregular network; (iv) the horizontal resolution and vertical precision at which the elevation data is represented; (v) the topographic complexity of the landscape being represented: and (vi) the algorithms used to calculate different terrain attributes (Theobald 1989, Chang and Tsai 1991, Florinsky 1998).

Measurement of errors in DEMs is often impossible because the true value for every geographic feature or phenomenon represented in a geographic data set is rarely determinable (Goodchild *et al.* 1994, Hunter *et al.* 1995).Uncertainty, instead of error, should be used to describe the quality of a DEM or DEM-derived product. One critical aspect of any DEM uncertainty study is how the resolution of the DEM impacts the accuracy of information derived from the DEM (Tang Guoan 2000). Commonly, the higher resolution DEMs have more information, and can describe the surface of the Earth more precisely; but on the other hand, as the resolution of the DEM increases, the volume of spatial data increases sharply, data redundancy increases also, and the cost of analysis will become greater (Tang Guoan 2001). It is especially difficult if a large area is investigated for slope farmland extraction. Remote sensing technologies are used to obtain the land use information of large areas; LANDSAT TM-5 images were used to classify the larger study site, for example the Loess Plateau of Northern Shaanxi Province, into different land use types, and the slope farmland areas were calculated employing land use data and DEMs. The object of this chapter mainly focuses on the uncertainty of the process through which the slope farmlands are calculated by slope gradient derived from different resolution DEMs and land use information classified from the TM-5 data with a spatial resolution of 30 m.

The Loess Plateau, in the Northern part of Shaanxi, is an area of serious soil erosion and the ecological environment has been destroyed through the long-time activity of humans. It has been discovered that 90% of farmland in this area is slope farmland (Deng Fengdong 2002). To restructure the ecosystem of this area and realize sustainable development, the local government must take a set of measures. A project, named 'Return the farmland back to woodland or pasture', was put into practice.

One object of this project is to decrease the areas of farmland with slope gradients higher than 25%. High resolution land use investigation is impossible for such a large area; therefore the extent of sloping farmland was calculated by land use data and DEMs were very helpful in this project.

Table 1. Major topographic variables of study site.

Area (km)	*Mean Elevation (m)*	*Gradient (°)*	*Curvature (°)*	*Ground Roughness*
10 × 10	1,127.69	27.10	31.27	1.1902

2 Research Sites and Data Source

The two 10 km × 10 km study sites were selected in the Loess Plateau of northern Shaanxi province (Figure 1). The altitude of the area is between 968 m and 1,336 m, maximum slope gradients are approximately 73% (Table 1).

Figure 1. The location of research sites.

Landuse data were classified into twelve different land use types by a series of procedure: image correcting, image enhancement, ground investigation, supervised classification, results verified. According to the result, approximate 27% of the area is farmland. The Landsat TM5 images were obtained in the summer of 2003, and the investigation of the whole Loess Plateau was finished. The accuracy of the result has been checked; the overall accuracy reached 85.47%, kappa = 0.817 (Figure 2).

DEMs data used in the research included three different scale level data: 1:10,000, 1:50,000, 1:250,000. The DEM of 1:10,000 scale was digitized from the topographic map of the same scale, and the contours were interpolated to improve the resolution of the DEM to 5m to satisfied the needs of analysis in this hilly region. DEMs of 1:50,000 and 1:250,000 scale levels were both obtained from the National Geomantic Centre of China (NGCC), the spatial resolution being 25 metres and 90 metres (Figure 3). The method of slope-derivation is the windows differential coefficient method researched by Burrough (1986). The method is integrated within the ARCGIS software. The method calculates the output slope value for each cell location based on the value from the DEM for the location and the values identified in a 3 x 3 cell rectangular shape called the neighbourhood; as the neighbourhood moves cell by cell, the DEM based slope extraction is processed. The slope was divided into a set of levels: 0°–3°, 3°–5°, 5°–8°, 8°–15°, 15°–25°, 25°–35°, 35°–45°, 45°–60°, 60°–90°, a total of nine levels. According to this principle, the slope maps, accordingly, could have been extracted from DEMs of different spatial scales (1:10,000, and 1:50,000, 1:250,000). Slope farmland extraction is based on the overlay analysis method. In the process, to make the data easy for statistical comparison, the DEMs was interpolated to have the same spatial resolution of 5 m, and then an overlay of the slope map and the land use map which was classified using the remote sensing images was used to calculate sloping farmland.

3 Accuracy of Land Use Classification

Two methods are frequently employed for assessing the accuracy of classifications: the method based on pixels and the method based on features. The first method samples the classified result map with equal intervals, then makes a comparison with the reference data to find the true land use type, to obtain the accuracy of the classification. The second method concentrates on the differences between various feature structures in the classification result and the real world feature structures to estimate the accuracy. Most classifications are based on pixel computing procedures, so the first method was always used to calculate the accuracy.

A confusion matrix lists the values for known types of reference data in the columns and for the classified data in the rows. The main diagonal of the matrix lists the correctly classified pixels. The overall accuracy can be calculated from the confusion matrix. If there are X types in the classification, then overall accuracy is:

$$overall\ accuray = \sum_{i=1}^{r} x_{ii} / N \tag{1}$$

where r is the count of classification type, N is the total number of samples.

Another accuracy assessing method is also based on the confusion matrix; it can assess the quality of the classifier. The Kappa coefficient is conventionally employed to assess the accuracy of classification results calculated by remote sensing. It takes the non-diagonal elements into account. Actually, every value in the confusion matrix is used in the Kappa coefficient. The Kappa coefficient can be calculated by:

$$\bar{k} = \frac{N \sum_{i=1}^{r} x_{ii} - \sum_{i=1}^{r} x_{i+} x_{+i}}{N^2 - \sum_{i=1}^{r} x_{i+} x_{+i}} \tag{2}$$

where

r = number of rows and columns in the error matrix,
N = total number of observations,
X_{ii} = observation in row i and column i,
X_{i+} = marginal total of row i, and
X_{+i} = marginal total of column i.

In practice, the follow equation is used to calculate the kappa coefficient:

$$\bar{k} = \frac{\theta_1 - \theta_2}{1 - \theta_2} \tag{3}$$

where:

$$\theta_1 = \sum_{i=1}^{r} x_{ii} / N\,;\ \ \theta_2 = \sum_{i=1}^{r} x_{i+} x_{+i} / N^2$$

647 sampling units were used to assess the accuracy of the map derived from remote sensing in this research; among these sampling units, 497 were obtained by random sampling and the others were selected from a QuickBird image (spatial resolution 61 cm). The confusion matrix is listed

below (Table 2). According to the confusion matrix, the overall accuracy reached 85.47%, and the kappa coefficient is 0.817.

4 Research Result and Analysis

(1) Slope extraction obviously changed as DEMs of different scales were used. The resolution of DEMs impacts the accuracy of slope extraction in three ways. First, in the zone of 0–5 degrees, the accuracy of slopes derived from DEMs changes little, with the results processed with the 1:10,000 DEM being close to the results processed with the 1:50,000 DEM. Secondly, in the zone of 5–25 degrees, slope had a definite change as the resolution of the DEM changed; the lower resolution of DEMs used, the larger the gradient derived from the DEMs. In the zone of 5–15 degrees, the change of slope gradient is great; there are two reasons for this: one is that the areas with 15–25 degree slopes were a great proportion of the total area of the research site. On the other hand, DEMs with lower spatial resolution will smooth the small relief features (for example, there are many valleys of 15–30 m in width in the Loess Plateau that can hardly be surveyed by the DEM with a spatial resolution of 25 m), leading to a greater area of slope being calculated in the zone. Thirdly, in the zone of slopes larger than 25 degrees, the area definitely decreased as the slope gradient increased (Figure 4).

Table 2. The confusion matrix of the classification result in the Loess Plateau.

Classified data	Reference data												Total
	V	FL	F	SH	O	G	B	I	P	C	SD	W	
Vegetable (V)	2	0	0	0	1	0	0	0	0	0	0	0	3
Farmland (FL)	1	205	0	2	2	4	0	2	0	0	1	2	219
Forest (F)	0	2	75	1	2	3	0	2	0	1	0	0	86
Shrub (SH)	0	1	12	20	3	5	2	0	0	0	0	0	43
Orchard (O)	0	1	4	2	49	2	2	3	0	0	0	1	64
Grass (G)	0	4	1	1	2	143	0	3	0	0	0	2	145
Bare rock (B)	0	0	0	0	0	2	4	0	0	0	0	0	6
Industry (I)	2	2	0	0	0	5	0	34	0	0	0	1	48
Paddy field (P)	0	0	0	0	0	0	0	0	2	0	0	0	2
Coniferous forest (C)	0	0	0	0	0	0	0	0	0	1	0	0	1
Sand (SD)	0	0	0	0	0	0	0	0	0	0	3	0	3
Water (W)	0	0	0	1	0	0	0	0	0	0	0	15	13
Total	5	215	92	27	59	164	8	44	2	2	4	25	647

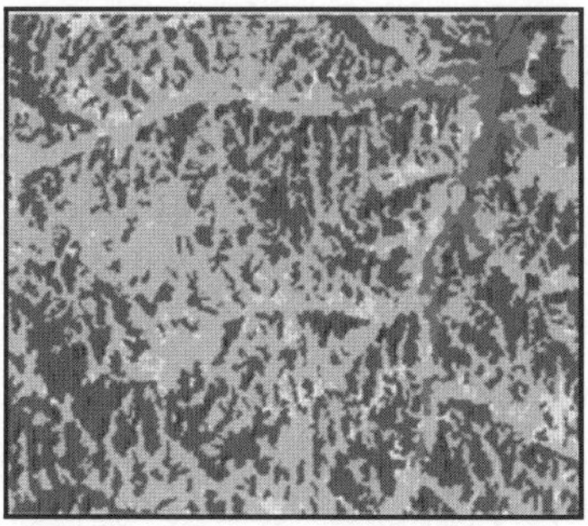

Figure 2. Land use map of study site.

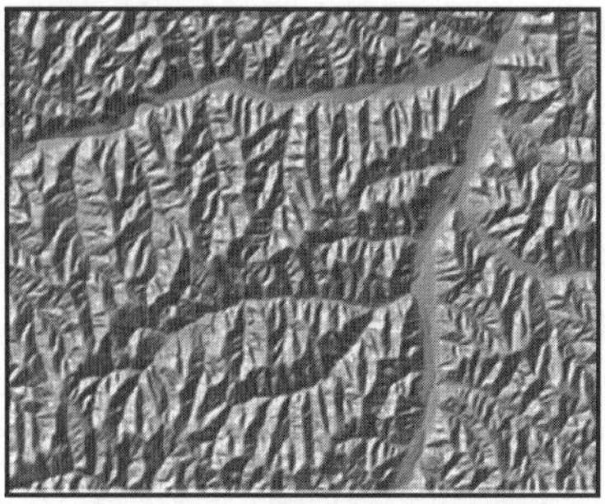

Figure 3. Hillshading with 5 m resolution

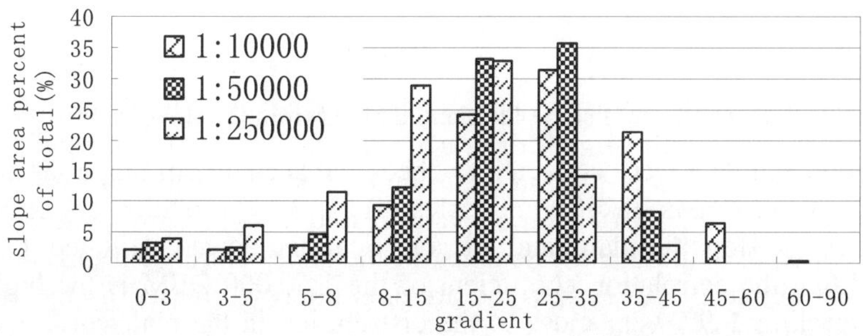

Figure 4. A comparison of gradient derived from DEMs of different resolution

(2) The impact of DEMs with different spatial resolutions on the slope farmland. According to the research results, it can be concluded that the error of slope farmland derived from the DEMs with the scale of 1:250,000 is too great to be used, and comparing the slope farmland derived from DEMs of the scale 1:10,000 with that of the scale 1:50,000, the main difference is seen if one focuses on the zone of 15–25 degrees and the zone of slopes larger than 35 degrees. 30% of the total farmland derived from the 1:10,000 DEM has slope gradients larger than 35 degrees, and also approximately 7% of total farmland is slope farmland with slope gradients larger than 60 degrees. But according to the ground investigations of the study site, the soil is very loose in this region and there is little farmland on slopes greater than 60 degrees because of the serious soil erosion. As the project 'Return the farmland back into woodland or pasture' was put in practice, the slope farmland with gradients larger than 25 degrees decreased sharply; there is not much farmland left and so there must be some errors in the slope farmland derived from the DEMs with a scale of

1:10,000. The error is mainly because the land use map with a resolution of 30 m cannot provide as many details about the surface of the Earth as DEMs with resolutions of 1:10,000 and there are many sub-pixels that cannot be classified by the land use map, but the distribution of real slopes has great impact on the real distribution of farmland (Figure 5).

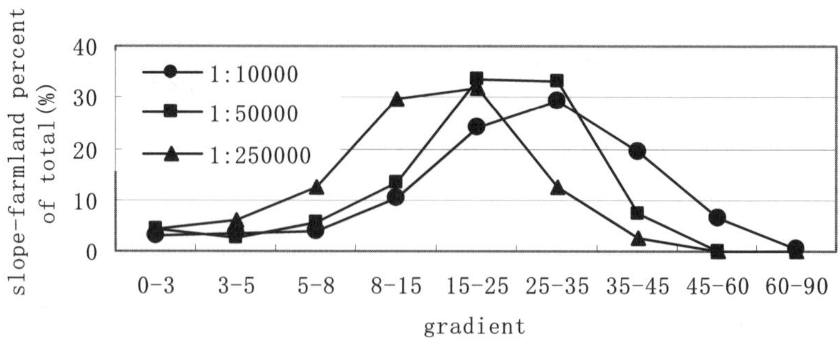

Figure 5. A comparison of slope-farmland derived from DEMs of different scale.

(3) Statistical data were collected to check the accuracy of slope farmland derived from DEMs of different scales (Table 3). Correlation coefficients between the slope farmland and the statistical data were calculated. It was found that the correlation coefficient of the 1:50,000 DEM is the highest one, reaching 0.92, very close to the distribution in the real world. Comparing the statistical data with calculated results, the area of slope farmland derived from the DEM of 1:50,000 scale is similar to the statistical data (Figure 6), and the area of slope farmland derived from the DEM of 1:10,000 scale is less precise than that derived from the DEM of 1:50,000 scale; the spatial resolution of the DEM of 1:250,000 is too low to calculate slope farmland.

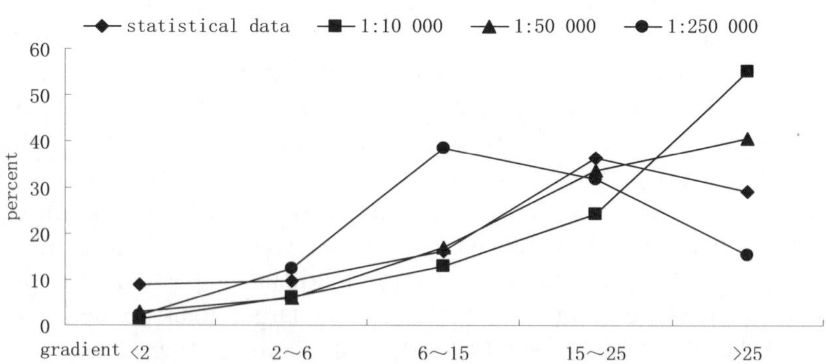

Figure 6. The ratio of slope farmland derived from different DEMs.

Table 3. The comparison of statistical data and slope farmland derived from DEMs.

	< 2	*2-6*	*6-15*	*15-25*	*> 25*	*correlation coefficient*
statistical data	8.82	9.73	16.21	36.19	29.04	1
1:10,000	1.40	6.32	12.84	24.19	55.24	0.73
1:50,000	2.90	6.03	16.82	33.62	40.64	0.92
1:250,000	2.20	12.45	38.42	31.71	15.22	0.42

5 Conclusion

Results showed that the DEM with a scale of 1:50,000 (with a resolution of 25 m) was basically enough for extracting slope fields with the required precision when Landsat TM5 was used for land use classification in the large areas of slope farmland. Although the accuracy of slope extracting with the scale of 1:10,000 is higher than other scale DEMs, the detailed land appearance descriptions did not show obvious improvements in the slope field precision because of the special land use of the studied area, especially the effect of land use type on plantation layout (plantation focused on Loess Mao with small slopes and large areas). Furthermore, with the constraints of using the resolution of 30 m for the land use, DEMs with the scale of 1:10,000 could not show its advantage of high precision, and with increased redundancy and computing load was not appropriate for obtaining slope farmland of large areas such as the Loess Plateau of northern of Shaanxi province. The DEM with a scale of 1:250,000 could not be used to extract slope fields effectively either because of its large generalization over the land layout of the studied area.

Although the DEMs with the scale of 1:50,000 can satisfy the needs of slope farmland investigations of large areas, to obtain more precise slope farmland maps, the land use result must be detailed and higher spatial resolution remote sensing images should be used to classify the land use types more precisely; for example, Quickbird images would be a good data source. If the resolution of the land use map can be improved to 5 m or better, higher DEM resolution is also needed, and in that case, the DEM with a scale of 1:10,000 will show its advantage of higher precision and more detailed features. On the other hand, it is feasible that DEMs with a scale of 1:10,000 could be used to improve the supervised classification.

References

Bolstad, P.V., Stowe, T., (1994), An evaluation of DEM accuracy - elevation, slope, and aspect, *Photogrammetric Engineering and Remote Sensing*, **(60)**11: 1327–1332.

Burrough, P. A. 1986. Principles of Geographical Information Systems for Land Resources Assessment, Oxford: Clarendon Press.

Deng Fengdong, (2002), The Application of Expert Classification System, Dynamic Monitor of Ecology Environment, *SPIE Third International Asia-Pacific Symposium*, Hangzhou.

Chang, K.T. and Tsai, B.W., (1991), The effect of DEM resolution on slope and aspect mapping, *Cartogr. Geogr. Inf. Systems,* **18**: 69–77.

Florinsky, I.V., (1998), Accuracy of local topographic variables derived from digital elevation models, *Int. J. Geogr. Inf. Sci.*, **12**: 47–61.

Goodchild, M.F., Buttenfield, B.P. and Wood, J., (1994), Introduction to visualizing data quality, In Hearshaw, H.M. and Unwin, D.J. (eds.): *Visualization in Geographic Information Systems,* New York: John Wiley and Sons: 141–149.

Hunter, G.J., Caetano, M. and Goodchild, M.F., (1995), A methodology for reporting uncertainty in spatial database products, *Journal of the Urban and Regional Information Systems Association,* **7**: 11–21.

Moore, I.D., Gessler, P.E., Nielsen, G.A. and Peterson, G.A., (1993), Soil attribute prediction using terrain analysis, *Soil Science Society of America Journal*, **57**: 443–452.

Tang Guoan, (2000), *A Research on the Accuracy of Digital Elevation Models*, Beijing and New York: Science Press.

Tang Guoan, (2001), Research on Accuracy of Slope Derived From DEMs of Different Map Scales. *Bulletin of Soil and Water Conservation,* **(21)**1: 53–56.

Theobald, D.M., (1989) Accuracy and bias issues in surface representation, In Goodchild, M., and Gopol, S. (eds.): *Accuracy of Spatial Databases*, Bristol, PA: Taylor and Francis: 99–106.

The Impact of Neighbourhood Size on Terrain Derivatives and Digital Soil Mapping

ZHU A-Xing, James E. BURT, Michael SMITH, WANG Rongxun and GAO Jing

Abstract

Slope gradient, slope aspect, profile curvature, contour curvature, and other terrain derivatives are computed from digital elevation models (DEMs) over a neighbourhood (spatial extent). This chapter examines the combined effect of DEM resolution and neighbourhood size on computed terrain derivatives and its impact on digital soil mapping. We employed a widely used regression polynomial approach for computing terrain derivatives over a user-specified neighbourhood size. The method first fits a least squares regression polynomial to produce a filtered (generalized) terrain surface over a user defined neighbourhood (window). Slope gradient, slope aspect, profile, and contour curvatures are then computed analytically from the polynomial. To examine the effects of resolution and neighbourhood, we computed terrain derivatives using various combinations of DEM resolution and neighbourhood size and compared those values with corresponding field observations in two Wisconsin watersheds. In addition, we assessed the effects of resolution and neighbourhood in the context of knowledge-based digital soil mapping by comparing soil class (series) predictions with observed soils. Our results show that a neighbourhood size of 100 feet produces the closest agreement between observed and computed gradient values, and that DEM resolution has little impact on the agreement. Both profile curvature and contour curvature are more sensitive to neighbourhood size than slope gradient, and sensitivity is much higher at small neighbourhood sizes than at large neighbourhood sizes. Because of the importance of terrain derivatives in the knowledge base, predictive accuracy using a digital soil mapping approach varies strongly with neighbourhood size. In particular, prediction accuracy increases as the neighbourhood size increases, reaching a maximum at a neighbourhood of 100 feet and then decreases with further increases in neighbourhood size. DEM resolution again does not seem to impact the accuracy of the soil map very much. This study concludes that, at least for knowledge-based soil mapping, DEM resolution is not as important as neighbourhood size in computing the needed terrain derivatives. In other words, assuming the DEM resolution is sufficient to capture information at the optimum neighbourhood size, there is no predictive advantage gained by further increasing DEM resolution.

Keywords: slope gradient, DEM, SoLIM, digital soil mapping, neighbourhood size, DEM resolution.

1 Introduction

Terrain derivatives, such as slope gradient, slope aspect, profile curvature, and contour curvature computed from digital elevation models (DEM), are among the key inputs to many geographic analyses using geographic information systems (GIS). These terrain derivatives are computed over a neighbourhood (spatial extent) and are certainly impacted by the size of neighbourhood used. The most popular method for computing terrain characteristics is perhaps the 3 x 3 roving window method used in popular software such as ESRI ArcGIS and TAPES (Moore 1992). It computes terrain attribute values over a 9 pixel square area with 3 pixels on each side; thus the spatial extent over which the values are computed is tied to the resolution of the DEM and changes as the resolution changes. As the resolution increases, the neighbourhood size decreases. Thus, terrain attribute values derived using such methods will change when the DEM resolution changes. The effect of DEM resolution (as a surrogate to neighbourhood size) on the spatial pattern of terrain derivatives has been studied by many authors (Chang and Tsai 1991, Wolock and Price 1994, Zhang and Montgomery 1994, Gao 1997, Goyal *et al.* 1998, Chaplot *et al.* 2000, Schoorl *et al.* 2000, Wilson *et al.* 2000, Thompson *et al.* 2001, McMaster 2002). These studies have generally concluded that as cell size increases, slope gradients tend to decrease, ranges in curvatures decrease, flow-path lengths tend to decrease and the accuracy of terrain attributes at particular locations tends to decrease.

There is an obvious problem with tying neighbourhood size directly to DEM resolution (Hodgson 1995, Wood 1996). There is no physical process based significance behind using a 3 x 3 window or any other particular number of cells for computing terrain attributes. For example, when the resolution of the DEM is 0.5 metre, there is no *a priori* reason for the slope gradient and aspect to be computed over a 1.5 by 1.5 metre area. Furthermore, when field natural scientists (such as soil scientists, geomorphologists and ecologists) are asked to define the slope gradient at a location for a particular phenomena, the slope gradient is often estimated over some characteristic distance, which the field scientist believes to be significant to the process under study, and in effect smoothes short-scale terrain complexity. Thus, it may be a mistake to compute terrain attributes over a neighbourhood tied to DEM resolution because the neighbourhood used by domain experts for slope gradient may be very different from that of the DEM resolution. Wood (1996) proposed calculating terrain derivatives using a varying window size that can be changed based on scale of interest. In other words, for a given DEM resolution, various neighbourhood sizes can be used to compute terrain derivatives depending on the scale of interest. Subsequently, several researchers have examined the effects of neighbourhood size on computed terrain derivatives and the application of varying neighbourhood size as a means of studying geomorphic features at multiple scales (Fisher *et al.* 2004, Schmidt and Hewitt 2004, Schmidt and Andrew 2005, Smith *et al.* 2006).

The work reported here investigates two suppositions. First, we believe that when terrain derivatives are used in GIS-based analysis, one needs to match the neighbourhood size, not the DEM resolution, with the scale of the physical process or phenomenon of interest. For example, if one has a 1-metre DEM, one should not necessarily compute derivatives on a metre scale. If the relevant physical processes are known to operate on a scale of 100 metres, the optimum neighbourhood will be of that size, not 1 metre. Second, we expect that if the neighbourhood is set appropriately, similar terrain values will be obtained from DEMs of any resolution up to the neighbourhood size. Thus if the appropriate neighbourhood is 100 metres, one would hope that DEM resolutions of 1 m, 5 m, 10 m … 100 m would give approximately the same results. Both of these conjectures (an optimum neighbourhood size and insensitivity to DEM resolution) are examined in what follows. Our goals are: (1) to examine the effect of neighbourhood size and DEM resolution on terrain derivatives computed from a DEM; and (2) to evaluate the impact of neighbourhood size on digital soil prediction as an example of an application that relies heavily on digital analysis of terrain. Section 2 describes the method used for computing the terrain derivatives, which is followed by the experimental design for our study. Section 4 describes the study area and the data sets used. Section 5 presents the results and discussion of these results. Section 6 summarizes the findings of this study.

2 Methods

For this study, we employed a least squares regression polynomial approach for computing terrain derivatives over a user-specified neighbourhood size as described in Wood (1996). The method first creates a least squares regression polynomial to produce a filtered (generalized) terrain surface over a user defined neighbourhood (window) (see Shary *et al.* (2002) and Schmidt *et al.* (2003) for discussions on polynomial methods). As is standard practice, a 2nd degree polynomial is used here (Evans 1980):

$$z = rx^2 + ty^2 + sxy + px + qy + u \tag{1}$$

The coefficients *p,r,s ...u* are found by moving a window of user-specified size across the DEM and minimizing the squared difference between the polynomial and the elevation values within this window (or neighbourhood area). This procedure is repeated for every elevation point, and thus z is considered a local polynomial. At every point, the polynomial is differentiated analytically to obtain slope, curvatures, and any other required values. This technique suppresses short-range variation at spatial scales smaller than the neighbourhood size, regardless of DEM resolution. Permitting the user to specify the neighbourhood size provides for control over the amount of short-scale variation in the analysis. In this implementation, the neighbourhood size is defined as the distance from the centre of the central pixel to the window edge (in some way it is like a radius, not a diameter). Other studies have shown that this method produces more accurate terrain derivatives than other common methods (e.g. Florinsky 1998).

3 Experimental Design

To investigate the effect of neighbourhood size and DEM resolution on terrain derivatives, we computed the slope gradient, profile curvature, and contour curvature using the above method at different neighbourhood sizes (from 10 ft to 300 ft) for DEM resolutions of 10 ft, 20 ft, 30 ft, 40 ft and 50 ft. Obviously, the DEM resolution defined the smallest possible neighbourhood for that DEM. Thus 20 ft neighbourhoods were possible only with DEM resolutions of 10 and 20 ft; the 30 ft and larger resolutions do not permit a neighbourhood of 20 ft. Field measurements of slope gradient for 81 sites were made by local soil scientists as part of another study. The field measurements were compared to the computed slope gradient values at these sites to examine the effect of neighbourhood size on computed terrain derivatives.

The computed terrain attributes (slope gradient, profile curvature, and contour curvature) were used together with other environmental variables related to soils (such as geology) as inputs to a knowledge-based digital soil mapping approach, the Soil Land Inference Model (SoLIM) (Zhu *et al.* 2001), for soil mapping. For a given location, the SoLIM approach combines information on soil environmental conditions (e.g. terrain derivatives, remote sensing data, and geology information) for a location with the knowledge of soil-environment relationships to predict the soil conditions at the location. To examine the impact of neighbourhood size on digital soil mapping, we held constant both the knowledge-base and the non-terrain data. We changed only neighbourhood size, which in turn gave varying terrain derivatives. Thus with each neighbourhood size for each DEM resolution, we obtain a version of the soil map using the SoLIM approach based on terrain that has been generalized at that neighbourhood size. For this exercise, we employed neighbourhood sizes ranging from 10 ft to 180 ft for DEM resolutions of 10 ft, 15 ft, and 30 ft. Field soil samples were collected on a hillslope in Dane County, Wisconsin to assess the accuracy of SoLIM-predicted soils under each neighbourhood size. In this way we obtained information about the impact of neighbourhood size on the accuracy of digital soil mapping.

The current implementation of SoLIM requires knowledge of the soil-environment relationships to be manually translated into a digital representation (Smith *et al.* 2006, Liu and Zhu accepted). The translated knowledge needs to be verified, which is achieved through subjective verification of preliminary inference results by soil scientists. If changes are suggested by soil scientists, the translated knowledge is revised based on their suggestions. The purpose of this validation is to ensure that the knowledgebase used for prediction agrees with the scientists' conception of soil-environment relations. This process continues until the soil scientists are satisfied with the preliminary inference result. In this study, we used results from the 10 ft DEM with 10 ft neighbourhood size for verification. Although the field validation data set is not used in the verification process, it is expected that predictions using the 10 ft DEM will be more accurate than those based on other DEM resolutions.

4 Study Area and Data

The study was conducted at two sites for which field data were available. Both sites are located in Dane County, Wisconsin, USA (Figure 1). The sites are in the "Driftless Area" of Wisconsin. The Driftless Area is part of

Wisconsin that did not experience direct glacial till deposition during Laurentide Ice Sheet advances. The landscape of the Driftless Area is characterized by plateaus or erosional remnants of dolostone (the Galena Formation) overlying sandstone scarps (the St. Peter Formation), which together form a branching network of valleys and ridges (Clayton and Attig 1997). Elevations over these sites range from 250 m in the bottomland drainage-ways to about 350 m on the upland ridge tops. Slope gradient ranges from 0–1% for summits and bottomlands to near 25% on the back slopes. Watershed 58, about 12,000 acres, is undergoing a detailed validation test for soil survey at 1:24,000 scale. Ultimately, about 500 field measurements of slope gradient will be collected. At the time of this study, 81 points were already available and form the basis for the analysis that follows. The slope gradients measured in this watershed were compared with the computed slope gradient values. A detailed soil survey (about 1:12,000 scale) using the SoLIM approach was completed for the Thompson Farm (about 160 acres) and soil information at 43 field sites was obtained for validation. Unfortunately, the slope gradient values at these 43 sites were not accurately measured. Thus, this site was used only for examining the impact of neighbourhood size on the accuracy of digital soil mapping.

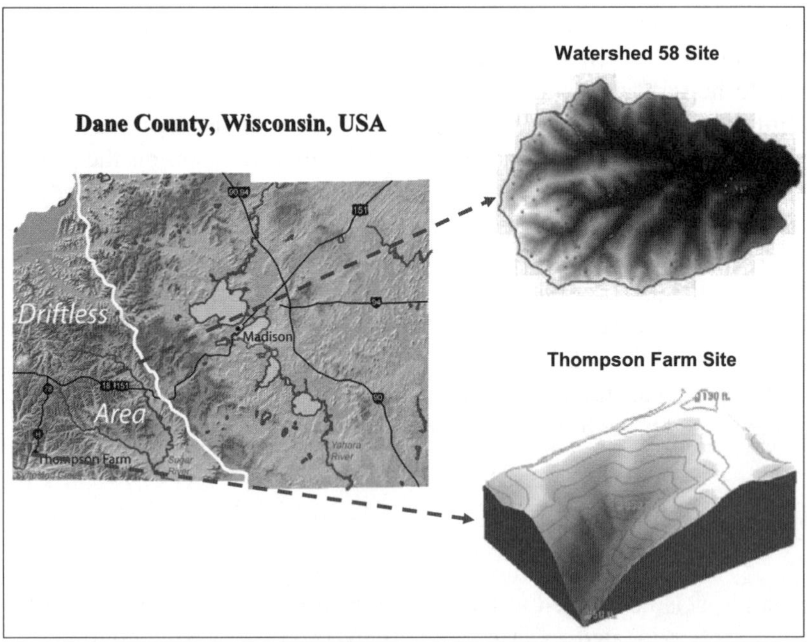

Figure 1. Locations of the study sites.

The 10 ft DEMs for both sites were derived from air photography of Dane County flown in 2000. The 15 ft DEM was created by resampling the 10 ft DEM using the nearest neighbour approach while the 20 ft, 40 ft and 50 ft DEMs were derived by resampling the 10 ft DEM through averaging. The 30 ft DEM was derived from 1995 air photography of Dane County. The quality of the 30 ft DEM is much better than the 10 metre DEM produced by the U.S. Geological Survey owing to the addition of extra spot elevations in the DEM generation process.

5 Results and Discussion

5.1 Comparison with field observation

In examining the effect of neighbourhood size on the computed terrain derivatives, we compute the RMSE (root mean squared error) between the computed slope gradient values and the observed slope gradient values at the 81 sites for each neighbourhood size. We then plot the RMSE against the corresponding neighbourhood size for each DEM resolution as seen in Figure 2. Two important points can be made from this figure. First, the RMSE is smallest for a neighbourhood size between 100 ft and 110 ft. This suggests that in investigating soil properties and conducting soil survey in the Driftless area, soil scientists seem to use a neighbourhood size about 100 ft. This confirms our supposition that for a given application one should not blindly accept DEM resolution as the best choice for neighbourhood size. The optimum value can be quite different, as seen in the figure. The optimum value is no doubt application-specific. Table 1 shows the ME (Mean Error) and RMSE for the slope gradient computed using ArcGIS (neighbourhood size tied to resolution) and for slope gradient computed using a neighbourhood size close to the optimum. It is clearly seen that fixed neighbourhood approaches (such as that used in ArcGIS) for computing slope gradient over-estimate the slope gradient values for applications at 1:24,000 scale.

We must emphasize that this comparison implies that physically meaningful slope gradient values are scale-specific and are sensitive to neighbourhood size. We are not concluding that using a neighbourhood size based on resolution will always produce “wrong” slope gradient values, nor are we implying that the optimum neighbourhood is always 100 ft. Rather, we argue that tying neighbourhood size to DEM resolution cannot guarantee that computed slope gradients are optimum. In this example, for soil mapping at 1:24,000 scale, the optimum value is about 100 ft over this area with moderate relief, but other applications at other scales might have very different optima.

The second point is that the resolution of the DEM does not seem to have much impact on the RMSE for the resolutions examined. This could have significant implication for digital soil survey because this suggests that the resources used to obtain finer DEMs may not be cost-effective after achieving resolution able to resolve features at the optimum neighbourhood scale. For example, if processes occurring at scales smaller than 100 ft do not affect the phenomenon of interest, there is no advantage in using a DEM with 10 ft, 20 ft or other similarly high resolution.

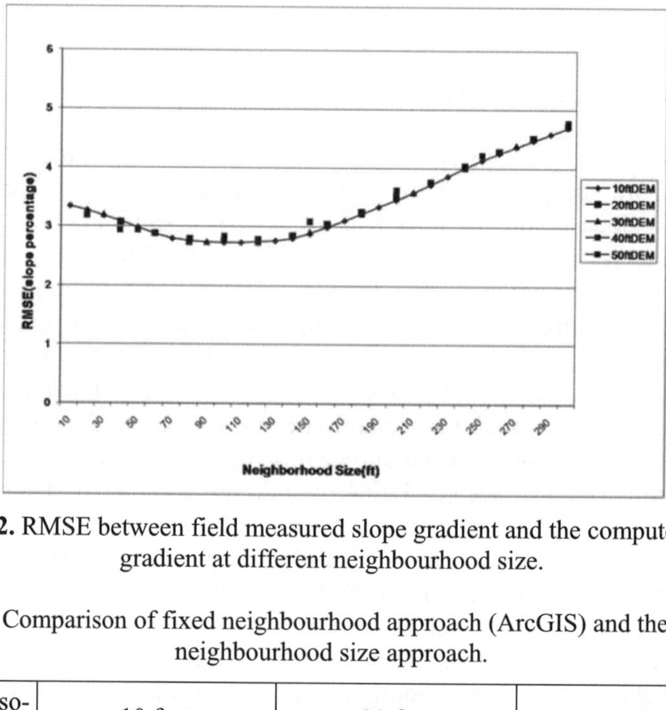

Figure 2. RMSE between field measured slope gradient and the computed slope gradient at different neighbourhood size.

Table 1. Comparison of fixed neighbourhood approach (ArcGIS) and the variable neighbourhood size approach.

DEM Resolution	10 feet		20 feet		30 feet	
Method	ArcGIS	Variable at 100 feet	ArcGIS	Variable at 100 feet	ArcGIS	Variable at 90 feet
ME	0.94	0.075	0.923	0.041	0.864	0.197
RMSE	3.422	2.726	3.188	2.764	3.28	2.739

5.2 Sensitivity to neighbourhood size

To further examine the neighbourhood size sensitivity and to compare the sensitivity among different variables (slope gradient, profile curvature, contour curvature), we will use two concepts: "standardized magnitude" and "relative change". The standardized magnitude concept measures individual deviations from the mean using the ratio of the individual value to the mean. For example, the standardized magnitude for slope gradient at a given point for a given neighbourhood size is the ratio of the gradient value at that point for the current neighbourhood size over the mean of the slope gradient values at that point for all neighbourhood sizes. Values of standardized magnitude far from unity mean greater deviation from the mean. Because standardized magnitude is dimensionless, values for one terrain variable can be compared with those of other terrain variables, which provides a way to assess relative sensitivity of different terrain derivatives to neighbourhood size. We must point out that standardized magnitude may not be appropriate for locations where the mean value approaches zero. The "relative change" concept measures the difference in standardized magnitude between two neighbourhood sizes at a point. It therefore provides an alternative way to characterize the sensitivity across neighbourhood size. In other words, it allows us to identify neighbourhood sizes to which the terrain conditions are more or most sensitive.

Figure 3 shows the standardized magnitudes for the three terrain variables across the neighbourhood sizes examined at two field points. This figure clearly shows that curvature measures are much more sensitive to neighbourhood size than slope gradient. At some neighbourhood sizes, the computed curvature values are a few times more than the overall mean while the computed gradient values are about the same as the overall mean. Although there is variability in sensitivity from location to location, the fact remains that curvature measures are more sensitive to neighbourhood size than slope gradient.

Terrain variables are more sensitive to neighbourhood size at small neighbourhood sizes than at large neighbourhood sizes. This is seen in Figure 4, which shows the relative change across different neighbourhood sizes for the two field points. The figure has two important features. The first is that the sensitivity is much stronger at small neighbourhood sizes and generally decreases as neighbourhood size increases. The second is that relative change for the two curvature variables fluctuates much more than that for slope gradient across neighbourhood size. However, the general pattern of sensitivity across neighbourhood size is that terrain variables are more sensitive to neighbourhood size when neighbourhood size is small and less sensitive when neighbourhood size is large. This makes the neighbourhood size selection for applications that require terrain information over small spatial scale (large map scale) much more critical. To obtain a good approximation of terrain information at a small spatial scale, one almost needs an exact match between the neighbourhood size used and the desired spatial scale. At large spatial scales, this exact match may not be necessary.

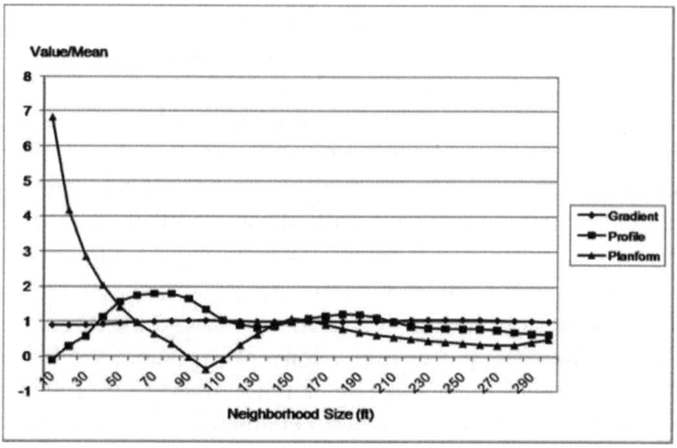

Figure 3(a)

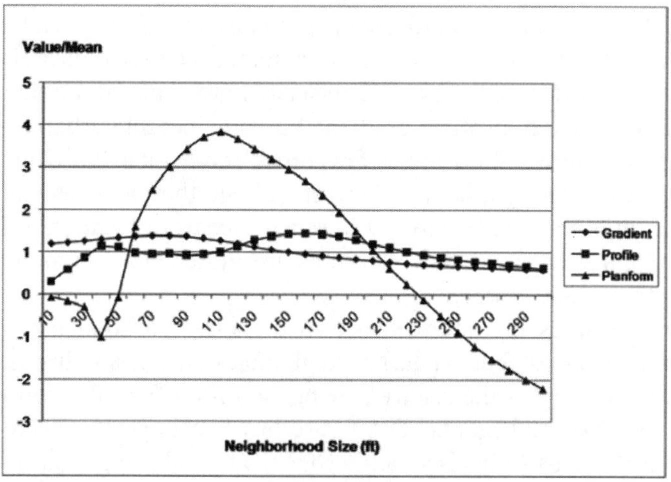

Figure 3(b)

Figure 3. Standardized magnitude across neighbourhood size (with DEM resolution fixed at 10 ft): (a) for Field Point 49; (b) for Field Point 105.

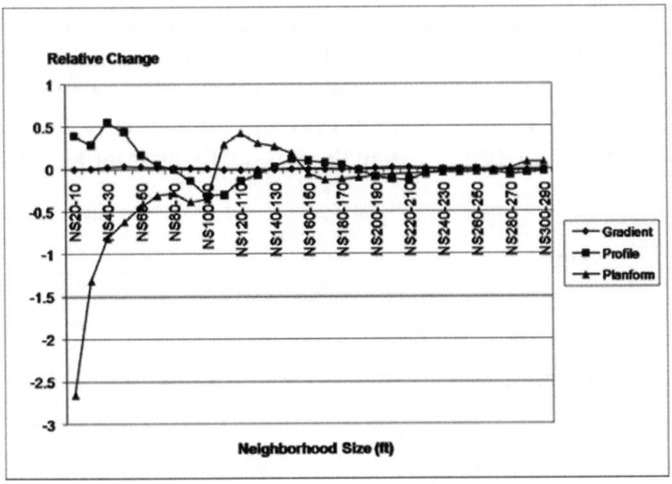

(a)

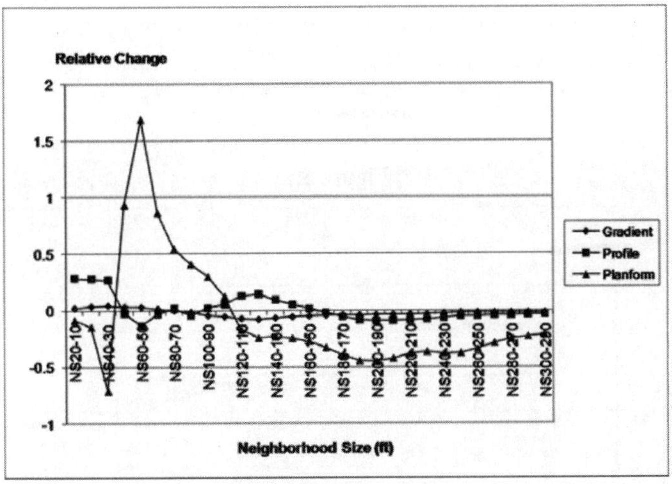

(b)

Figure 4. Relative change in standardized magnitude across neighbourhood size (with DEM resolution fixed at 10 ft): (a) Field Point 49; (b) Field Point 105.

With a variable window approach, terrain variables are less sensitive to DEM resolution. Figure 5 shows the computed slope gradient and curvature values at neighbourhood size about 150 ft across DEM resolution. As can be seen from Figure 5, the standardized magnitude is much closer to unity across all resolutions than what was observed with neighbourhood size (Figure 4). As before, we see that the deviation for the curvature variables is higher than for slope gradient. Another observation is that there is seemingly no consistent trend in how the standardized magnitude changes with DEM resolution. This, along with the evidence seen in Figure 2, leads us to believe that when a varying window approach is used for computing terrain derivatives, DEM resolution does not play a major role as long as the resolution of the DEM is within the physically appropriate scale.

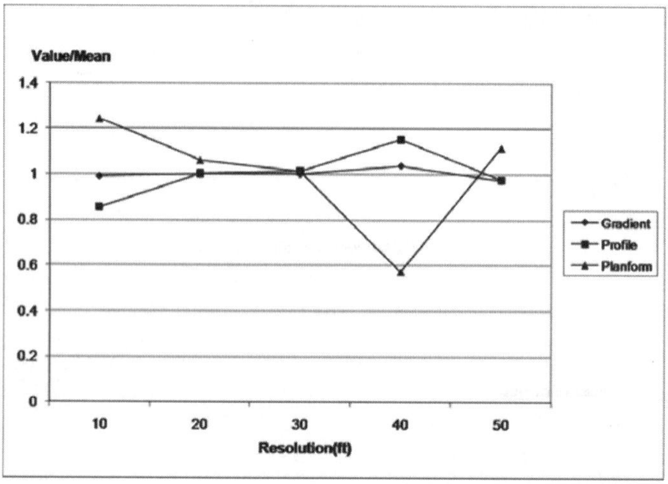

Figure 5(a)

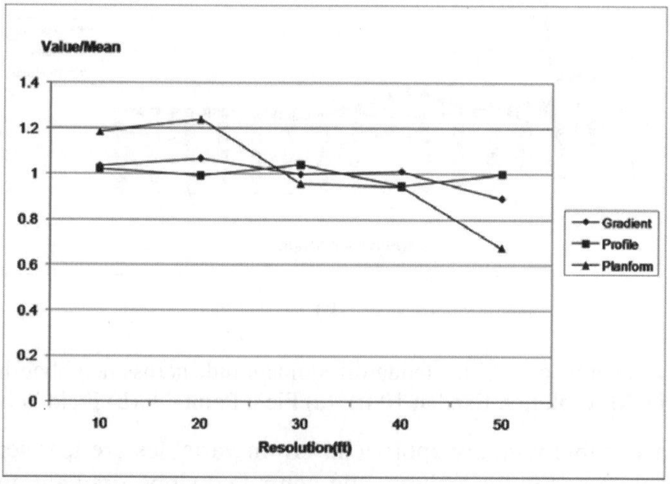

Figure 5(b)

Figure 5. Standardized magnitude values across DEM resolution (with neighbourhood size controlled around 150 ft): (a) Field Point 49; (b) Field Point 105.

5.3 Impact on Digital Soil Mapping

The relationship between the accuracy of digital soil predictions and neighbourhood size is shown in Figure 6. It is clear that neighbourhood size has profound impact on the accuracy of the soil map. The difference in accuracy between different neighbourhood sizes can be quite substantial, with the accuracy at one neighbourhood size perhaps double that at another. It is important to note that the most accurate soil map is not obtained at the smallest neighbourhood size. The accuracy peaks at the neighbourhood size around 100 feet. Note that the somewhat high accuracy for the 10 ft DEM at the 10 ft neighbourhood size is related to the verification process. Still, this accuracy is lower than that using a 100 ft neighbourhood size. The finding here further suggests that removing certain fine scale variations in the DEM is important for digital soil mapping because these fine scale details do not contribute to the differentiation of soil at the scale of interest to the soil scientists. This finding confirms what has been reported by Smith *et al.* (2006).

The difference in accuracy among the different DEM resolutions is very small. This suggests that DEM resolution does not have significant impact on the accuracy of soil maps. This can be easily understood from what was discovered above. As shown in the slope gradient analysis, DEM resolution does not have much impact on the slope gradient value and it therefore cannot have significant impact on the accuracy of the soil map. A minor observation can be made about the pattern present in Figure 6. Although the DEM does not seem to have significant impact on the accuracy of the soil map, the coarse DEM seems to have higher accuracy than the finer DEM for neighbourhood sizes below the optimal neighbourhood size. The reason is that the coarse DEM smoothes out the fine details, which are not important to soil formation at the scale under study.

6 Conclusions

This study examines the effect of neighbourhood size on terrain derivatives computed from DEMs and its impact on the accuracy of digital soil mapping. Although the study is conducted at two different sites with different foci, similar findings were obtained. First, slope gradient is less sensitive to neighbourhood size than the two curvature variables. Second, terrain derivatives are more sensitive to neighbourhood size when neighbourhood size is small and the sensitivity decreases as neighbourhood size increases. Third, terrain derivatives are less sensitive to DEM resolution than to neighbourhood size. Fourth, in the context of soil mapping, the optimal neighbourhood size for determining slope gradient in areas with moderate relief is about 100 feet. Fifth, the accuracy of soil predictions based on terrain derivatives is strongly related to the neighbourhood size used to compute these terrain derivatives. The impact of DEM resolution on the accuracy is rather insignificant as long as the DEM resolution is finer than the optimal neighbourhood size. These findings have important implication for digital soil mapping where terrain derivatives are the key inputs. Future studies should examine how the optimal neighbourhood size changes with respect to different landscape conditions (such as different levels of relief) and should include more DEM resolutions.

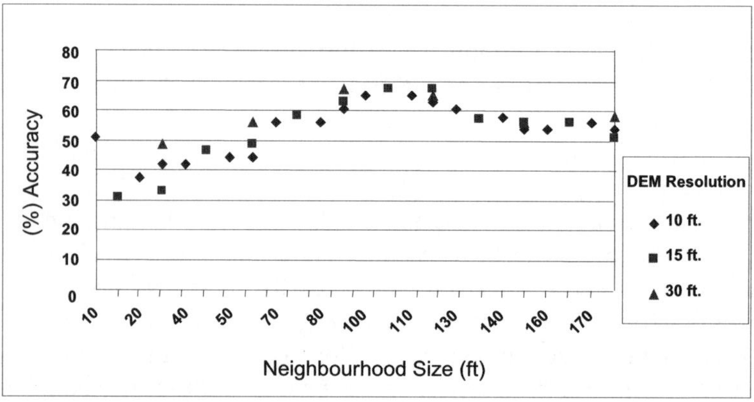

Figure 6. Accuracy of digital soil mapping and neighbourhood size.

Acknowledgements

This study was funded through support from the Chinese Academy of Sciences International Partnership Project "Human Activities and Ecosystem Changes" (Project No.: CXTD-Z2005-1) and the 'Hundred Talents' Program of Chinese Academy of Sciences. Support from the Natural Resources Conservation Service, U.S. Department of Agriculture is also greatly appreciated. We thank Amanda Moore for her assistance in collecting field data over the Thompson Farm site.

References

Chang, K. and Tsai, B., (1991) The effect of DEM resolution on slope and aspect mapping, *Cartography and Geographic Information Systems*, **18**: 69–77.

Chaplot, V., Walter, C. and Curmi, P., (2000) Improving soil hydromorphy prediction according to DEM resolution and available pedological data, *Geoderma*, **97**: 405–422.

Clayton, L. and Attig, J.W., (1997), Pleistocene Geology of Dane County, Wisconsin, *Wisconsin Geologic and Natural History Survey Bulletin #95*: 3.

Evans, I.S., (1980), An integrated system of terrain analysis and slope mapping, *Zeitschrift fur Geomorphologie*, Suppl. Bd. Vol. 36: 274–295.

Fisher, P., Wood, J. and Cheng, T., (2004), Where is Helvellyn? Multiscale morphometry and the mountains of the English Lake District, *Transactions of the Institute of British Geographers,* **29**: 106–128.

Florinsky, I.V., (1998), Accuracy of local topographic variables derived from digital elevation models, *International Journal of Geographical Information Science*, **12**: 47–61.

Gao, J., (1997), Resolution and accuracy of terrain representation by grid DEMs at a micro-scale, *International Journal of Geographical Information Science*, **11**: 199–212.

Goyal, S.K., Seyfried, M.S. and O'Neill, P.E., (1998), Effect of digital elevation model resolution on topographic correction of airborne SAR, *International Journal of Remote Sensing*, **19**: 3075–3096.

Hodgson, M.E., (1995), What cell size does the computed slope/aspect angle represent? Photogrammetric Engineering and Remote Sensing, **61**: 513–517.

Liu, J. and Zhu, A.X., accepted. Automated Soil Mapping Using Linguistic Soil-landscape Knowledge, *Soil Science Society of America Journal.*

McMaster, K.J., (2002), Effects of digital elevation model resolution on derived stream network positions, *Water Resources Research*, **38**: 1–9.

Moore, I.D., (1992), Terrain analysis program for environmental sciences (TAPES), *Agricultural Systems and Information Technologies*, **(4)**2: 37–39.

Schmidt, J., Evans, I.S. and Brinkmann, J., (2003), Comparison of polynomial models for land surface curvature calculation, International Journal of Geographical Information Science, **17**: 797–814.

Schmidt, J. and Andrew, R. (2005), Multi-scale landform characterization, *Area,* **(37)**3: 341–350.

Schmidt, J. and Hewitt, A.E., (2004), Fuzzy land element classification from DTMs based on geometry and terrain position, *Geoderma,* **121**: 243–256.

Schoorl, J.M., Sonneveld, M.P.W. and Veldkamp, A., (2000), Three-dimensional landscape process modeling: the effect of DEM resolution, *Earth Surface Processes and Landforms*, **25**: 1025–1034.

Shary, P.A., Sharaya, L.S. and Mitusov, A.V., (2002), Fundamental quantitative methods of land surface analysis, *Geoderma*, **107**: 1–32.

Smith, M.P., Zhu, A.X., Burt, J.E. and Styles, C., (2006), The Effects of DEM Resolution and Neighbourhood Size on Digital Soil Survey, *Geoderma*, (in press).

Thompson, J.A., Bell, J.C. and Butler, C.A., (2001), DEM resolution: effects on terrain attribute calculation and quantitative soil-landscape modelling, *Geoderma*, **100**: 67–89.

Wilson, J.P, Repetto, P.L. and Snyder, R.D., (2000), Effect of data source, grid resolution, and flow routing method on computed topographic attributes, In: Wilson, J.P. and Gallant, J.C. (eds.): *Terrain Analysis: Principles and Applications,* New York: John Wiley & Sons: 133–161.

Wolock, D.M. and Price, C.V., (1994), Effects of digital elevation model map scale and data resolution on a topography based watershed model, *Water Resources Research,* **30**: 3041–3052.

Wood, J., (1996), Scale-based characterisation of digital elevation models, In Parker, D. (ed.): *Innovations in GIS 3*, London: Taylor & Francis: 163–75.

Zhang, W.H. and Montgomery, D.R., (1994), Digital elevation model grid size, landscape representation and hydrologic simulations, *Water Resources Research*, **30**: 1019–28.

Zhu, A.X., Hudson, B., Burt, J., Lubich, K. and Simonson, D., (2001), Soil mapping using GIS, expert knowledge, and fuzzy logic, *Soil Science Society of America Journal,* **65**: 1463–1472.

The Impact of DEM Error on Predictive Vegetation Mapping

Brian G. LEES, HUANG Zhi, Kimberley VAN NIEL
and Shawn W. LAFFAN

Abstract

Digital elevation models (DEM) are one of the most important data sources for Land Use-Land Cover (LULC) and Predictive Vegetation Mapping (PVM). A number of indices are derived from DEMs and their use depends on the nature of the classification problem and the tool being employed. In some cases it is the practice to pre-classify these prior to modelling. This chapter examines the impact of doing this on the production of a LULC classification, and on the production of a surface, or field, prediction of a single species. Secondly, the error in classification resulting from error in the original DEM is examined to give some comparison. We show that, contrary to widespread practice, leaving the input variables in an unprocessed form is clearly better than almost any of the 'improvements' usually made. This applied to both classification of LULC and to the prediction of a surface, or field, representing a single species. As expected, forest type mapping is likely to be quite sensitive to some level of DEM error. We can see that the DEM error has an uneven impact on the different forest types. Importantly, when increasing the level of DEM error, we found a non-linear decrease in classification performance.

1 Introduction

1.1 Land use and land cover classification

Land use and land cover classification (LULC) is a common practice. Land cover is a critical component of climate and global change models (Brown *et al.* 1993, Wilson and Henderson-Sellers 1985) and land cover changes may be as important as climate change when assessing total human biome impacts (Chapin *et al.* 2000). As a result, global land cover data and databases have been developed (Defries and Townshend 1994, Skole 1994). Digital Elevation Models (DEMs) are a fundamental data source for land use and land cover classification, with DEM derived topographic indices often being used in conjunction with remotely sensed data. The numerous ways of extending point or line data using predictive modelling are reviewed in the comprehensive work by Guisan and Zimmerman (2000). Most use spatial estimates of physical cnvironmcnt variables, based on the assumption that the physical environment controls, at some level, land cover distributions. Many of these predictor variables are either elevation itself, or are derived from a DEM, including slope, aspect, net solar radiation, temperature, and precipitation (see Franklin (1995) for a listing of commonly used measures).

These variables are modelled transformations, derived from data that have some degree of error. We know that this error will be propagated through any subsequent data development and modelling process such as predictive land cover classification and vegetation modelling (Goodchild 1989, Richards 1993, Unwin 1995, Holmes *et al.* 2000, Van Niel *et al.* 2004). However, the way in which these variables are transformed before inclusion in the modelling process also generates an error that is propagated to the subsequent classification or prediction (Lees 1996).

1.2 Aim

In this chapter we look at both of the cases described above, using the same data set to give some comparison of the effects. This study tries to answer the questions: (1) while categorizing input data may reduce error in some cases, what is the overall impact of this process on the models? (2) How sensitive is the same land cover classification to the error and uncertainty of input data?

We looked at three things:

1. The impact of transforming or categorizing a single DEM-derived variable on a decision tree model of forest types is investigated.
2. The impact of categorizing a DEM-derived variable for modelling a single forest tree species using GAM and GLM analysis is explored.
3. The impact of varying degrees of DEM error on a decision tree model of forest types.

2 Analyses

For all analyses, the DEM used was part of the Kioloa GLCTS Pathfinder Site data set (Lees 1999). This is a complex land cover data set of 15.75 km by 15.75 km, with 275,000 pixels (Lees and Ritman 1991). The DEM was developed by digitizing elevation contours (10 m interval), stream lines and spot heights from 1:25,000 scale topographic maps of the region then interpolating the data using IDW. The resolution is 30 m. Van Niel *et al.* (2004) determined that the level of error in the DEM was comparable to DEM error levels found in USGS and British Ordnance Survey DEMs.

The classification scheme consists of seven forest types and two other land cover classes (cleared and ocean) (Table 1). Details of the data set and study location are given in Lees and Ritman (1991) and Huang and Lees (2004; 2005).

2.1 Impact of categorical data on categorical outcomes

The first analysis examines the impact on a forest type decision tree model of transforming or categorizing predictor variables developed from a DEM. The predictor variable chosen was Flow Length, a measure generally used as a surrogate for soil wetness. Each test of the model used variations in this single input variable along with a well-tried suite of companion variables; TM Bands 2, 4 & 7, Elevation, Geology, Slope, Aspect, and Flow Accumulation. Flow Length is roughly analogous to 'position on slope', and Flow Accumulation is roughly analogous to 'catchment'; both have been estimated using the D8 algorithm implemented in ArcMap 9.0.

Table 1. Land use class types and frequencies in the base forest type data set.

Class	*Forest type*	*Cell count*
1	Dry Sclerophyll	76,315
2	E. botryoides	7,375
3	Lower slope wet forest	6,588
4	Wet E. maculata	46,364
5	Dry E. maculata	15,501
6	Rainforest Ecotone	11,568
7	Rainforest	14,063
8	Cleared land	20,213
9	Ocean	77,638

The histogram of Flow Length has large numbers of cells at low wetness values with low numbers of cells at high values. This is usually interpreted as a transition from hillslope drainage to in-channel drainage. Intuitively, it would seem sensible to compress this long tail of high values as it would appear likely to have little impact on our analysis, possibly even improving our discrimination of the hillslope values. In this case, a histogram stretch would be the normal procedure before grey scaling the data for analysis with values assumed to be 'in-channel' saturated at the point at which the histogram levels.

The forest types were then modelled using a decision tree with the predictor variables described above along with Flow Length: (1) without modification, (2) stretched and saturated at 1000 m, (3) stretched and saturated at 300 m, (4) split into six categories defined using equal intervals, and (5) with a histogram stretch and then split into six categories defined using equal intervals. Table 2 shows that all of the changes to Flow Length described above decrease the accuracy of the models.

Table 2. Using the Lees & Ritman (1991) LX930 dataset in a DTA with field data automatically broken into 10 intervals. Correlates DEM and FLOWACC are present. Training and test data sets held constant. Training 'n' = 1000, Test 'n' = 600. Stopping 'n' = 40. The effect of stretching the data can be seen to degrade the accuracy from 81.8% to 72.3%. Using categorical descriptions of 'position on slope' has a similar negative impact.

Degree of pre-processing	*Accuracy (%)*
No modification	81.8
Flowlength saturated at 1,000 m	72.3
Flowlength saturated at 300 m	72.3
Flowlength as 6 classes	73.3
Flowlength (stretched) as 6 classes	73.0

This demonstrates that the form in which the DEM derivatives are presented to the LULC classification model is very important to model outcomes, and that further processing of the predictor variable in every case reduced the model accuracy.

2.2 Impact of categorical data on field models

The question arising from the previous test is, of course, is this effect algorithm specific? Would we find a similar effect using GAM and GLM models of a single tree species that produces a field output (field surface)?

Again, Flow Length is categorized and tested against the field data to compare modelling outcomes. In this analysis, we model the presence/absence response of a single tree species (*Acmena smithii*) in the study area using a generalized additive model (GAM). There is a strong relationship between topographic position and *Acmena smithii*, a Gondwanan relic occurring in gullies.

The categories were: (1) Flow Length without modification, as in the control used in the first test, (2) split into three categories, (3) split into five categories, and (4) split into seven categories. The categories were defined using equal intervals as before, natural breaks and quartiles (Zimmermann 2000). In GAMs, classified data should be modelled as factors because it is not continuous. However, in the literature, indices in GAMs are often modelled with smoothing splines so, despite our reservations about the procedure, we ran both to show the differences. Models were built using all of these alternatives to demonstrate the effect of the variable classification procedures.

70% of the data were used for the model development and a randomly selected 30% were reserved for testing. Analyses were conducted in SPlus.

The results (Table 3) once again show that there is a decreasing effectiveness of the variable with increasing generalization. When modelled appropriately (as factors rather than as splines), accuracy stays the same or degrades. Despite one classified dataset (NB 7) having a better percentage correctly classified (%CC) and area under the curve (AUC), it is clear that the best result is with the variable unclassified.

These results make sense when one considers that, on topography with a typical length of slope of 300 metres for example, the 'crest' might occupy ~100 metres, the 'upper slope' ~70 metres , the 'mid-slope' ~70 metres, the 'lower slope' ~70 metres, and the 'flat' ~100 metres. The actual transition between forest types here can be over a distance of less than 10 metres and so the 'real' uncertainty is magnified seven fold by the use of a categorical hill slope position. Clearly, the solution is to simply use downhill distance in metres.

2.3 Field data for categorical outputs

Finally, we conducted a sensitivity analysis to see how the degree of error in the original DEM data impacts the results. We used the C4.5 decision tree algorithm (Quinlan 1993). In the classification process, 80% of samples were used for training. This tree was used for all the sensitivity analyses in this study. The remaining 20% of samples were used for testing classification accuracy of the base data set and the results of sensitivity analyses. The overall accuracy of the base data set is 0.651.

Table 3. Examining the impact of categorizing the DEM-derived predictor variable topographic position on GAM and GLM models of a single tree species. Again, the variable in question is subjected to different categorization processes, and tested against the field data to compare modelling outcomes. Four tests were carried out. Column three shows the change in model deviation, column four is chi-squared, column five is Kappa, column six is the percentage correctly classified and column seven is the area under the curve. We should emphasize that it is bad practice to process classified data in a spline, but as this appears in the literature we show it.

Model		*Change in model Deviance*	*P(Chi)*	*p-kappa*	*% Correctly Classified*	*AUC*
Unclassified		94.9	0.00001	0.615	87.1	0.869
Smoothing spline	Equal interval 7	87.8	0.00001	0.458	76.1	0.808
	Equal interval 5	79.1	0.78050	0.553	83.9	0.782
	Natural breaks 7	77.4	0.00576	0.691	90.3	0.897
	Natural breaks 5	85.0	0.00074	0.558	85.8	0.811
	Quantile 7	94.9	0.00001	0.553	83.9	0.842
	Quantile 5	99.4	0.00001	0.484	79.4	0.801
As factors	Equal interval 7	87.8	NA	0.458	76.1	0.808
	Equal interval 5	79.2	NA	0.553	83.9	0.782
	Natural breaks 7	78.6	NA	0.615	87.1	0.821
	Natural breaks 5	85.1	NA	0.558	85.8	0.811
	Natural breaks 3	76.5	NA	0.536	82.6	0.775
	Quantile 7	96.5	NA	0.553	83.9	0.842
	Quantile 5	99.4	NA	0.484	79.4	0.801
	Quantile 3	79.4	NA	0.372	70.3	0.738

In this study, we assumed the DEM error is stochastic in nature but with a spatially autocorrelated distribution (Van Niel *et al.* 2004). Also, the DEM error was assumed to be directly associated with (therefore a small percentage of) the nominal elevation value. For an error model, a Monte Carlo simulation approach was used as it is more readily applied to the C4.5 decision tree classifier than mathematical methods such as the Taylor series (e.g. Crosetto e*t al.* 2000, Crosetto and Tarantola 2001, Canters *et al.* 2002, Davis & Keller 1997). It is not possible to build a mathematical error model for a land cover classification process that is not continuously linearly differentiable so we chose the one-at-a-time design (Hamby 1994, Crosetto *et al.* 2000). The one-at-a-time approach is not only simple but also very appropriate for this study.

The approach used in a Monte Carlo analysis is to perturb one or more of the input datasets at some chosen level of random error to generate a different realization of the original dataset. We used 1000 iterations to ensure convergence on a stable solution (Heuvelink 1998), while random values were generated using the Mersenne Twister pseudo-random number generator, as recommended by Van Niel and Laffan (2003).

The sensitivity analysis in this study was conducted by perturbing the DEM using five levels of spatially autocorrelated random error. The five levels are ±2%, ±4%, ±6%, ±8% and ±10% of individual elevation values. The general Monte Carlo procedure used involves the following steps.

1. Extract the elevation values of 1000 randomly selected pixels from the DEM data (this is an average spacing of approximately 17 cells for this data set).
2. Randomly generate error values for the 1000 selected pixels, calculated as a percentage of each elevation value. The percentage is within the chosen error level (e.g. ±2%).
3. Generate a spatially autocorrelated error surface from the 1000 pixels using Kriging interpolation.
4. Generate a perturbed realization of the DEM by adding the error surface to the original DEM.
5. Generate derivative data sets (slope and aspect) from the perturbed DEM.
6. Generate a perturbed forest type map using the previously trained C4.5 decision tree with the perturbed DEM, slope and aspect, and the original geology and Landsat datasets.

Three criteria were used to evaluate the results. We first assessed the overall test accuracy difference between each of the 1000 forest type maps and the base forest type map (the accuracy criterion), and secondly we assessed the number of pixels in the perturbed classification that changed their classes when compared to the base forest type map (the pixels changed criterion). Thirdly, a spatial assessment of the change in pixels was also generated by calculating the frequency with which each pixel changed across all iterations for each error level. To assess the convergence on a stable solution, the mean and standard deviation of the accuracy criterion were assessed as the number of iterations increased.

Figure 1 and Table 4 clearly indicate the magnitude of the mean overall accuracy reduction and that its standard deviation increases with increasing error levels. However, the effect is not linear and becomes more prominent when the uncertainty level increases. Of note, the convergence line of the uncertainty level of ±10% is clearly separate from those of the other four uncertainty levels (Figure 1A and Figure 2). The range of the mean differences over all error levels is less than 0.01.

Figure 3 and Table 5 show the proportion of pixels that change their initial class assignments because of the errors. As expected, the magnitude of the mean number of pixels and its standard deviation increases with increasing error levels, although the mean change is close to 10% for each error level.

Table 4. Summarized results of the sensitivity analyses.

DEM Error Level	*±2%*	*±4%*	*±6%*	*±8%*	*±10%*
Mean difference of overall test accuracy	-0.0263	-0.0265	-0.0268	-0.027	-0.0281
STD of overall test accuracy difference	0.0046	0.0053	0.0061	0.0067	0.0071
Mean number of pixels changed	26925	27016	27414	28038	28807
STD number of pixels changed	371	368	403	461	556

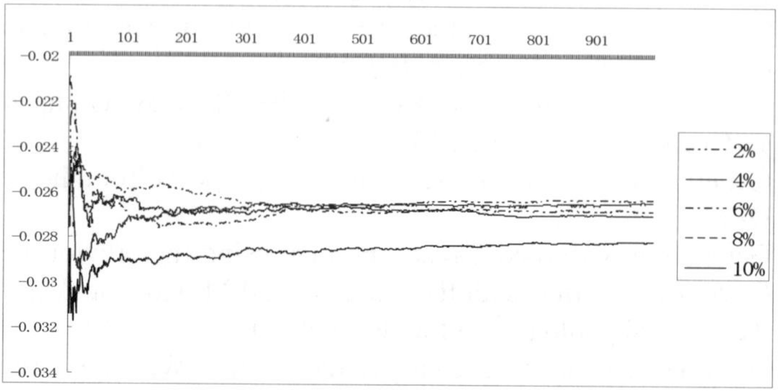

(A)

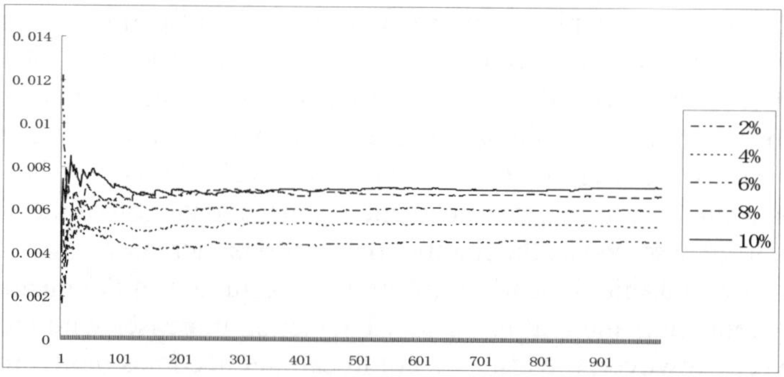

(B)

Figure 1. Convergence of the sensitivity analysis for the test accuracy criterion over the 1000 iterations. (A) Mean difference, (B) STD.

As for the accuracy criterion, the effect is not linear, and becomes more prominent when the uncertainty level increases. The convergence lines in Figure 1A and box plots in Figure 2 show that the uncertainty levels of ±2% and ±4% are very similar, but that there is an increasing difference as the uncertainty level increases to ±10%.

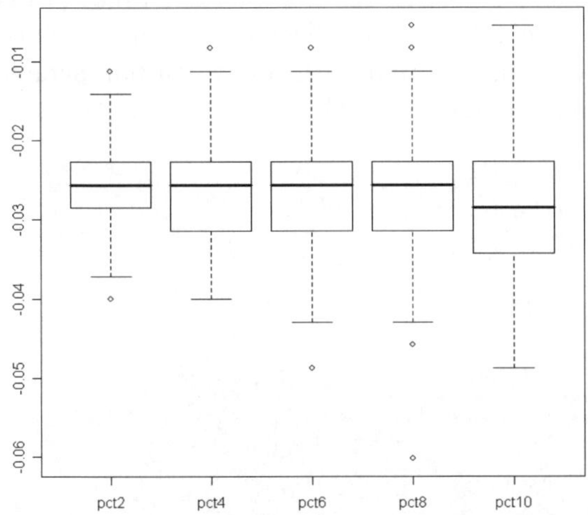

Figure 2. Boxplots for the accuracy criterion.

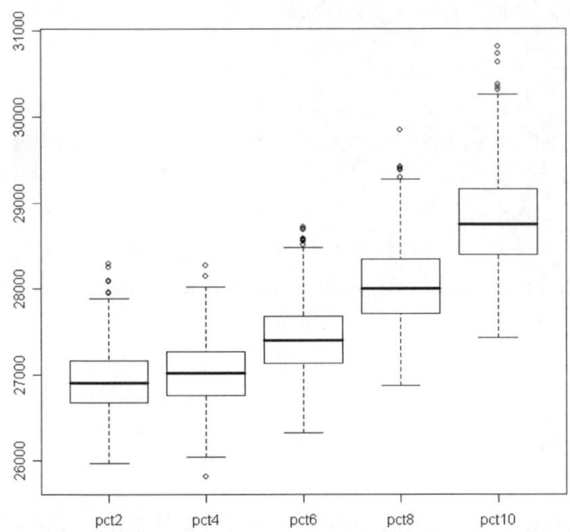

Figure 3. Boxplots for the pixels changed criterion.

While the impact of random DEM error on the results of forest type mapping is not surprising, we do now have an assessment of the extent to which this occurs. For example, the DEM error we used caused approximately 10% of all pixels to change their initial class assignment in each perturbed classification. Additionally, for the ±10% level of error, 36% of all pixels changed at least once across all perturbations, 16% changed 100 or more times, and 13% changed 200 or more times (Table 5). A spatial assessment of the pixel changes indicates that their distribution is primarily within the forest classes (Figure 4), for which non-parametric classifiers like decision trees are most needed.

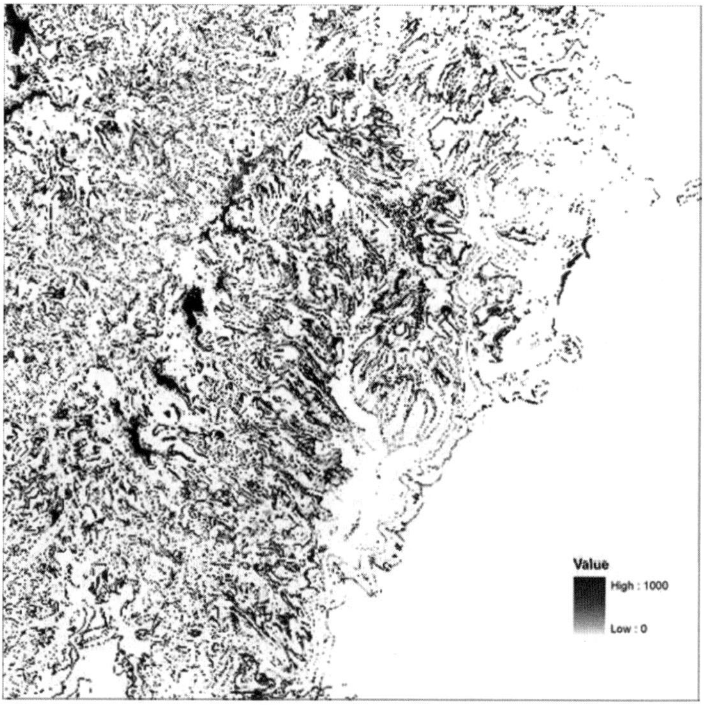

Figure 4. Spatial assessment of pixels changed with the ±10% DEM uncertainty level.

3 Discussion

Error is generated through the models both by the way the elevation data is presented and by the propagation of error in the elevation data itself. The nature of error propagation in an analysis depends both on the primary data set error and on the type of analysis employed. We have shown how a single variable—topographic position—can be a source of error depending upon the way it is presented to the model.

Table 5. Number of class changes for cells in each land cover class when using the ±10% DEM uncertainty level, summarized by class.

Class	*Min*	*Max*	*Mean*	*STD*	*Cells changed once (%)*	*Cells changed >100 times (%)*	*Cells changed >200 times (%)*
1	0	1,000	203	346	67.0	30.4	24.4
2	0	1,000	566	417	89.7	69.1	65.1
3	0	1,000	348	442	63.1	42.3	39.0
4	0	1,000	36	149	34.6	6.8	4.7
5	0	1,000	108	257	55.9	18.8	14.6
6	0	1,000	167	315	50.7	26.6	21.6
7	0	1,000	74	167	41.7	19.0	13.4
8	0	867	7	60	3.9	1.7	1.3
9	0	542	5	50	1.8	1.1	1.0

The use of categorical labels for phenomena with monotonically varying characteristics is an obstacle to both classification and (machine) learning. A little thought shows that such a mismatch between data model and data structure is clearly a potential source of error. For a more obvious example, the relationships between North, South, East and West are not inherently apparent to a machine learning system or classifier. However, recoding the directions as degrees makes the relationships clear. It also allows intermediate values to be fitted into the schema. This is one of the considerable advantages that the ratio and interval data types have over the nominal data type in learning systems. However, merely reclassing nominal data to ratio or interval data is not a solution.

The nominal data type tends to be associated with the entity data model. This often means that real spatial variation within the bounds of the entity has been suppressed in order to simplify the representation. Reclassing the nominal label to a ratio value using some external source of knowledge cannot recover the suppressed variation. The solution to this starts in the field. If field data collection procedures allow the observer to subjectively classify the data, prior to recording it, everything that follows is degraded as a result. If instead of, say, recording the distance down-slope in metres and the local slope in degrees, the field data collector recorded the position as 'mid-slope', and the slope as 'moderately inclined', the field notes could be deciphered using Speight (1984) as being located 'not adjacent above a flat or depression' and 'not adjacent below a crest or flat', and between 5o45' and 18o slope. However, the degree of uncertainty that has been introduced already is significant. Certainly, treating these as fuzzy values in later analyses would help, but this chapter suggests that avoiding the creation of uncertainty is probably better.

Decision tree classifiers claim to be more error tolerant than traditional statistical models in classification (Quinlan 1986). Nevertheless, the results of the sensitivity analysis indicate that the decision tree forest type mapping is quite sensitive to DEM error. With as small as a ±2% DEM error level, the overall test accuracy was reduced by more than 2%. More significantly, the same uncertainty level has caused nearly 10% of the study area to change its initial class assignment. The forest type mapping became more sensitive with the increase of the DEM uncertainty level, and the effect is not linear. The *pixels changed* criterion and the *accuracy* criterion have different implications for critical uncertainty levels. However, the *accuracy* criterion is not as reliable as the *pixels changed* criterion, as it is based on only a limited number of test samples dispersed across the study area. The assessment of changing class assignment indicates that most of the pixel changes occurred within particular forest classes. This confirmed the findings of Huang and Lees (2004) that Forest type 2 and Forest type 3 are more difficult to classify in this study site.

4 Conclusions

From the above analyses of the results, we can derive the following general conclusions. Firstly, contrary to widespread practice, leaving the DEM derivatives input to a model in an unclassified form was clearly better than almost any of the 'improvements' made. This applied to both classification of LULC and to the prediction of a surface, or field, representing a single species. This is an important message.

Secondly, as expected, forest type mapping is likely to be quite sensitive to some level of DEM error. We can see that the DEM error has an uneven impact on the different forest types. However, when increasing the level of DEM error, we can expect to see a non-linear decrease in classification performance.

References

Brown, J.F., Loveland, T.R., Merchant, J.W., Reed, B.C. and Ohlen, D.O., (1993), Using multisource data in global land–cover characterization: concepts, requirements, and methods, *Photogrammetric Engineering and Remote Sensing*, **(59)**6: 977–987.

Canters, F.W., De Genst, W. and Dufourmont, H., (2002), Assessing effects of input uncertainty in structural landscape classification, *International Journal of Geographical Information Science*, **(16)**: 129–149.

Chapin, F.S. III, Zavaleta, E.S., Eviner, V.T., Naylor, R.L., Vitousek, P.M., Reynolds, H.L., Hooper, D.U., Lavorel, S., Sala, O.E., Hobbie, S.E., Mack, M.C. and Diaz, S., (2000), Consequences of changing biodiversity, *Nature*, **405**: 234–242.

Crosetto, M., Tarantola, S. and Saltelli, A., (2000), Sensitivity and uncertainty analysis in spatial modelling based on GIS, *Agriculture, Ecosystems, and Environment*, **81**: 71–79.

Crosetto, M., and Tarantola, S., (2001), Uncertainty and sensitivity analysis: tools for GIS–based model implementation, *International Journal of Geographical Information Science,* **15**: 415–437.

Davis, T. and Keller, C.P., (1997), Modelling uncertainty in natural resource analysis using fuzzy sets and Monte Carlo simulation: slope stability prediction, *International Journal of Geographical Information Science*, **11**: 409–434.

Defries, R.S. and Townshend, J.R.G., (1994), NDVI-derived land cover classifications at a global scale, *International Journal of Remote Sensing,* 15: 3567–3586.

Franklin, J., (1995), Predictive vegetation mapping: Geographic modelling of biospatial patterns in relation to environmental gradients, *Progress in Physical Geography*, **19**: 474–499.

Goodchild, M.F., (1989), Modeling error in objects and fields, In Goodchild, M.F. and S. Gopal (eds.): *Accuracy of Spatial Databases,* Chapter 10, New York: Taylor & Francis.

Guisan, A. and Zimmerman, N., (2000) Predictive habitat distribution models in ecology, *Ecological Modelling*, **135**: 147–186.

Hamby, D.M., (1994), A review of techniques for parameter sensitivity analysis of environmental models, *Environmental Monitoring and Assessment*, **32**: 135–154.

Heuvelink, G.B.M., (1998), *Error Propagation in Environmental Modelling with GIS*, London: Taylor & Francis.

Holmes, K., Chadwick, O. and Kyriakidis, P., (2000), Error in a USGS 30-meter digital elevation model and its impact on terrain modelling, *Journal of Hydrology*, **233**: 154–173.

Huang, Z and Lees, B.G., (2005), Representing and reducing error in natural resource classification, *International Journal of Geographic Information Science,* **(19)**5: 603–621.

Huang, Z and Lees, B.G., (2004), Combining Non–Parametric models for multi-source predictive forest mapping, *Photogrammetric Engineering and Remote Sensing*, **(70)**4: 415–427.

Lees, B.G., (1996), Improving the spatial extension of point data by changing the data model, In Goodchild, M. et al., (eds): *Proceedings of the Third International Conference on Integrating GIS and Environmental Modeling*, (Santa Fe, New Mexico, National Centre for Geographic Information and Analysis, WWW;CD).

Lees, B.G., (1999), *ANU Kioloa/NASA Pathfinder GLCTS Site: Kioloa DTM*, Available online at: http://sres–associated.anu.edu.au/pathfinder/dtm.html

Lees, B. and Ritman, K., (1991), Decision-tree and rule-induction approach to integration of remotely sensed and GIS data in mapping vegetation in disturbed or hilly environments, *Environmental Management*, **15**: 823–831.

Quinlan, J.R., (1986), Induction of Decision Trees, *Machine Learning*, **1:** 81–106.

Quinlan, J.R., (1993), *C4.5: Programs for Machine Learning*, Morgan Kauffman.

Richards, J.A., (1993) *Remote Sensing Digital Image Analysis,* Berlin: Springer-Verlag.

Skole, D.L. (1994), Data on global land–cover change: acquisition, assessment and analysis, In Meyer, W.B. and Turner II, B.L. (eds.): *Changes in land use and land cover: a global perspective,* Cambridge: Cambridge University Press: 437–471.

Speight, A., (1984), Landform, In McDonald, R.C. *et al.* (eds): *Australian Soil and Land Survey Field Handbook,* Melbourne: Inkata Press.

Unwin, D.J., (1995), Geographical Information Systems and the Problem of error and uncertainty, *Progress in Human Geography*, **19**: 549–558.

Van Niel, K.P. and Laffan, S.W., (2003), Gambling with randomness: the use of pseudo-random number generators in GIS, *International Journal of Geographical Science*, **17**: 49–68.

Van Niel, K., Laffan, S.W. and Lees, B.G., (2004), Error and uncertainty in environmental variables for ecological modelling: Sensitivity of variables to error in source elevation data, *Journal of Vegetation Science*, **15**: 747–756.

Wilson, M.F. and Henderson-Sellers, A., (1985), A global archive of land cover and soils data for use in general circulation climate models, *Journal of Climatology*, **(5)**22: 119–143.

Zimmermann, N., (2000), *Tools for analyzing, summarizing and mapping of biophysical variables*, Available online at: http://www.wsl.ch/staff/niklaus.zimmermann/progs.html (Accessed: 4/20/2001).

Section 5: Applications of Terrain Analysis

Global Lineaments: Application of Digital Terrain Modelling

Igor V. FLORINSKY

Abstract

In the past few decades, there have been proposals suggesting that hidden global linear (helical) structures exist, which are tectonically and topographically expressed. In this study, this hypothesis was checked using digital terrain modelling. The study was based on a 30 arc-minute gridded global digital elevation model. Eighteen topographic variables were for the first time calculated and mapped for the entire surface of the Earth. Digital terrain analysis provided support for the existence of global lineaments: on maps of specific catchment area, it was possible to detect five mutually symmetrical pairs of helical structures encircling the Earth from pole to pole. The structures are topographically expressed by patterns of the global ridge network. They are apparently associated with traces of the torsional deformation of the planet: two double helices are in reasonable agreement with theoretically predicted traces of shear fractures, while another two double helices are in reasonable agreement with ideal traces of cleavage cracks. Geological phenomena observed along the structures are discussed (i.e. fracturing, faults, crystal, and ore deposits). It is probable that double helices are relict structures similar to a planetary network of helical lineaments on Venus.

Keywords: tectonics, geological structure, catchment area, helix, planet.

1 Introduction

Lineaments are commonly recognized as linear surface manifestations of geological features of various origin, age, depth, and size (Hobbs 1904, O'Leary *et al.* 1976, Makarov 1981). They are usually associated with fracture zones, faults, folds, other linear features, and their sequences. Lineaments, as a rule, are topographically expressed, and can be observed on maps and remotely sensed images at a wide range of scales. At a regional scale, one can treat lineaments geometrically as planar straight lines. At continental and global scales, they should be considered as spatial curves.

Much attention has been paid to planetary systems of lineaments. There have been three groups of investigations: (a) detection of regularities in the global distribution and direction of lineaments by analysing physiographic and geological maps and then developing models for the regularities observed (Chebanenko 1963, Moody 1966, Katterfeld and Charushin 1973, Besprozvanny *et al.* 1994); (b) development of a physical-mathematical model of a global tectonic process causing an ideal planetary network of lineaments and then comparison of the ideal and actual lineament networks (Vening Meinesz 1947, Dolitsky and Kiyko 1963, Chebanenko and Fedorin 1983); and (c) laboratory simulation of a global lineament network using rotatable spheres (Knetsch 1965, Cherednichenko *et al.* 1966). The origin of global lineament systems was usually associated with rotation-derived forces.

Rance (1968) developed a physical-mathematical model of the torsional deformation of a sphere. The torsion was attributed to an action of possible mantle convection currents on the crust. According to the model, there are two systems of traces of torsional failure surfaces on the surface of the sphere: shear fractures and cleavage cracks (Rance 1967). Geometrically, traces of torsional deformation constitute two systems of double helices encircling the sphere from pole to pole (Figure 1a). The traces vary in inclination at the equator: pairs of mutually symmetrical helices tracing shear fractures are inclined at 15° to 18° and 165° to 162°, and other pairs of helices tracing cleavage cracks are inclined at 56° to 62° and 124° to 118°. A search for actual global helical tectonic features resulted in the detection of several relatively small lineaments referring to faults, trenches, ridges, fracture zones, and seamount chains in basins of the Pacific and Indian Oceans (Rance 1967, Rance 1969).

O'Driscoll (1980) detected two global topographically and tectonically expressed double helical zones by a visual analysis of physiographic maps. The zones had the same inclination at the equator: about 32° and 160° (Figure 1b). O'Driscoll believed that these are fundamental structural belts governing the global deformation network and the planetary evolution. Volkov (1995) reported six global double helical structures also detected by a visual analysis of physiographic maps. At the equator, three of them were inclined at about 12° and 168°, and the other three structures were inclined at about 22° and 158° (Figure 1c). Volkov presumed that these are traces of tidal effects within the Earth-Moon resonance system of past ages.

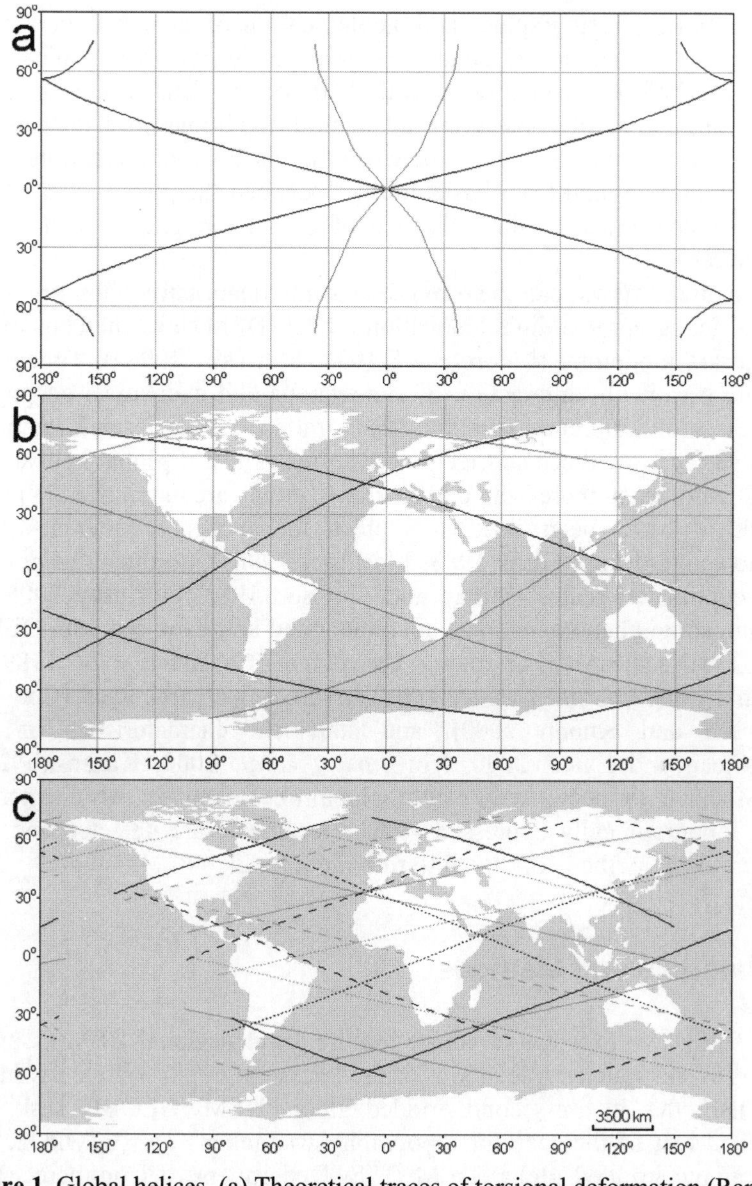

Figure 1. Global helices. (a) Theoretical traces of torsional deformation (Rance 1967): shear fractures (black) and cleavage cracks (grey). (b) Axes of double helical zones (O'Driscoll 1980). (c) Six double helices (Volkov 1995): lines of different style show different structures. The Mercator projection was originally used. The Plate Carrée projection is used here.

Although much attention has been paid to global lineaments, their existence is still questionable. This is because of (1) the qualitative character of topographic, physiographic, and geological maps analysed in previous works; (2) inaccurate presentation of seafloor bathymetry on those maps produced before reasonably accurate bathymetric data became available; (3) impossibility of considering all natural conditions in a mathematical model; (4) obvious differences between the Earth's rotation and its laboratory simulation; and (5) a basic conflict between the plate tectonic theory and the possibility of the existence of global topographic and tectonic structures.

Technical flaws can be obviated using quantitative descriptions of global topography, a digital elevation model (DEM), and methods of digital terrain modelling (Moore *et al.* 1991, Florinsky 1998a). Topography, resulting from the interaction of endogenous and exogenous geophysical processes of different spatial and temporal scales, carries information on both surface processes and tectonic features. Thus, if global helical structures really exist, there is a good chance that they are manifested in relief.

DEMs have been used to detect lineaments at regional scales (Schowengerdt and Glass 1983, Florinsky 1996, Chorowicz *et al.* 1999) and continental scales (Moore and Simpson 1983, Florinsky 2005). Although some phenomena have been modelled and explored with DEMs at the global scale, such as mantle convection (Cazenave *et al.* 1989), the Earth's crust (Mooney *et al.* 1998), hydrological processes (Coe 1998, Renssen and Knoop 2000), and statistical characteristics of relief (McClean and Evans 2000, Vörösmarty *et al.* 2000, Kazanskii 2005), DEMs have not been used to study global lineaments. In this chapter, hidden, topographically expressed global lineaments are detected and interpreted using methods of digital terrain modelling.

2 Materials and Methods

The study was based on a 30 arc-minute gridded global DEM assembled from several sources. Elevations of the land topography were derived from GLOBE, the 30 arc-second gridded global DEM (GLOBE Task Team 1999). Most of the seafloor topography was taken from ETOPO2, the 2 arc-minute gridded global DEM (U.S. Department of Commerce 2001). Bathymetry of the Antarctic Continental Shelf, Caspian Sea, and some large lakes was digitised using topographic maps. The DEM consisted of 721 columns by 361 rows. For Antarctica and Greenland, GLOBE includes elevations of ice surfaces rather than subglacial topography (Hastings and Dunbar 1998). These areas were included in the DEM to retain a united configuration of data.

Global DEMs include high frequency noise (Coe 1998, Hastings and Dunbar 1998, Arabelos 2000) leading to the derivation of useless, noisy digital models and unreadable maps of secondary topographic variables (Florinsky 2002). The problem can be partially resolved by DEM smoothing. To denoise the DEM, one, two, and three iterations of smoothing were applied to the DEM using a 3 × 3 kernel with linear inverse distance weights.

This study was the first application of digital terrain modelling to reveal lineaments at the global scale. Topographic attributes *a priori* 'effective' for this purpose were unknown. It was reasonable to use a representative set of variables. DTMs of the following local topographic attributes were derived from the smoothed DEMs: twelve curvatures (i.e. horizontal, vertical, accumulation, difference, ring, minimal, maximal, mean, Gaussian, unsphericity, horizontal excess, and vertical excess curvatures), slope steepness, slope aspect, rotor, and a model of accumulation zones. DTMs of two regional topographic variables were derived from the smoothed DEMs: specific catchment area and specific dispersive area. Definitions, formulae, and interpretations of the variables can be found elsewhere (Shary *et al.* 2002). Local variables were calculated by the method designed for a spheroidal trapezoidal grid (Florinsky 1998b). Regional variables were derived by a single flow direction algorithm including preliminary filling of sinks (Martz and de Jong 1988) adapted to a spheroidal trapezoidal grid.

The parameters of the Krassovsky ellipsoid, as well as formulae for lengths of meridian and parallel arcs and area of a spheroidal trapezium (Morozov 1979), were employed to calculate changing sizes and area of a spheroidal trapezoidal moving window (Florinsky 1998b) during the DEM smoothing and the derivation of topographic variables. The DEM was processed as a virtually closed spheroidal matrix of elevation values. As a result, there was not the usual loss of border columns and rows due to the application of moving windows. All DTMs produced had a resolution of 30 arc-minutes, and consisted of 721 columns by 361 rows.

To gain a better representation and understanding of global patterns of topographic attributes (Figure 2), their values were transformed as follows (Shary *et al.* 2002):

$$T' = \mathrm{sign}(T) \cdot \ln(1 + 10^m |T|), \tag{1}$$

where T is an attribute, $m = 0$ for slope steepness, aspect, and regional variables, $m = 16$ for accumulation, ring, and Gaussian curvatures, and $m = 9$ for other variables. Specific catchment and dispersive areas were also mapped classifying their values into two levels (Figure 3). The Plate

Carrée projection was used to map topographic variables. DTM treatment was accomplished with LandLord 4.0 (Florinsky *et al.* 1995).

Maps of topographic variables were visually examined in detail. Attention was paid to lineaments running over the entire globe or a hemisphere. Contrary to many lineaments of regional and continental scales, global lineaments are not manifested as uninterrupted linear patterns of the map image or sequences of such lines. A global lineament may be visually detected due to traits of the image *texture* strung out along some direction along a line running over the Earth (Figure 4). Structures detected were mapped using the Plate Carrée projection for the entire Earth, and polar stereographic projections for the Northern and Southern hemispheres (Figure 5) with ArcView GIS 3.0 (© ESRI, 1992–1996).

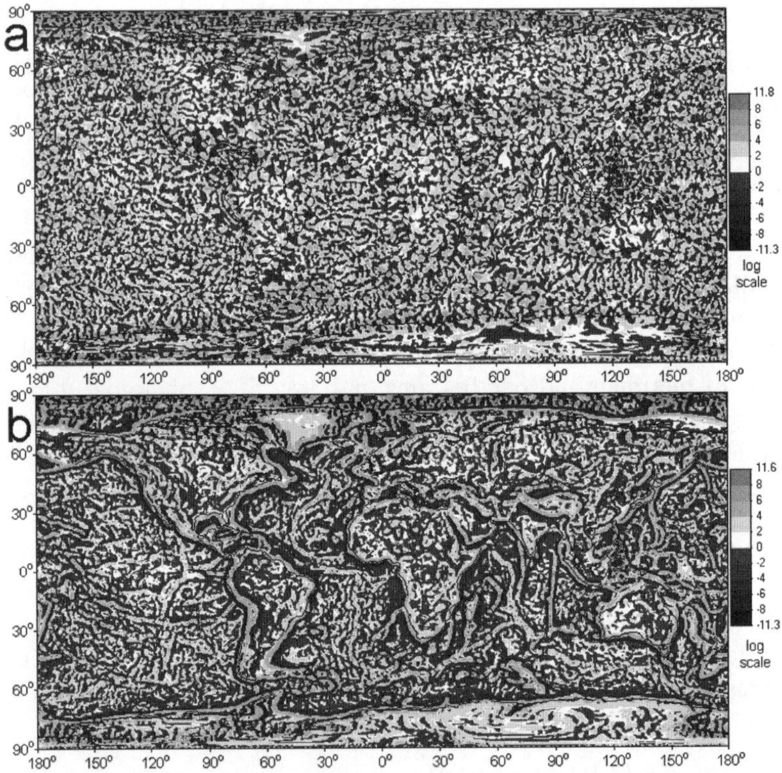

Figure 2 (a-b). Global maps of topographic variables derived from the 3-times smoothed DEM. (a) Horizontal curvature. (b) Vertical curvature.

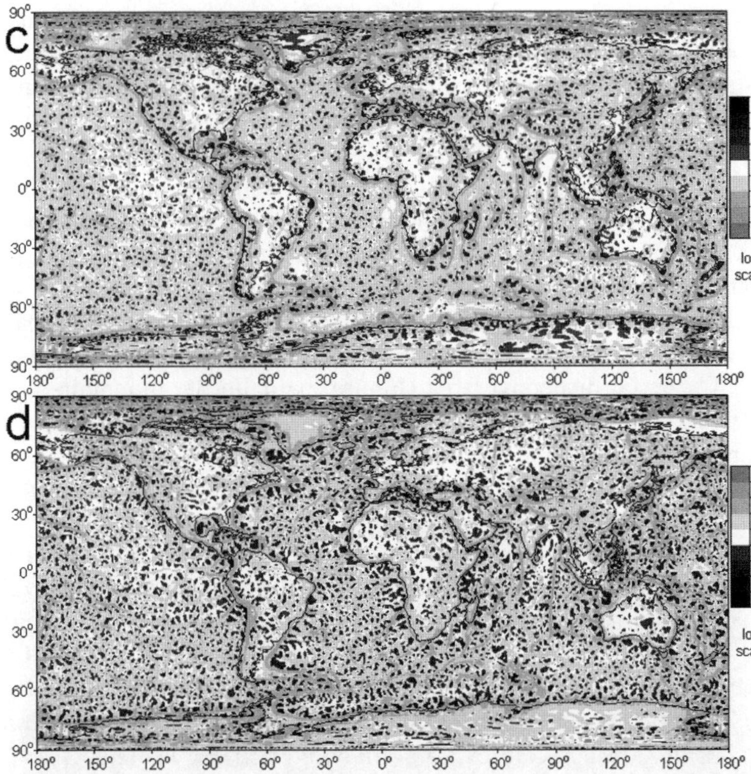

Figure 2 (c-d). Global maps of topographic variables derived from the 3-times smoothed DEM. (c) Minimal curvature. (d) Maximal curvature.

3 Results and Discussion

Global maps of topographic variables represent peculiarities of the Earth's mega-relief in different ways, according to the physical and mathematical sense of a particular variable. For example, horizontal curvature (Figure 2a) delineates areas of flow divergence and convergence (positive and negative values, respectively. These areas correspond to spurs of valleys and ridges (dark and light coloured patterns of the map image, respectively), which form so-called flow structures. At this generalization level, flow structures are most pronounced in ocean basins. Vertical curvature (Figure 2b) is a measure of relative acceleration and deceleration of flows (positive and negative values, respectively). Among other features, the map of vertical curvature shows 'mega-scarps', such as edges of continents and mountains. Low negative values of minimal curvature delineate valleys and troughs, while positive values delineate 'mega-hills' (Figure 2c). High positive values of maximal curvature show ridges, while negative values show 'mega-depressions' (Figure 2d). Catchment area measures an upslope area, which is potentially drained through a given point on the land surface. At the global scale, low values of specific catchment area (Figure 2e) delineate land and oceanic ridges as black lines (e.g. the Andes, Alps, mid-ocean ridges), while its high values show land and oceanic valleys as white lines and depressions as light coloured areas (e.g. the Mediterranean Sea, Gulf of Mexico, Angola Basin). Dispersive area measures a downslope area, which may be potentially exposed by flows passing through a given point on the land surface. At this generalization level, high values of specific dispersive area (Figure 2f) delineate mountain systems and highlands as light coloured areas (e.g. the Himalayas, Urals, Ethiopian Highlands) as well as land and oceanic ridges as white lines.

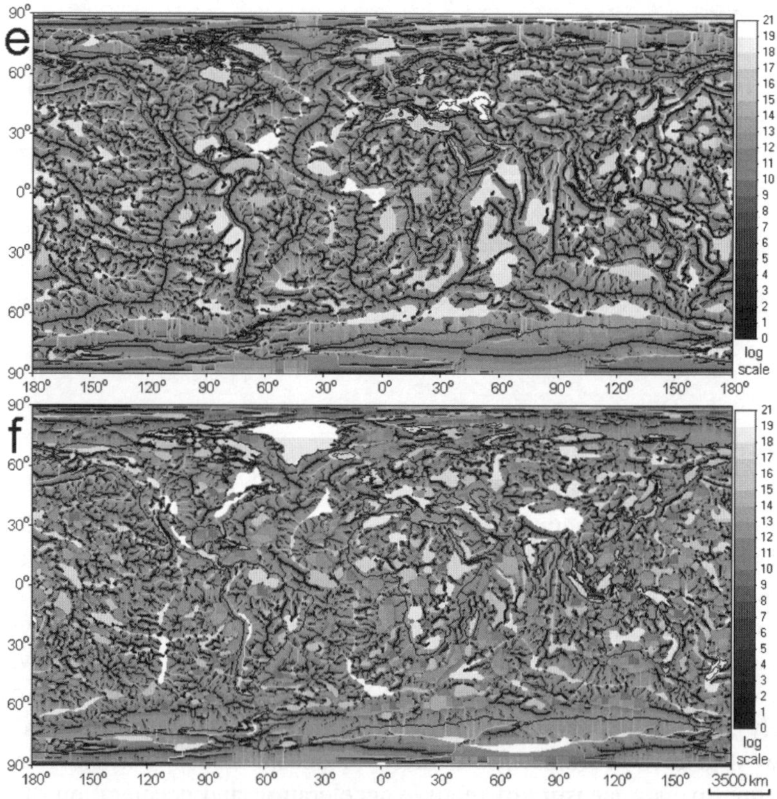

Figure 2 (e-f). Global maps of topographic variables derived from the 3-times smoothed DEM. (e) Specific catchment area. (f) Specific dispersive area.

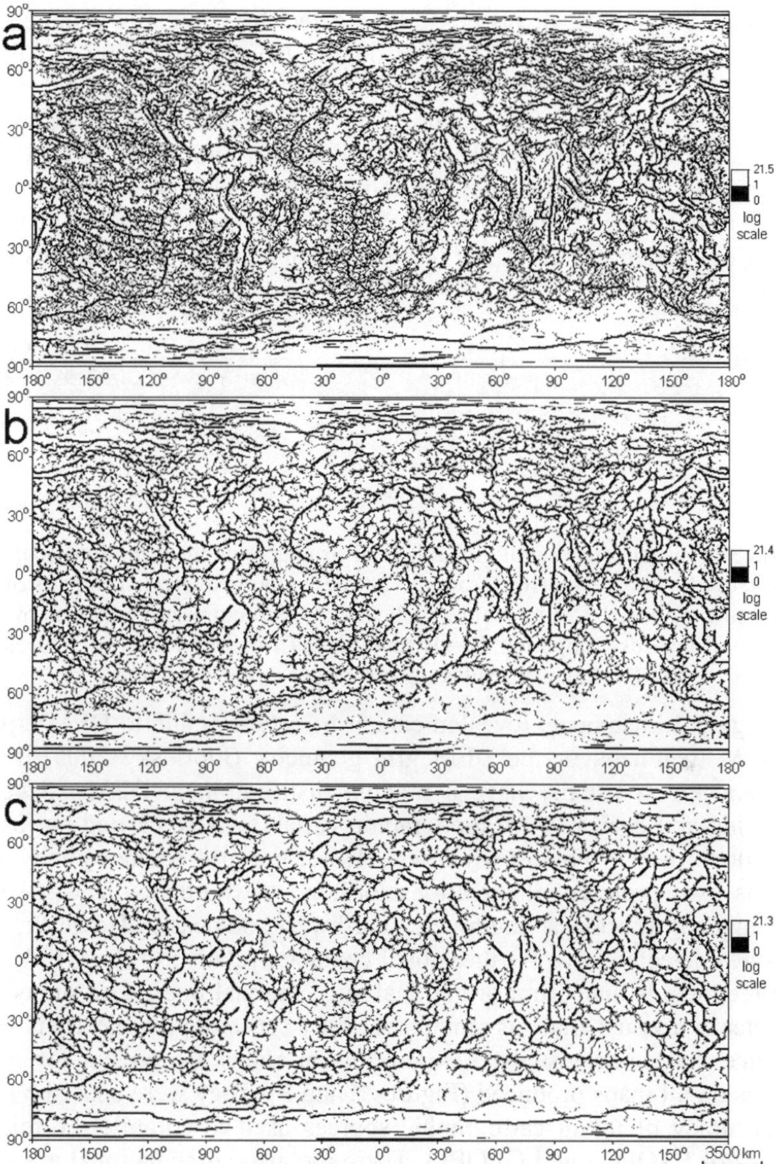

Figure 3. Global maps of catchment area classified into two levels. (a) 1-time smoothing. (b) 2-times smoothing. (c) 3-times smoothing.

Maps of specific catchment area with values classified into two levels (Figure 3) were best suited to detect global lineaments. These maps display the ridge network of the planet. The greater the number of DEM smoothing, the more generalized picture of the network is mapped. Analysing these maps (Figure 4), it was possible to detect five mutually symmetrical pairs of global lineaments, viz. five double helices encircling the Earth from pole to pole (Figure 5). The structures revealed are helical zones rather than simply lines. Each double helix is named after the area(s) of intersection(s) of its arms (Table 1). Arms running clockwise upward and counter-clockwise upward (dextral and sinistral helices) are called right and left arms, respectively. Each helical zone transgresses plate boundaries and regions dissimilar in respect to their tectonic origin, rock composition, and age.

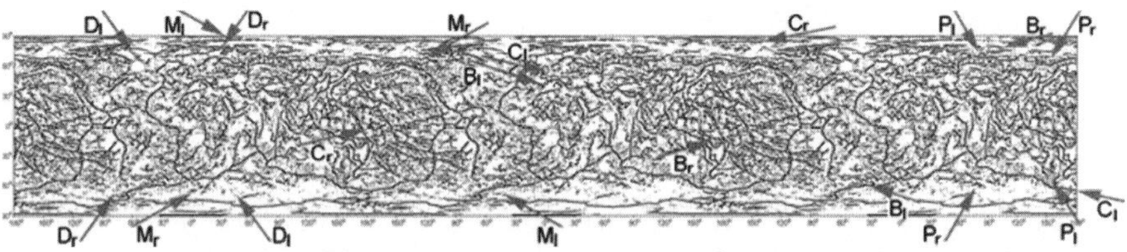

Figure 4. Visual detection of global helical structures on the three-fold map of specific catchment area derived from the 2-times smoothed DEM. Ten pairs of arrows show proposed positions of ten helical arms. Structures are indicated by labels: C = Caucasus-Clipperton, B = Biscay-Santa Cruz, M = Marcus, D = Dakar, and P = Palawan. Subscripts: r = right arm, l = left arm.

The global lineaments revealed cannot be artefacts due to DEM errors, the DEM treatment, or the DEM grid geometry (Florinsky 2005). First, noise and errors usually have a random distribution in DEMs. Second, smoothing and derivation of topographic variables were carried out using local filters (n × n moving windows). Third, the grid geometry may amplify its own preferential directions: orthogonal (north-south, east-west) and diagonal (northeast-southwest, northwest-southeast). However, the structures detected have (a) the global character relative to the DEM; and (b) directions distinct from orthogonal and diagonal ones. The subjective character of a visual analysis remains the only cause of possible artefacts.

Notice that some visually recognizable artefacts are typical for Polar Regions on all maps produced (Figures 2 and 3). They were caused by too low accuracy of initial cartographic sources used to compile the related portions of ETOPO2 and GLOBO. There are also computational artefacts on maps of specific catchment and dispersive areas, which are manifested as straight parallel lines located predominantly in the Polar Regions (Figure 2e, f). They are well known artefacts of single flow direction algorithms common for flat slopes. However, the artefacts did not influence the detection of global helical structures since the artefacts are situated within limited zones of the maps.

Table 1. Parameters of the global topographic helices.

Structure	Left arm		Right arm		Geograph-ical coordinates of the arm intersec-tion(s)
	Lengths (km)	Inclination at the equa-tor (°)	Lengths (km)	Inclination at the equator (°)	
Caucasus-Clipperton	55,800	167.5	31,500	12.5	46.4°N, 44.81°E; 5.9°N, 134.7°W
Biscay-Santa Cruz	39,600	162.2	29,800	17.5	44.4°N, 7.3°W; 12.9°S, 171.4°E
Marcus	26,500	150.6	24,900	29.7	21.4°N, 157.5°E
Dakar	17,700	126.9	17,200	53.3	14.9°N, 16.0°W
Palawan	15,400	121.3	15,300	59.5	9.9°N, 119.1°E

The Caucasus-Clipperton double helix (Figure 5a) coincides with one of the structures reported by Volkov (1995) (Figure 1c). The left arm of the Biscay-Santa Cruz structure (Figure 5b) partly agrees with the left arm of one of the helices detected by O'Driscoll (1980) (Figure 1b). A comparison of inclination angles of theoretical traces of torsional deformation (Section 1, Figure 1a), and that of the double helices revealed (Table 1) shows that the Caucasus-Clipperton and Biscay-Santa Cruz structures can be assigned to traces of shear fractures, while the Dakar and Palawan structures can be assigned to traces of cleavage cracks. The mean deviation of inclination angles of the structures from the theoretical values is 2.8°. Of the five double helical structures detected, four have inclination angles fitting theoretical values. This suggests that one may consider topographically expressed helical structures as fracture traces of global torsion. There are several deviations from the theory, such as: (a) arms of each double helix meet off the equator; and (b) there is the Marcus double helix with 'abnormal' inclination. The discrepancies might be in part attributed to the deviation of the Earth's shape from a sphere, as assumed in the model of Rance (1968).

The polar stereographic map of the Northern hemisphere presents the Caucasus-Clipperton structure (Figure 5a) as a nearly ideal plane two-arm Archimedean spiral (Fikhtengolts 1966, p. 520). Away from the pole, it is approximated by the equation:

$$r = a\theta \tag{2}$$

where r is a radius, θ is an angle, and a is a constant; $a = 1.01$. The left arm of this structure is explicitly described as a spherical Archimedean spiral (Klíma *et al.* 1981):

$$\sigma = R\psi_0\lambda/(2\pi) \tag{3}$$

where σ is a length of a meridian arc between the pole and a point on the spiral (ψ, λ), R is a radius of the sphere, ψ_0 is a value by which ψ changes if λ changes by 2π, $\psi = 0.5\pi - \varphi$, φ is the latitude, and λ is the longitude; ψ_0 = 79.55°. Volkov (2006, personal communication) supposed that all lineaments delineated by him (Figure 1c) are spherical Archimedean spirals. This issue needs further investigation.

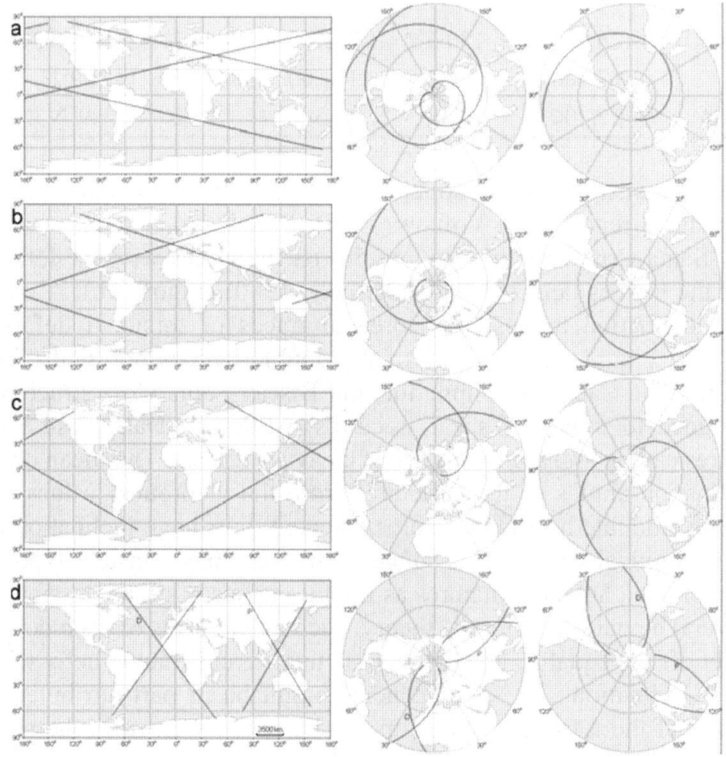

Figure 5. Double helical structures for the entire Earth and the Northern and Southern hemispheres. (a) Caucasus-Clipperton. (b) Biscay-Santa Cruz. (c) Marcus. (d) Dakar (D) and Palawan (P).

Literature provides evidence for shears and increased fracturing along double helices. In particular, the left arm of the Caucasus-Clipperton structure (Figure 5a) within Europe was associated with a strike-slip fault system (Moody 1966). Within the Pacific Ocean basin, the left arm of the Biscay-Santa Cruz structure (Figure 5b) was also interpreted as a system of strike-slip faults (Moody 1966). This helix is coaxial with a zone of planetary fracturing within Eurasia (Miroshnichenko *et al.* 1984). Along this zone, there is a belt of intensive rock fracturing 3,000 km wide, including right-lateral strike-slip faults along the northern border of the belt (Poletaev 1986). Relatively small strike-slip faults southward of Madagascar and northward of New Guinea (Moody 1966) seem to be parts of the right arm of the Marcus structure (Figure 5c). Fragments of the right arm of the Dakar structure (Figure 5d) were associated with shear systems (Vening Meinesz 1947), strike-slip faults (Moody 1966), and fracture zones (Miroshnichenko *et al.* 1984). The left arm of the Dakar double helix is coaxial with a zone of planetary fracturing (Miroshnichenko *et al.* 1984). Between Indonesia and the Philippines, a fragment of the right arm of the Palawan double helix (Figure 5d) was interpreted as a system of strike-slip faults (Chebanenko 1963, Moody 1966). In Australia, the left arm of the Palawan structure and the right arm of the Biscay-Santa Cruz structure were also associated with strike-slip faults (Chebanenko 1963).

Intensive crystallo- and metallogeny are observed along both arms of the Caucasus-Clipperton helix (Figure 5a). Evseev (1989) reported finds of large crystals of viterite and strontianite along a line coaxial with the left arm of the structure in Europe, from Scotland to the Caspian Sea. Along the right arm of the structure in Siberia, between 70°E and 170°W, there were numerous finds of unique large crystals including amethyst, aquamarine, axinite, azurite, beryl, calomel, calcite, charoite, diopside, fluorite, garnet, hematite, magnetite, pegmatite, smoky and green quartz, spinel, spherulite, topaz, tourmaline, etc. (Evseev 1993). Large iron and gold ore deposits of endogenic and metamorphic origin are located along both arms of this helix in Eurasia, Central and South America (Volkov 1995). It is common knowledge that lineaments of the regional and continental scales may control ore deposits since an increased fracturing of the crust along lineaments, especially at sites of their intersection, is favourable to magmatic intrusions (Favorskaya 1977, O'Driscoll 1986). This may explain the occurrence of crystal and ore deposits along helical structures.

Global helical topographic structures are not unique to the Earth. Slyuta *et al.* (1989) discovered a dense, regular network of dextral and sinistral spiral structures on radar scenes of the Northern hemisphere of Venus (Figure 6). These planetary structures are wound around the axis of rotation of Venus. They are topographically manifested as troughs, scarps, and depressions. Slyuta *et al.* (1989) believed that strong rotational forces had formed the network during the deceleration of Venus's rotation. They suggested that the helical network is a relict feature, an 'imprint' of ancient rotational stress fields, because the current rotation velocity of Venus is quite slow. The low intensity of erosion has allowed relict helical structures to persist on the Venusian surface. This author supposes that double helices of the Earth are also relict features. Strong erosion led to their hidden manifestation in modern topography (Figure 4).

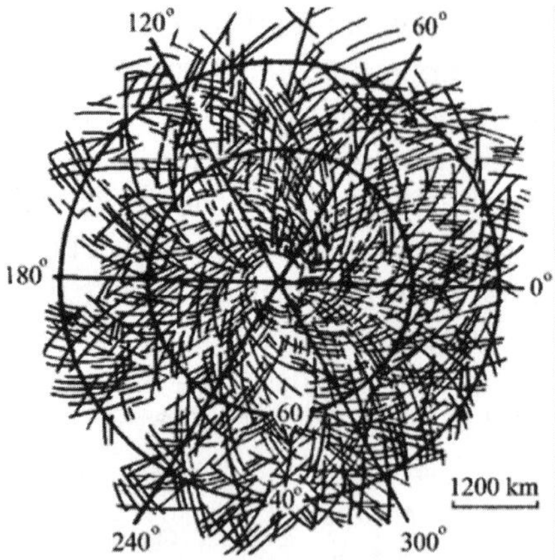

Figure 6. Spiral structures of the Northern hemisphere of Venus (after Slyuta et al. 1989, Figure 2, © Plenum Publishing Corporation, 1990; reproduced with kind permission of Springer Science and Business Media).

Contrary to regional lineaments and some structures of the continental scale clearly recognizable on the maps of specific catchment area (e.g. mountain chains, mid-ocean ridges, and ocean troughs), double helical structures are not in harmony with the plate-tectonic theory. Notice that the existence of transcontinental and planetary lineaments was one of the main geological and geomorphic facts contradicting the plate-tectonic ideas (Favorskaya 1977, Pavlenkova 1995, Pratt 2000, Smoot 2001). To link these facts with possible continental drift, seafloor spreading, and subduction, some modifications of the plate-tectonic hypothesis have been proposed. For instance, Moody (1966) proposed that continental drift is a movement of crustal blocks along major lineament zones with relatively stable mantle roots. Besprozvanny *et al.* (1994) suggested that regularities in the global lineament network are caused by dissipative structures of the upper core. Once again, our results set one to think about a tectonic paradigm more adequately depicting the actual structure and evolution of the Earth.

4 Conclusions

The application of digital terrain modelling provided support for the hypothesis for the existence of double helical structures of the Earth, which are topographically expressed and possibly associated with the torsional deformation of the planet. To understand fully the origin and properties of helical structures, comprehensive studies should be conducted. A quantitative analysis of the global topography may be essential to give a clearer insight into the planetary evolution.

Eighteen topographic variables were for the first time calculated and mapped for the entire surface of the Earth, including both land and seafloor topography. To produce readable and interpretable global maps of topographic attributes, DEM denoising was the key step in data processing. This paper focused on lineaments, but global maps of topographic variables can be useful to study other problems of tectonics and geophysics (ring structures, lithospheric strain, etc.). These maps can be integrated into virtual geological and geomorphic globes (Rundquist *et al.* 2002, Tooth 2006).

Acknowledgements

The author thanks V.G. Trifonov (Geological Institute, Russian Academy of Sciences, Moscow, Russia), A.E. Fedorov (Annual Seminar 'The System of the Planet Earth: Non-Conventional Problems of Geology', Moscow, Russia), Yu.V. Volkov (Research Computing Center, M.V. Lomonosov Moscow State University, Moscow, Russia), and H. Rance (Queensborough Community College, City University of New York, Bayside, USA) for discussions, as well as P.A. Shary (Institute of Physical, Chemical, and Biological Problems of Soil Science, Russian Academy of Sciences, Pushchino, Russia), G. Ventura (Istituto Nazionale di Geofisica e Vulcanologia, Roma, Italy), and an anonymous referee for useful criticism. The study was partly supported by RFBR grant 06-05-74882.

References

Arabelos, D., (2000), Intercomparisons of the global DTMs ETOPO5, TerrainBase and JGP95E, *Physics and Chemistry of the Earth (A)*, **25**: 89–93.

Besprozvanny, P.A., Borodzich, E.V. and Bush, V.A., (1994), Numerical analysis of ordering relations in the global network of lineaments, *Physics of the Solid Earth*, **30**: 150–159.

Cazenave, A., Souriau, A. and Dominh, K., (1989), Global coupling of Earth surface topography with hotspots, geoid and mantle heterogeneities, *Nature*, **340**: 54–57.

Chebanenko, I.I., (1963), Principal Regularities of Fault Tectonics of the Earth's Crust and Its Problems, Kiev: Ukrainian Academic Press (in Russian).

Chebanenko, I.I. and Fedorin, Ya.V., (1983), On a new type of rotation-tectonic lines in the Earth's lithosphere, *Doklady Akademii Nauk SSSR*, 270: 406–409 (in Russian).

Cherednichenko, A.I., Burmistenko, V.M., Tokovenko, V.S. and Chebanenko, I.I., (1966), Attempt of laboratory simulation of planetary faults (lineaments) of the Earth, *Dopovidi Academii Nauk Ukrainy*, 10: 1333–1336 (in Ukrainian, with English abstract).

Chorowicz, J., Dhont, D. and Gündoğdu, N., (1999), Neotectonics in the eastern North Anatolian fault region (Turkey) advocates crustal extension: mapping from SAR ERS imagery and digital elevation model, *Journal of Structural Geology*, **21**: 511–532.

Coe, M.T., (1998), A linked global model of terrestrial hydrologic processes: simulation of modern rivers, lakes, and wetlands, *Journal of Geophysical Research*, **D103**: 8885–8899.

Dolitsky, A.V. and Kiyko, I.A., (1963), On causes of deformation of the Earth's crust, In Nalivkin, D.V. and Tupitsin, N.V. (eds.): *Problems of Planetary Geology,* Moscow: Gosgeoltekhizdat: 291–312 (in Russian).

Evseev, A.A., (1989), Regularity in the distribution of discoveries of large crystals, *New Data on Minerals*, **36**: 53–67 (in Russian).

Evseev, A.A., (1993), Siberia's crystals and symmetry in the distribution of occurrences of minerals, *World of Stones*, 1: 11–20.

Favorskaya, M., (1977), Metallogeny of deep lineaments and new global tectonics, *Mineralium Deposita*, **12**: 163–169.

Fikhtengolts, G.M., (1966), A Course in Differential and Integral Calculus, Vol. 1, 6th ed., Moscow: Nauka (in Russian).

Florinsky, I.V., (1996), Quantitative topographic method of fault morphology recognition, *Geomorphology*, **16**: 103–119.

Florinsky, I.V., (1998a), Combined analysis of digital terrain models and remotely sensed data in landscape investigations, *Progress in Physical Geography*, **22**: 33–60.

Florinsky, I.V., (1998b), Derivation of topographic variables from a digital elevation model given by a spheroidal trapezoidal grid, *International Journal of Geographical Information Science*, **12**: 829–852.

Florinsky, I.V., (2002), Errors of signal processing in digital terrain modelling, *International Journal of Geographical Information Science*, **16**: 475–501.

Florinsky, I.V., (2005), Artificial lineaments in digital terrain modelling: can operators of topographic variables cause them? *Mathematical Geology*, **37**: 357–372.

Florinsky, I.V., Grokhlina, T.I. and Mikhailova, N.L., (1995), LANDLORD 2.0: the software for analysis and mapping of geometrical characteristics of relief, *Geodezia i Cartografia*, 5: 46–51 (in Russian).

GLOBE Task Team, (1999), *The global land one-kilometer base elevation (GLOBE) digital elevation model, version 1.0,* Boulder: NOAA, National Geophysical Data Center, Available online at: http://www.ngdc.noaa.gov/mgg/topo/globe.html (accessed 22 October 2005).

Hastings, D.A. and Dunbar, P.K., (1998), Development and assessment of the global land one-km base elevation digital elevation model (GLOBE), *ISPRS Archives*, **32**: 218–221.

Hobbs, W.H., (1904), Lineaments of Atlantic Border region, *Geological Society of America Bulletin*, **15**: 483–506.

Katterfeld, G.N. and Charushin, G.V., (1973), General grid systems of planets, *Modern Geology*, **4**: 253–287.

Kazanskii, B.A., (2005), Calculation of the Earth's topography-related potential energy from digital data, *Izvestiya, Physics of the Solid Earth*, **41**: 1023–1026.

Klíma, K., Pick, M. and Pros, Z., (1981), On the problem of equal area block on a sphere, *Studia Geophysica et Geodaetica*, **25**: 24–35.

Knetsch, G., (1965), Über ein Structur-Experiment an einer Kugel und Beziehungen zwischen Gross-Lineamenten und Pol-Lagen in der Erdeschichte, *Geologische Rundschau*, **54**: 523–548.

Makarov, V.I., (1981), Lineaments: problems and trends of studies by remote sensing techniques, *Izvestiya Vuzov, Geologia i Razvedka*, 4: 109–115 (in Russian).

Martz, L.W. and de Jong, E., (1988), CATCH: a Fortran program for measuring catchment area from digital elevation models, *Computers and Geosciences*, **14**: 627–640.

McClean, C.J. and Evans, I.S., (2000), Apparent fractal dimensions from continental scale digital elevation models using variogram methods, *Transactions in GIS*, **4**: 361–378.

Miroshnichenko, V.P., Berezkina, L.I. and Leontieva, E.V., (1984), *Planetary Fracturing of Sedimentary Cover of the Lithosphere from Remotely Sensed Data*, Leningrad: Nedra (in Russian).

Moody, J.D., (1966), Crustal shear patterns and orogenesis, *Tectonophysics*, **3**: 479–522.

Mooney, W., Laske, G. and Master, T., (1998), CRUST 5.1: A global crustal model at 5x5, *Journal of Geophysical Research*, **B103**: 727–747.

Moore, I.D., Grayson, R.B. and Ladson, A.R., (1991), Digital terrain modelling: a review of hydrological, geomorphological and biological applications, *Hydrological Processes*, **5**: 3–30.

Moore, R.F. and Simpson, C.J., (1983), Image analysis – a new aid in morphotectonic studies. In *17th International Symposium on Remote Sensing of Environment*, 9–13 May 1983, Ann Arbor, USA, Vol. 3, Ann Arbor: Environmental Research Institute of Michigan: 991–1002.

Morozov, V.P., (1979), *A Course in Spheroidal Geodesy. 2nd enl. and rev. ed.*, Moscow: Nedra (in Russian).

O'Driscoll, E.S.T., (1980), The double helix in global tectonics, *Tectonophysics*, **63**: 397–417.

O'Driscoll, E.S.T., (1986), Observations of the lineament–ore relation, Philosophical Transactions of the Royal Society of London, Series A: Mathematical and Physical Sciences, **317**: 195–218.

O'Leary, D.W., Friedman, J.D. and Pohn, H.A., (1976), Lineament, linear, lineation: some proposed new standards for old terms, *Geological Society of America Bulletin*, **87**: 1463–1469.

Pavlenkova, N.I., (1995), Structural regularities in the lithosphere of continents and plate tectonics, *Tectonophysics*, **243**: 223–229.

Poletaev, A.I., (1986), *Seismotectonics of the Main Kopetdag Fault Zone*, Moscow: Nauka (in Russian).

Pratt, D., (2000), Plate tectonics: a paradigm under threat, *Journal of Scientific Exploration*, **14**: 307–352.

Rance, H., (1967), Major lineaments and torsional deformation of the Earth, *Journal of Geophysical Research*, **72**: 2213–2217.

Rance, H., (1968), Plastic flow and fracture in a torsionally stressed planetary sphere, *Journal of Mathematics and Mechanics*, **17**: 953–974.

Rance, H., (1969), Lineaments and torsional deformation of the Earth: Indian Ocean, *Journal of Geophysical Research*, **74**: 3271–3272.

Renssen, H. and Knoop, J.M., (2000), A global river routing network for use in hydrological modelling, *Journal of Hydrology*, **230**: 230–243.

Rundquist, D.V., Ryakhovsky, V.M., Gatinsky, Yu.G. and Chesalova, E.I., (2002), GIS-project 'The geodynamic globe, scale 1:10,000,000' for global monitoring of various geological processes, *Proceedings of the All-Russian Scientific Conference 'Geology, Geochemistry, and Geophysics on the Boundary of the 20th and 21st Centuries'*, 8–10 Oct. 2002, Moscow, Russia, Vol. 1 ,Moscow: Svyaz-Print: 87–88 (in Russian).

Schowengerdt, R.A. and Glass, C.E., (1983), Digitally processed topographic data for regional tectonic evaluations, *Geological Society of America Bulletin*, **94**: 549–556.

Shary, P.A., Sharaya, L.S. and Mitusov, A.V., (2002), Fundamental quantitative methods of land surface analysis, *Geoderma*, **107**: 1–32.

Slyuta, E.N., Kudrin, L.V. and Sinilo, V.P., (1989), Preliminary data on the nature of a planetary system of lineaments observed in radar images of Venus (data from Venera-15 and -16), *Cosmic Research*, **27**: 786–797.

Smoot, N.C., (2001), Earth geodynamic hypotheses updated, *Journal of Scientific Exploration*, **15**: 465–494.

Tooth, S., (2006), Virtual globes: A catalyst for the re-enchantment of geomorphology? *Earth Surface Processes and Landforms*, **31**: 1192–1194.

U.S. Department of Commerce, NOAA, National Geophysical Data Center, (2001), *2–minute gridded global relief data (ETOPO2),* Available online at: http://www.ngdc.noaa.gov/mgg/fliers/01mgg04.html (accessed 21 October 2005).

Vening Meinesz, F.A., (1947), Shear patterns of the Earth's crust, *Transactions of the American Geophysical Union*, **28**: 1–61.

Volkov, Y.V., (1995), Loxodromy and minerageny (the influence of astronomic resonances in the Earth-Moon system on the origin of ore deposits in the Earth's crust), *Bulletin of Moscow Society of Naturalists, Geological Series*, **(70)**6: 90–94 (in Russian, with English abstract).

Vörösmarty, C.J., Fekete, B.M., Meybeck, M. and Lammers, R.B., (2000), Geomorphometric attributes of the global system of rivers at 30-minute spatial resolution, *Journal of Hydrology*, **237**: 17–39.

Modelling Channelling and Deflection of Wind by Topography

John B. LINDSAY and James J. ROTHWELL

Abstract

Topographic indices have been used extensively in the past to model wind-related phenomena such as soil erosion, snow redistribution, and atmospheric deposition of contaminants. A new index of exposure/sheltering to wind, the channelling/deflection index (CDI), is presented and evaluated in this chapter. Unlike existing windiness indices, most of which are based on ray-tracing algorithms, the CDI is capable of modelling channelling and deflection of flowlines, as well as wind shadowing. The CDI is a measure of how much more or less windy a location is as a result of the influence of upwind topography. The method for calculating the CDI applies a flow routing algorithm to model airflow patterns. Like all topographic indices involving the analysis of extended neighbourhoods, the CDI is affected by edge contamination; the location of digital elevation model (DEM) edges can significantly influence the CDI pattern estimated for an area. Analyses showed that edge effects can be diminished by using an appropriately sized buffer of elevation data in the windward direction. A 4 km buffer was found to be sufficient to accurately model the CDI in five study areas of varying topography, based on DEM data with a 10 m grid resolution.

Keywords: wind modelling, exposure, sheltering, topographic index, DEM.

1 Introduction

The partial control that topography has over many environmental phenomena largely occurs for two reasons. Firstly, slope gradient, slope orientation, and relative landscape position affect the flow paths taken by runoff and therefore the abundance of near-surface water at a location. Secondly, these topographic attributes also influence climatic variables such as temperature, solar radiation, and exposure to wind. Each of these environmental conditions (i.e. the abundance of water, energy, heat, wind, etc.)

affects soil development, the distributions of flora and fauna, and the flux of fine sediment, essential nutrients, and contaminants in catchments. This chapter is concerned with the association between topography and wind exposure. Topography affects local windiness by deflecting and channelling moving air (Böhner and Antonić in press). In extreme cases of wind deflection, topography can act as an obstacle or barrier resulting in wind sheltering, also known as wind shadowing. The degree to which a location is exposed or sheltered influences evaporation processes, and therefore the abundance of soil water and vegetation patterns, as well as the potential for aeolian transportation and deposition of fine sediment, soil, snow, and particulate contaminants.

Computational fluid dynamics (CFD) can be used to map the pattern of airflow over a surface by solving the Navier-Stokes equations, a series of differential equations used to describe fluid motion, for a given boundary condition (Girault and Raviart 1986). Although this is an extremely computationally demanding task requiring specialized modelling software and expert knowledge (Chaderjian *et al.* 2006), it is possible to create an airflow model for complex topography (Uchida and Ohya 1999). For applications requiring continuous, instantaneous airflow data, the physically based approach of CFD is needed. However, for applications that require information about the spatial variation in longer-term site windiness, other simpler approaches may be more suited. Terrain analysis and the derivation of simple topographic indices of exposure from DEMs have been used in the past to measure longer-term site windiness. Terrain analysis is the field of study or analytical approach that uses the direct and indirect control that topography has on process functioning to gain understanding of various phenomena in a landscape. This approach takes advantage of the fact that often terrain attributes, such as slope, aspect, and landscape position, are much easier to measure than the spatial and temporal distributions of other controlling factors (Wilson and Gallant 2000). Terrain analysis relies on the use of DEMs to represent topography. Considerable work has taken place over the last two decades to develop simple topographic indices derived from DEMs to act as surrogates for other more difficult to measure phenomena (e.g. Moore *et al.* 1991, Wilson and Gallant 2000). Researchers have proposed several DEM-derived indices for evaluating the spatial pattern of wind exposure/sheltering (e.g. Murakami and Komine 1983, Lapen and Martz 1993, Antonić and Legović 1999, Winstral and Marks 2002). These indices are by their nature simplifications of a complex dynamic system, removed from the governing process equations and other controlling factors such as roughness due to land cover. Nonetheless, exposure/sheltering indices have been used successfully to model the spatial patterns of snow depths (Lapen and Martz 1993, Winstral and Marks 2002,

Erickson *et al.* 2005), soil loss (Chappell 1996), vegetation patterns (Huang 2002), and concentrations of atmospherically-deposited contaminants (Antonić and Legović 1999). In this chapter, we (1) evaluate the suitability of existing DEM-derived exposure/sheltering indices, and (2) present a new index that incorporates topographic channelling and deflection of wind.

2 Topographic Indices of Exposure to Wind

Numerous topographic indices of site windiness exist. Relative terrain aspect is perhaps the simplest measure of topographic exposure, taking into account land-surface orientation only. Relative terrain aspect is the angle between the land-surface aspect and the wind direction (Antonić and Legović 1999, Böhner and Antonić in press). It is calculated respecting the circularity of angular measures and is bound between 0°, indicating a location that is oriented in a windward direction (i.e. exposed), and 180°, indicating a leeward orientation (i.e. sheltered). In application, the specified wind azimuth is usually assumed to represent a regional wind direction. Figure 2A shows the pattern of relative terrain aspect for a part of the Bleaklow plateau in the southern Pennines, UK, based on a 10 m DEM (Figure 1) derived from laser altimetry (LiDAR). As relative terrain aspect is calculated solely using information about topography within a small neighbourhood of 3 × 3 grid cells, local-scale topographic variation is well represented in Figure 2A. It is apparent from this image, however, that relative terrain aspect does not account for distant topographic obstacles that cause sheltering effects. For example, a site oriented towards the wind flux that is behind a wind-sheltering hill would be considered to be exposed in a relative terrain aspect image despite the obstacle. This is clearly a limitation of the index.

There are several topographic indices that use ray-tracing, i.e. querying elevations along a search path in a specified direction, to account for the influence of distant topographic obstacles on exposure/sheltering. For example, horizon angle (Antonić and Legović 1999), also referred to as maximum upwind slope (Winstral *et al.* 2002, Erickson *et al.* 2005), is commonly used as an exposure/sheltering index (Figure 2B), although it is also frequently applied to solar radiation modelling (Dozier *et al.* 1981). The usefulness of horizon angles for assessing local windiness was recognized long before automated estimation from DEMs was possible, e.g. the field-based, empirical TOPEX index refined by Wilson (1984). Horizon angle is defined as the maximum elevation angle along a ray in the direction of the hypothetical wind flux (Antonić and Legović 1999). Winstral and Marks (2002) refer to the grid cell with the maximum elevation angle in a search path as the shelter-defining cell. The distance between a grid cell and its shelter-defining cell is usually small enough that Earth curvature can be ignored in estimating the horizon angle. An assumption is made that large positive values of horizon angle indicate areas that are relatively sheltered from wind in a specified direction (Böhner and Antonić in press). Locations with negative horizon angles (declinations) are located above their horizon and are therefore relatively exposed. Antonić and Legović (1999) recommend setting all declinations to zero, recognizing that the exposure of a site located above its horizon is often affected to a greater extent by altitude. This same approach to handling declinations was also used to derive the TOPEX index.

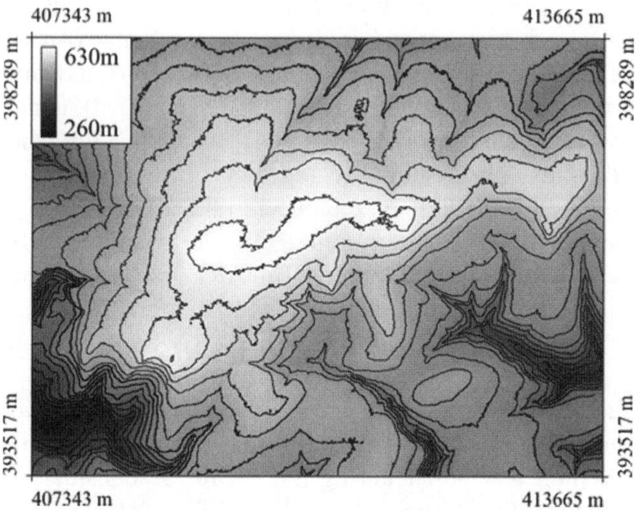

Figure 1. A 10 m resolution LiDAR digital elevation model of the Bleaklow plateau, southern Pennines, UK. Contours are drawn at a 20 m interval.

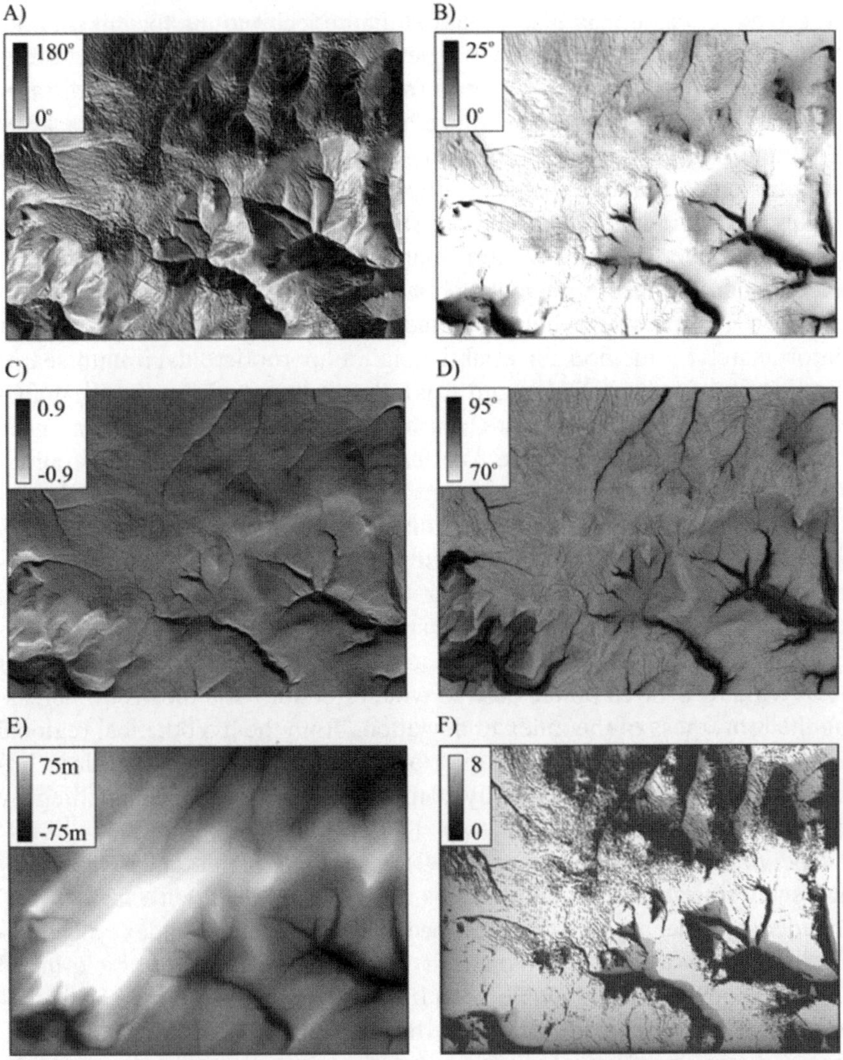

Figure 2. Various topographic indices of exposure/sheltering derived for the LiDAR DEM, including (A) relative terrain aspect, (B) horizon angle, (C) exposure towards the wind flux, (D) openness, (E) directional relief, and (F) fetch. Each image is based on a hypothetical wind direction of 225° (i.e. a wind from the southwest), except for openness (D), which is derived from data in all eight main compass directions. Distance values for the fetch image (F) have been log-transformed to enhance visualization.

Horizon angle is frequently calculated using a maximum search distance. This can significantly reduce the computational effort required to calculate the index, something that can be problematic when analysing large DEMs. There may also be sound theoretical reasons for this practice in addition to the computational benefits. Böhner and Antonić (in press) argue that whilst for solar radiation modelling an infinite search distance (implying that the ray-tracing procedure always terminates at the DEM edge) is preferred, for modelling exposure/sheltering to wind, large search distances can ignore the adaptability of airflow to terrain, i.e. topographic deflection of winds. In their study of snow redistribution depths in a head-water catchment in south-western Idaho, Winstral *et al.* (2002) also concluded that a distance-limited estimation of horizon angle is preferred. This results in greater weight being applied to the effects of local topography. Unfortunately, a method for establishing an appropriate maximum search distance remains a challenge and it is often determined through trial-and-error optimization. Furthermore, not all researchers have found the need for a maximum search distance. Wörlen *et al.* (1999), for example, found a strong relationship between measured wind speeds and horizon angles estimated without a maximum search distance. In addition to the ambiguity involved in selecting an appropriate search distance, there is also the problem that, being based on ray-tracing, horizon angle does not account for topographic deflection of wind. Winstral and Marks (2002) suggest averaging horizon angle values across a range of directions (e.g. every 5° over a 30° window centred on the desired wind direction) as a means of increasing the robustness of the index to deviations from the hypothetical regional wind direction. Although averaging over a range of directions does improve results, it does not actually compensate for upwind channelling by topography.

Openness (Yokoyama et al. 2002) is a topographic index of exposure/sheltering that is related to horizon angle (Figure 2C). To the authors' knowledge, openness has never been applied as an index of exposure/sheltering to wind, although it is very similar to the field-based TOPEX index described by Wilson (1984). Openness is defined as the average zenith angle (i.e. 90° minus the horizon angle) in the cardinal and diagonal directions along a distance-limited search path. Therefore, unlike other DEM-derived exposure indices, openness is directionally independent. This characteristic, however, means that openness, like the TOPEX index, is perhaps less suited to measuring wind exposure in areas where there is a dominant wind direction or local channelling of air (Quine and White 1994).

In addition to relative terrain aspect and horizon angle, Antonić and Legović (1999) and Böhner and Antonić (in press) identify a third DEM-derived topographic index for modelling wind exposure/sheltering. *Exposure towards the sloped wind flux* (Figure 2D) combines relative terrain aspect and horizon angle in a single index. This terrain attribute accounts for land-surface orientation relative to the wind and the shadowing effects of distant topographic obstacles. Exposure towards the sloped wind flux (cosα) can be conceptualized as the angle between a plane orthogonal to the wind and a plane that represents the local topography at a grid cell (Böhner and Antonić in press) and is calculated as follows:

$$\cos\alpha = \cos\mu \cdot \sin\beta + \sin\mu \cdot \cos\beta \cdot \cos(\delta - \gamma) \tag{1}$$

where μ is the terrain slope, γ is the terrain aspect, δ is the azimuth of the wind flux, and β is the horizon angle in the wind direction (Antonić and Legović 1999). Notice that cos(δ - γ) is equivalent to the relative terrain aspect. If the horizon angle is set to zero, Equation (1) yields the exposure toward the horizontal component of the wind flux. Equation (1) is commonly used for topographic solar radiation modelling (Böhner and Antonić in press) where α is the solar illumination angle for a given surface, defined by μ and δ. Exposure towards the wind flux has also been found to be one of the most useful indices for explaining spatial variability in various atmospherically-deposited contaminants (Antonić and Legović 1999).

In an early paper on the subject, Lapen and Martz (1993) described two DEM-derived measures of wind exposure/sheltering: directional relief and fetch. Directional relief (Figure 2E), like the horizon-angle based indices described above, is a measure of the degree to which a site is located above or below its surroundings in a specified direction. The main difference, however, is that directional relief is not an angular measure but rather an altitudinal difference. It is calculated by subtracting the elevation of a DEM grid cell from the average elevation of the grid cells that lie between it and the edge of the DEM in a specified direction (Lapen and Martz 1993). Thus, positive values indicate that a grid cell is lower than the average elevation of the grid cells in a direction (i.e. relatively sheltered) and a negative directional relief value indicates that the grid cell is higher (i.e. relatively exposed). The calculation of directional relief is therefore similarly based on a ray-tracing procedure.

Figure 2F shows the Lapen and Martz's (1993) measure of fetch, i.e. the distance of unobstructed airflow, for the Bleaklow LiDAR DEM. The Lapen and Martz (1993) fetch algorithm searches each grid cell along a ray in a specified direction until either the DEM edge is encountered or the following condition is met:

$$Z_{test} \geq Z_{core} + D \cdot I \tag{2}$$

where Z_{core} is the elevation of the grid cell at which fetch is being determined, Z_{test} is the elevation of the grid cell being tested as a potential topographic obstacle, D is the distance between the two grid cells in metres, and I is the height increment in m·m-1 (notice I is effectively unitless). If the search does not identify an obstacle grid cell before the edge of the DEM is reached, the distance between the DEM edge and Z_{core} is entered. Based on the Lapen and Martz (1993) procedure, edge distances are assigned negative values to differentiate between these artificially truncated fetch values and those for which a valid topographic obstacle is identified. For the purpose of effective visualization, Figure 2F shows the logarithm-transformed, absolute fetch values rather than the original distances measured by the fetch algorithm. In Equation (2), I is essentially the minimum tangent of the slope between Z_{core} and Z_{test} needed for the test grid cell to be considered a significant topographic barrier to wind. Lapen and Martz (1993) suggest values for I in the range of 0.025 to 0.1 based on their study of snow distributions in low-relief agricultural landscapes of the Canadian Prairies. Fetch analysis, based on Equation (2), has been applied to the study of patterns of tree species in Taiwan (Huang 2002).

Most of the existing exposure/sheltering indices focus on identifying areas of wind shadow that result from topographic obstacles. A common problem with these indices is their inability to incorporate the channelling/deflection of wind by topography. This is a limitation of the ray-tracing procedure on which most of these indices are based. The main drawback to using ray-tracing in wind modelling applications is that moving air and light do not behave similarly. Whereas a ray of light ends when it encounters a barrier (except for back-scatter), a flowline of air will be deflected around the obstacle, altering its flow direction. That is, wind is capable of being deflected from its path by topographic features. Evidently, the channelling and deflection of near-surface winds by topography is not adequately modelled by existing wind exposure/sheltering indices. In the following section, we describe a simple topographic index that can be used to simulate channelling and deflection of winds by topography.

3 The Channelling/Deflection Index

DEM-based flow routing algorithms have been used extensively to model the flow of surface and near-surface water (e.g. O'Callaghan and Mark 1984, Freeman 1991, Tarboton 1997). These algorithms are used to simulate the spatial patterns of flow directions and contributing area (A), or flow accumulation, for a given surface. Water and air are both fluids. One of the most significant differences between these two fluids is the fact that airflow can occur in an uphill direction whilst water only flows downhill. We propose a method, described below, for modelling wind channelling/deflection that is based on applying a flow routing algorithm to a surface that combines topography and information about regional wind speed and wind direction. This technique effectively compensates for the downhill-only nature of DEM-based flow algorithms by altering the DEM. The upwind source area that results from this analysis is the area from which the wind flux passing through a location originates. The channelling/deflection index (CDI) can be calculated as follows:

$$CDI_{ij} = \frac{A_{ij}(C)}{A_{ij}(P)} \tag{3}$$

where A_{ij} is the upwind source area for grid cell (i, j), P is a grid of a planar surface representing the wind strength and direction, and C is a grid derived by the combination of the streamlined DEM, S, and a planar surface representing the hypothetical wind characteristics. C can be conceptualized as the surface that results from streamlining the DEM and tilting it downwards in the direction of the wind flux by an amount that is proportional to wind speed. C and P are calculated as:

$$C_{ij} = S_{ij} + P_{ij} \tag{4}$$

$$P_{ij} = \tan\lambda \cdot \sin\delta \cdot X_i + \tan\lambda \cdot \cos\delta \cdot Y_j + k \tag{5}$$

where S_{ij} is the elevation of grid cell (i, j) in the streamlined DEM, λ is the gradient of the plane in the wind direction δ, X_i is the x co-ordinate of the ith column in the grid, Y_j is the y co-ordinate of the jth row in the grid, and k is a constant. It should be noted that the contributing area function, A_{ij}, is dependent on both slope and aspect of the surface to which it is applied and not on the actual values of the surface at a specific location. The actual values of C and P therefore have no effect on A_{ij}. Thus, the value of k does not influence CDI_{ij} and can be set to zero or any other desired value. For example, k could be set to a high positive or negative number if there were an algorithmic limitation on the range of values contained in C and P—the X_i and Y_j terms in Equation (4) can otherwise yield very large positive or negative values.

The DEM is not directly used to model the spatial pattern of wind channelling and deflection. Instead a streamlined version of the terrain model is applied. This accounts for the fact that (1) a zone of reduced wind speed and turbulence extends for some distance in the leeward direction of an obstacle, and (2) air starts to rise some distance before it reaches an obstacle due to a wedge of high pressure located in the windward direction. A streamlined terrain model, S, which incorporates these wind zones, can be calculated as follows:

$$S_{ij}\left(\beta_{ij}, I\right) = \begin{cases} Z_{ij} & \tan\beta_{ij} < I \\ Z_{cd} - D \cdot I & \tan\beta_{ij} \geq I \end{cases} \quad (6)$$

where grid cell (c, d) is the first grid cell encountered in a ray extending from grid cell (i, j) in direction δ that satisfies the condition $\tan\beta_{ij} < I$. I is the same height increment described in Equation (2) and is specified by the user. Equation (6) is first used to model the wind shadow in the leeward direction. In a second step, horizon angle is calculated using the leeward streamlined DEM and a wind direction of δ -180. Equation (6) is then used a second time to create the final streamlined DEM, i.e. the terrain model that incorporates both the leeward wind shadow and the windward high-pressure 'ramp'. The leeward effects of an obstacle on airflow extend over a much greater distance than the windward effects, implying that $I_{leeward} < I_{windward}$. Research suggests that the effect of an obstacle on wind patterns can be observed for a distance of 10 to 40 times the obstacle height on the leeward side and approximately two times the obstacle height on the windward side (Lapen and Martz 1993, Huang 2002). This implies values of $I_{leeward}$ and $I_{windward}$ of 0.025–0.1 and 0.5, respectively, although these are approximations. Figure 3 shows the effect of using Equation (6) to streamline the Bleaklow LiDAR DEM. An algorithm for streamlining terrain based on Equation (6) has been implemented in TAS GIS, a freely distributed software package for spatial analysis and environmental modelling (Lindsay 2005).

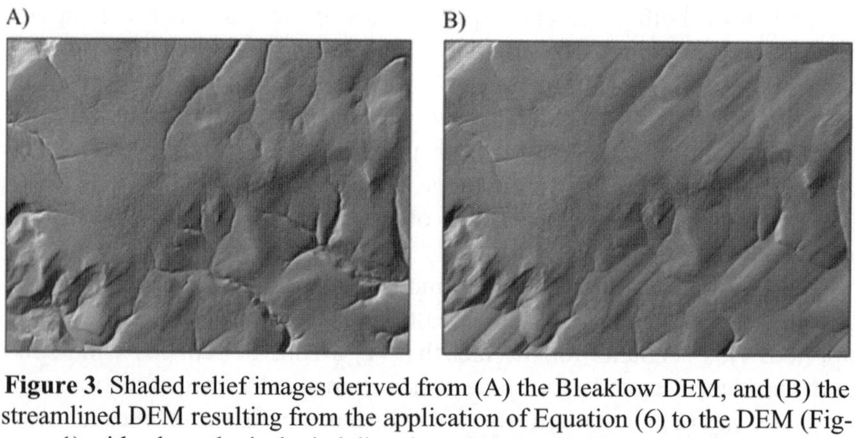

Figure 3. Shaded relief images derived from (A) the Bleaklow DEM, and (B) the streamlined DEM resulting from the application of Equation (6) to the DEM (Figure 1) with a hypothetical wind direction of 225° and Ileeward and Iwindward values set to 0.067 (i.e. a slope of 1:15) and 0.5, respectively.

Equation (4) describes the local balance between the force of gravity pulling a parcel of air downslope, and the horizontal pressure gradient force pushing the parcel in the direction of the regional wind. *P* can be thought of as a plane describing the spatial pattern (gradient and aspect) of atmospheric pressure. Thus, λ is directly proportional to the horizontal pressure gradient force and therefore also to wind speed. Increasing λ, effectively increasing wind speed, reduces the relative influence of topography, or gravitational force, on the airflow pattern resulting from Equation (3) (Figure 4). This implies that stronger winds are deflected, in the horizontal plane, by topographic obstacles to a lesser extent than gentler winds, due to increased momentum. In actuality the horizontal pressure gradient force is usually much smaller than the force of gravity. In fact, if a typical value for the horizontal pressure gradient (a tangent slope of approximately 0.01 Pa·m-1) were used for λ, Equation (3) would be heavily weighted towards the influence of the gravitational force. The resulting modelled airflow pattern would suggest that air drained from a landscape towards the nearest downslope topographic low. Clearly this is unrealistic. The answer to this problem lies in the fact that Equation (4) represents the relative balance of all the forces acting on an air parcel. Although the gravitational force is relatively large, it is severely dampened by the nearly equivalent, though variable, vertical pressure gradient force. This can be accounted for either by including a vertical pressure gradient term in Equation (4) by the addition of a weighting parameter, or equivalently, by ensuring that λ is sufficiently large to give the necessary relative weighting to the horizontal pressure gradient. We prefer the latter approach because it provides a simpler model (recalling that the actual values of *P* need not represent realistic values of atmospheric pressure), and because in most applications we are less concerned with the airflow pattern at a specific wind speed than we are with the pattern resulting from a range of typical wind speeds. This can be achieved by averaging CDI_{ij} over a range of λ values (λ_{min} to λ_{max}), yielding the pattern of wind channelling/deflection over a range of wind speeds. As guidance, it is reasonable that $\tan\lambda_{min} \geq I_{windward}$, implying that air is capable of flowing over obstacles to an extent. At very steep gradients (e.g. $\lambda > 80°$) the effective wind speed is so high that the pathways of individual flowlines are hardly influenced by topography. Thus the value of λ_{max} can be set accordingly.

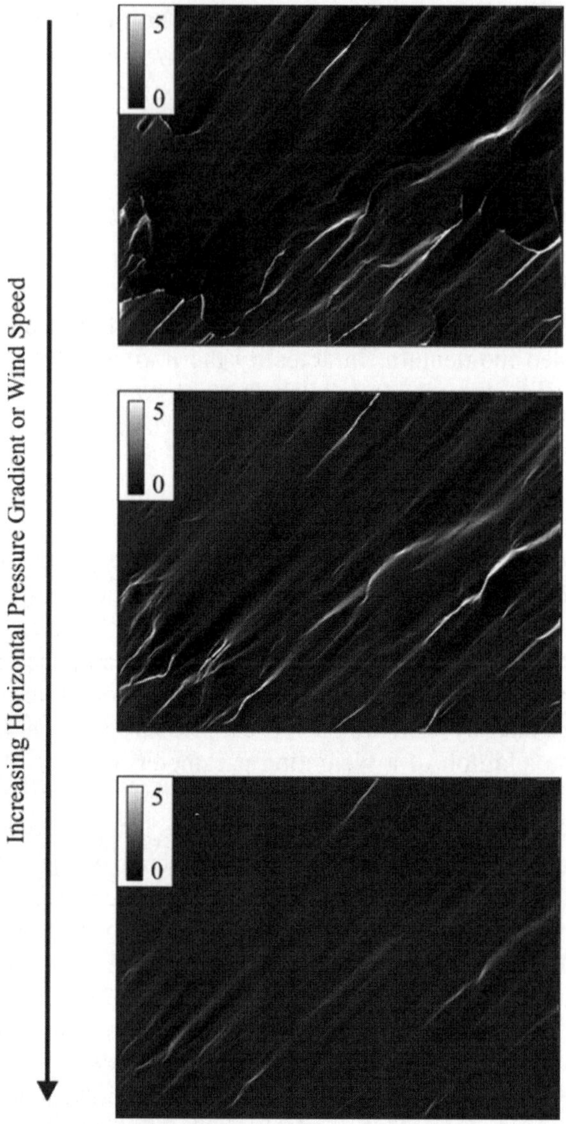

Figure 4. The implications of increasing the plane gradient, λ, (i.e. increasing wind speed) on the modelled pattern of the topographic deflection index (CDI). The regional wind direction is constant for each simulation and assumed to be 225° (i.e. a wind from the south-west).

It may not be readily apparent why $A_{ij}(C)$ is normalized by $A_{ij}(P)$ in Equation (3). Edge effects are not usually problematic for surface water applications of flow routing algorithms so long as the DEM edges do not intersect significant catchment divides. However, in the airflow model described above, all flowlines continue without terminating from one edge of the DEM to the opposite edge; it is impossible to define meaningful flow boundaries from the monotonic surfaces *C* and *P*. Therefore, estimates of A_{ij} are severely impacted by edge contamination and the pattern of upwind source area is heavily influenced by the extent of the DEM. Dividing $A_{ij}(C)$ by $Aij(P)$ normalizes the pattern of upwind source area, effectively removing much of the edge contamination (Figure 5). Normalization compensates for the fact that individual flowlines do not have start points, or rather their start points are unlikely to coincide with the windward edges of the DEM. Even after normalisation, however, locations nearest the edge in the windward direction will suffer from a degree of edge contamination. As such, in estimating the pattern of the CDI for a site, it would be advisable to buffer the area with digital elevation data in the upwind direction. The sensitivity of the CDI to edge contamination is examined further in Section 4.

The normalization of the upwind source area (Equation (3)) also provides a convenient interpretation of the CDI: grid cells with CDI values less than unity are predicted to experience sheltering by upwind topography due to deflection and values greater than unity experience upwind channelling, i.e. relative exposure (Figure 5C). CDI_{ij} can therefore be conceptualized as a measure of how much larger or smaller the upwind source area is for a location as a result of the influence of topography. Also note that the CDI is a unitless index. Because the CDI is intended to be an index of near-surface exposure/sheltering, it is useful to modify Equation (3) to account for areas where the streamlined DEM and the original DEM are not equal (i.e. wind shadows in Figure 6), such that:

$$CDI_{ij} = \begin{cases} 0 & S_{ij} \neq Z_{ij} \\ \dfrac{A_{ij}(C)}{A_{ij}(P)} & S_{ij} = Z_{ij} \end{cases} \tag{7}$$

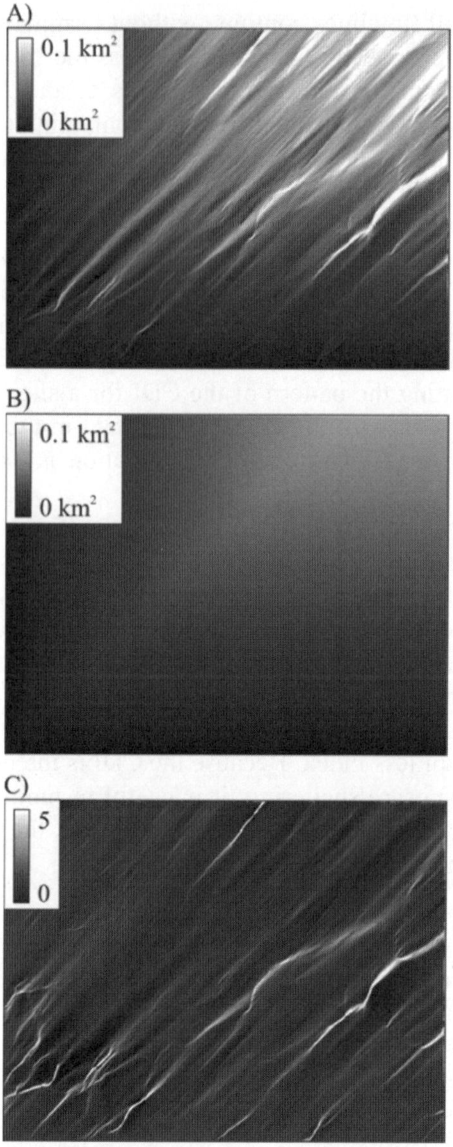

Figure 5. (A) The spatial pattern of upwind contributing area derived from a surface combining the LiDAR DEM and regional wind characteristics, (B) the pattern of upwind contributing area derived from a plane describing wind characteristics, and (C) the pattern of the topographic deflection index (CDI) resulting from the ratio of (A) to (B). The regional wind direction is assumed to be 225° (i.e. a wind from the south-west).

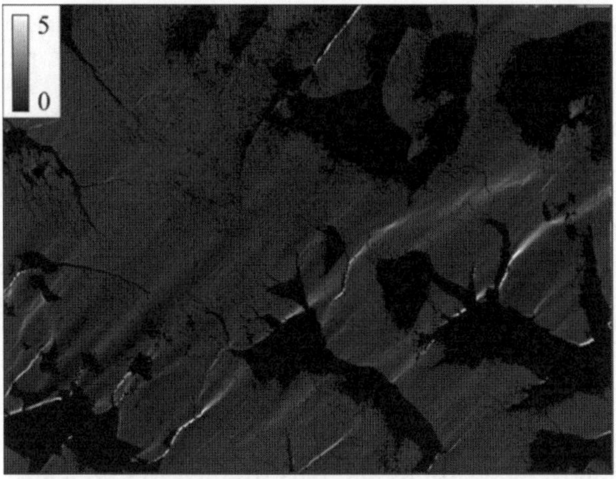

Figure 6. The CDI image that results from the use of the script (described in text) to derive the pattern of CDI for the Bleaklow study area. The CDI has been averaged across a range of λ values from 20° to 70°.

The choice of an appropriate algorithm to estimate A is an important consideration. There are at least six different flow routing algorithms that are commonly used to estimate contributing areas. Details on individual algorithms can be found in the excellent review by Gallant and Wilson (2000). Each flow algorithm differs in essentially two ways: (1) in the method used to calculate flow direction, and (2) in the scheme used to divide the flow entering a grid cell to one or more of its neighbours (Lindsay 2003). Probably the most fundamental difference among flow routing algorithms is the way in which they handle flow divergence. Algorithms that send all of the flow entering a grid cell to one neighbour are sometimes called single flow direction (SFD) algorithms. In contrast, algorithms that are capable of representing flow divergence are often referred to as multiple flow direction (MFD) algorithms because they disperse the flow entering a grid cell to two or more neighbouring cells (Wolock and McCabe 1995). Wind is dispersive and therefore it is necessary to use an MFD type flow routing algorithm to estimate the CDI (Equation (3)). Several MFD flow algorithms exist including FD8 (Freeman 1991, Quinn *et al.* 1991), DEMON (Costa-Cabral and Burges 1994), and D∞ (Tarboton 1997). Each of these algorithms can be used to estimate the CDI. D∞, the least dispersive of the MFD algorithms, is perhaps the most widely available because of theoretical advantages, its robustness against directional biases caused by the grid structure of data, and its relative simplicity. With greater flow dispersion than D∞, controlled by a user-specified dispersion parameter, FD8 can also produce realistic flow patterns. The flow-tube based approach used in DEMON is well suited to modelling airflow; however, the algorithm's complexity is such that very few analysis packages have implemented the scheme.

In addition to flow routing algorithm considerations, it is also necessary to select a method of removing depressions and forcing small gradients across any flat areas in *C*. This is a pre-processing step that ensures the proper functioning of the flow algorithm. Flow algorithms are incapable of modelling flow paths out of depressions and across flat areas; modelled flowlines terminate in these features (Jenson and Domingue 1988). Notice that this issue is only relevant when λ is sufficiently small to allow depressions to occur in the combined surface, *C*. That is, when the gradient of *P*, which does not contain depressions, is much larger than the gradients of *S*, depressions are unlikely to occur in *C*. Several algorithms have been developed to remove depressions and flats from DEMs, which can be categorized as depression filling, depression breaching, and combination algorithms (Lindsay and Creed 2005). For this application, depression filling methods (e.g. the efficient algorithm described by Planchon and Darboux 2001) are probably the most appropriate. It should be noted that depressions in *C* are likely areas of flow separation.

The CDI can be calculated in most GIS packages that possess an appropriate flow routing algorithm. The following code is a TAS GIS script that can be used to estimate an average CDI image from a DEM over a range of λ values:

```
Total CDI='S'*0 //Initialise a grid the same size
as the DEM
startLamda='S'*0+20 //Initialise a second grid
REPEAT 6 TIMES
   tanLamda=tan((LOOPCOUNT-1)*10+'startLamda')
   P='tanLamda'*sin(225)*'X'+'tanLamda'*cos(225)*'Y'
   C='S'+'P'
   C filled=DEPFILL('C',1,true)
   A1=CONTAREA('C filled',Dinf,SCA,1,true,false)
   A2=CONTAREA('P',Dinf,SCA,1,true,false)
   CDI='A1'/'A2'
   Total CDI='Total CDI'+'CDI'
END REPEAT
CDI 225=('Total CDI'/7)*('S'='DEM')
```

In the above script, δ is 225° (a south-west regional wind direction) and λ ranges from 20° to 70°, calculated using a 10° step. The final image

'CDI 225' (Figure 6) is averaged over the range of λ values and locations within the wind shadows are assigned null values.

4 Sensitivity to Edge Contamination

Edge contamination occurs in the CDI for two main reasons. First, topographic features beyond the windward edges of the study area can cause wind shadows that extend into the site, i.e. the streamlined DEM could fail to incorporate important distant features. The likelihood of this occurring is largely dependent on the local relief characteristics. Secondly, in the method described above there is an implicit assumption that grid cells along the windward edges of the study site have equal upwind source areas. In actuality, however, the upwind source areas of edge grid cells will be preconditioned by distant, unmodelled topography. Therefore, airflow can be channelled before it enters the site, with implications for the within-site airflow pattern. This unknown boundary condition will not be incorporated into the modelled airflow pattern. As an MFD flow algorithm is needed to estimate A_{ij}, however, distant topography will impact CDI_{ij} to a lesser degree than nearer topography. This characteristic will reduce edge contamination in a CDI image.

So then, how large of an upwind buffer is needed to ensure that the CDI image is sufficiently unaffected by edge contamination? To evaluate the sensitivity of the pattern of the CDI to edge contamination British Ordnance Survey Land-Form PROFILE 1:10,000 10 m DEMs (interpolated from contours) were obtained for five randomly selected 3 km × 3 km areas in the UK. The five study areas demonstrated a range of topographic settings with varying relief from 117 m to 351 m and average slopes among the sites varying from 7.5° to 12.9°. The CDI was first estimated for each study area using a random wind direction. The D∞ routing algorithm was used to estimate upwind source areas and $I_{leeward}$ and $I_{windward}$ were set to 0.067 and 0.5, respectively. The CDI estimates were repeated another 12 times per site, increasing the windward buffer of elevation data by 1 km with each model run. Therefore, the maximum buffer width that was used was 12 km beyond the windward edges of each study area. The pattern of the 12 km buffered CDI image served as the reference image for comparisons. The 12 CDI images with buffer widths ranging from 0 km (i.e. the DEM of the study site with no buffer) to 11 km were then correlated with the reference image. Comparisons were restricted to the 9 km^2 study areas and did not include data within buffer areas.

Figure 7 shows the relation between buffer width and the correlation coefficient, *r*, measured between the CDI image derived using a particular buffer width and the reference image. The findings illustrate that for each site *r* increased rapidly with buffer widths from 0 km to 4 km and levelled off with values above 0.9 for buffer widths greater than this threshold. Therefore, the pattern of the CDI for an area derived using a buffer width of 4 km was found to be very similar to the pattern that was calculated using a 12 km buffer width. Thus, the vast majority of the spatial variation in airflow patterns within a site can be accounted for by the influence of topography within 4 km in the upwind direction, with between 6% and 52% of the pattern accounted for by the topographic variation directly within the site (i.e. *r* values of images with no buffers ranged from approximately 0.24 to 0.73). Despite possessing a range of topographic characteristics, each of the five study areas demonstrated remarkably similar patterns, particularly with respect to the levelling off of *r* values with more extensive buffers. However, because topographic indices involving slope and aspect demonstrate a strong dependency on DEM grid resolution, the specific value for the buffer threshold (here interpreted to be approximately 4 km) may actually be dependent on the spatial resolution of the DEM used in the analysis. Nonetheless, this analysis demonstrates that (1) a buffer is required to minimize the effects of edge contamination on CDI images, and (2) the buffer needs not be excessively wide when compared to the overall extent of the site over which the CDI is being modelled.

5 Discussion and Conclusions

A review of DEM-derived indices of local exposure/sheltering to wind revealed that most existing indices adequately represent wind shadowing by barriers but are incapable of modelling topographic channelling and deflection of winds. This was viewed as a major limitation to the application of these topographic indices to the study of erosion of fine sediment and soil, snow accumulation, vegetation patterns, deposition of atmospheric contaminants, and other wind-related phenomena. A new exposure/sheltering index, the channelling/deflection index (CDI), was proposed to compensate for the shortcomings of existing measures of local windiness. The CDI accounts for topographic shadowing, through the incorporation of a 'streamlined' DEM, as well as the influences of channelling and deflection by topographic obstacles, through application of the same flow routing methods that are commonly used to simulate the overland flow of water. The topographic index is a relative measure of how windy a location is expected to be, over a range of atmospheric conditions focused in a particular direction, compared with a featureless surface. The CDI reduces a complex dynamic system to a simple index with only five significant adjustable parameters (δ, $I_{leeward}$, $I_{windward}$, λ_{min}, λ_{max}). With the exception of the derivation of a streamlined DEM, which required the development of a specialized (though publicly available) algorithm, the CDI is estimated using spatial analysis functions that are commonly available in GIS packages.

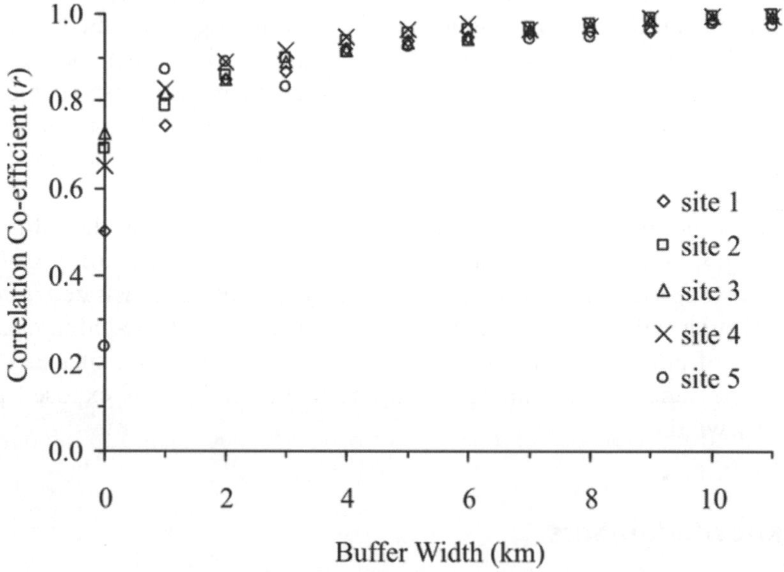

Figure 7. The relation between buffer width and the association (strength of the Pearson's correlation co-efficient, r) between the derived CDI pattern and the reference image, derived using a 12 km buffer width.

Because DEM edges are arbitrary with respect to wind flow boundaries, the CDI inherently suffers from a degree of edge contamination. Edge effects are somewhat diminished through normalization of the upwind source area by the equivalent value derived from a featureless surface, and by the need for a divergent flow routing algorithm. Some degree of edge contamination can however remain. As such, in application the CDI should always be estimated using an upwind buffer of elevation data. An analysis of five study sites possessing a range of topographic characteristics showed that a buffer of elevation data of approximately 4 km width in the upwind direction (based on 10 m grid resolution elevation data) was sufficient to accurately replicate the pattern of the CDI derived using a much more extensive buffer.

The method of calculating the CDI described in this chapter uses a planar surface to direct winds across a landscape (i.e. analogous to a pressure surface). A planar surface is suitable for relatively small areas but for larger extents the influence of the Coriolis Effect on wind directions would need to be taken into account. This could be accomplished by using a curved surface, instead of a plane. Earth curvature would also have to be accounted for when modelling flow patterns over more extensive areas, which could be challenging given that most existing GIS packages are limited to operations on rectangular co-ordinate systems. In light of the above discussion, the CDI can be thought of as a framework for incorporating channelling/deflection effects into a topographic index rather than a rigid procedure. In application, individual researchers may choose to modify various components of the method described here, e.g. in the use of a non-planar surface or in the incorporation of roughness effects in the flow routing method.

Whilst the method described in this chapter for modelling the influence of topography on site windiness is sensible and robust, the utility of CDI for analysing wind-related phenomena has yet to be demonstrated. Ongoing research will evaluate the usefulness of the CDI for modelling spatial patterns of atmospherically-deposited contaminants in moss species (bio-indicators) and for explaining variations in the erosion of exposed peat soils by wind.

Acknowledgments

The authors would like to thank Clive Agnew, Mark Hildebrandt, Laurice Magdalani, and Stuart Lane for their helpful advice, which greatly strengthened this chapter.

List of Symbols and Abbreviations

α	Sloped wind flux or solar illumination angle
β	Horizon angle in the wind direction
γ	Terrain aspect
δ	Azimuth of wind flux
λ	Gradient of a plane representing a wind force
λ_{min}	Lower range of plane gradients used to calculate the CDI
λ_{max}	Upper range of plane gradients used to calculate the CDI
μ	Terrain slope

A	Contributing area (flow accumulation) grid
c	Column of grid cell (c, d)
C	Grid derived by the addition of S and P
CDI	Channel/deflection index
CFD	Computational fluid dynamics
D	Distance between two grid cells
d	Row of grid cell (c, d)
DEM	Digital elevation model
i	Column of cell (i, j)
I	Height increment (m·m-1)
j	Row of cell (i, j)
k	Constant value used to calculate P
MFD	Multiple flow direction
P	A planar surface grid representing a hypothetical wind characteristics
S	Streamlined DEM
SFD	Single flow direction
X	The x-coordinate of a column in a grid
Y	The y-coordinate of a row in a grid
Z_{core}	Elevation of a grid cell at which fetch is measured
Z_{test}	Elevation of a grid cell being tested as a potential topographic obstacle to wind

References

Antonić, O. and Legović, T., (1999), Estimating the direction of an unknown air pollution source using a digital elevation model and a sample of deposition, *Ecological Modelling*, **124**: 85–95.

Böhner, J. and Antonić, O., (In press), Land surface parameter specific topoclimatology. In Hengl, T. and Reuter, H.I. (eds): *Geomorphometry: concepts, software, applications*, Office for Official Publications of the European Communities, Luxembourg, EUR 22670 EN: 147–171.

Chaderjian, N.M., Ahmad, J.U., Kramer, M.G. and Holst, T.L., (2006), Navier-Stokes simulation of local winds over the earth's topography, *NASA Advanced Supercomputing (NAS) Division Technical Report; NAS–06–015*.

Chappell, A. (1996), Modelling the spatial variation of processes in the redistribution of soil: digital terrain models and 137Cs in southwest Niger, *Geomorphology*, **17**: 249-261.

Costa-Cabral, M.C. and Burges, S.J., (1994), Digital elevation model networks (DEMON): A model for flow over hillslopes for computation of contributing and dispersal areas, *Water Resources Research*, **30**(6): 1681–1692.

Dozier, J., Bruno, J. and Downey, P., (1981), A faster solution to the horizon problem, *Computers and Geosciences*, **7**: 145–151.

Erickson, T.A., Williams, M.W. and Winstral, A., (2005), Persistence of topographic controls on the spatial distribution of snow in rugged mountain terrain, Colorado, United States, *Water Resources Research*, **41**, W04014, doi:10,1029/2003WR002973.

Freeman, T.G., (1991), Calculating catchment area with divergent flow based on a regular grid, *Computers & Geosciences*, **17**(3): 413–422.

Gallant, J.C. and Wilson, J.P., (2000), Primary terrain attributes, In Wilson, J.P. and Gallant, J.C. (eds.): *Terrain Analysis: Principles and Applications*, New York: John Wiley & Sons: 51–85.

Girault, V. and Raviart, P.A., (1986), *Finite element methods for Navier–Stokes equations: Theory and algorithms*, (Springer Series in Computational Mathematics. Volume 5), Berlin: Springer-Verlag.

Huang, K.Y., (2002), Evaluation of the topographic sheltering effects on the spatial pattern of Taiwan fir using aerial photography and GIS, *International Journal of Remote Sensing*, **(23)**10: 2051–2069.

Jenson, S.K. and Domingue, J.O., (1988), Extracting topographic structure from digital elevation data for geographical information system analysis, *Photogrammetric Engineering and Remote Sensing*, **(54)**11: 1593–1600.

Lapen, D.R. and Martz, L.W., (1993), The measurement of two simple topographic indices of wind sheltering-exposure from raster digital elevation models, *Computers & Geosciences*, **(19)**6: 769–779.

Lindsay, J.B., (2005), The Terrain Analysis System: A tool for hydro–geomorphic applications, *Hydrological Processes*, **(19)**5: 1123–1130.

Lindsay, J.B., (2003), A physically based model for calculating contributing area on hillslopes and along valley bottoms, *Water Resources Research*, **39**, 1332, doi:10.1029/2003WR002576.

Lindsay, J.B. and Creed, I.F., (2005), Removal of artefact depressions from digital elevation models: towards a minimum impact approach *Hydrological Processes*, **19**: 3113–3126.

Moore, I.D., Grayson, R.B. and Ladson, A.R., (1991), Digital Terrain Modelling: A Review of Hydrological, Geomorphological, Biological Applications, *Hydrological Processes*, **(5)**1: 3–30.

Murakami, S. and Komine, H., (1983), Prediction method for surface wind velocity distribution by means of regression analysis of topographic effects on local wind speed, *Journal of Wind Engineering and Industrial Aerodynamics*, **15**: 217–230.

O'Callaghan, J.F. and Mark, D.M., (1984), The extraction of drainage networks from digital elevation data, *Computer Vision, Graphics, & Image Processing*, **28**: 323–344.

Planchon, O. and Darboux, F., (2001), A fast, simple and versatile algorithm to fill the depressions of digital elevation models, *Catena*, **46**: 159–176.

Quine, C.P. and White, I.M.S., (1994), Using the relationship between rate of tatter and topographic variables to predict site windiness in upland Britain, *Forestry*, **67**(3): 245–256.

Quinn, P., Beven, K., Chevallier, P. and Planchon, O., (1991), The prediction of hillslope flow paths for distributed hydrological modelling using digital terrain models, *Hydrological Processes*, **5**: 59–79.

Tarboton, D.G., (1997), A new method for determination of flow directions and upslope areas in grid digital elevation models, *Water Resources Research*, **(33)**2: 309–319.

Uchida, T. and Ohya, Y., (1999), Numerical simulation of atmospheric flow over complex terrain, *Journal of Wind Engineering and Industrial Aerodynamics*, **81**: 283–293.

Wilson, J.D., (1984), Determining a TOPEX score, *Scottish Forestry*, **38**: 251–256.

Wilson, J.P. and Gallant, J.C., (2000), Digital terrain analysis, In. Wilson, J.P and Gallant, J.C. (eds.): *Terrain Analysis: Principles and Applications*, New York: John Wiley & Sons: 1–27.

Winstral, A. and Marks, D., (2002), Simulating wind fields and snow redistribution using terrain–based parameters to model snow accumulation and melt over a semi–arid mountain catchment, *Hydrological Processes*, **16**: 3585–3603.

Winstral, A., Elder, K. and Davis, R.E., (2002), Spatial snow modeling of wind-redistributed snow using terrain–based parameters, *Journal of Hydrometeorology*, **(3)**5: 524–538.

Wolock, D.M. and McCabe Jr., G.J., (1995), Comparison of single and multiple fow direction algorithms for computing topographic parameters in TOPMODEL, *Water Resources Research*, **(31)**5: 1315–1324.

Wörlen, C., Schulz, K., Huwe, B. and Eiden, R., (1999), Spatial extrapolation of agrometeorological variables, *Agricultural and Forest Meteorology*, **94**: 233–242.

Yokoyama, R., Shirasawa, M. and Pike, R.J., (2002), Visualizing topography by openness: a new application of image processing to digital elevation models, *Photogrammetric Engineering and Remote Sensing*, **(68)**3: 257–265.

Spatial Correlation of Topographic Attributes in Loess Plateau

ZHANG Ting, LI Jun, WANG Chun and ZHAN Lei

Abstract

As one of the important parts of Digital Terrain Analysis (DTA), spatial correlation analysis of topographic attributes (TAs) is an effective method of analysing the topographical environment. This chapter proposes a spatial correlation model for nine selected TAs, providing an effective method for quantitative DTA research and landform recognition. Forty seven different loess landforms were selected as test areas and their corresponding 5 m grid cell DEM data as test data. With grey correlation analysis, spatial correlations for these TAs were analysed and the TAs' correlation model built. Furthermore, the variations of the correlation curves are discussed. Results show that (1) TA correlation curves are similar to the spectrum, which provides a means of modelling the natural environment; (2) the correlation curve changes with the topographical area; and (3) the correlation curve reflects the landform and evolution pattern of the sample area.

Keywords: Loess plateau, DEM, topographic attribute, correlation.

1 Introduction

Terrain analysis is an important method of analysing the natural topographical environment. Traditional terrain analysis based on topographical maps, which deeply affect our life, is ineffective. DEM data, because of their availability and GIS-assisted processing, have taken the place of the topographical map in many aspects of terrain analysis. As the basic information source not only can DEMs be used for computing topographic attributes (TAs), but can also be employed in data mining at a profound level.

The analysis of TAs is the basis of digital terrain analysis based on a DEM. Correlation and variation laws of multiple TAs express the form and the evolution processes of the physical environment. In practice, the relationship between TAs appears to be of more importance than the TAs, particularly with reference to the surface.

At present, digital terrain analysis mainly includes algorithm research and quantitative analysis of TAs, extracting and recognizing topographical structure lines and landform units, research on soil erosion, the simulation of drainage basin characteristics, and geological calamity prevention in loess areas, etc. (Tang 1991, 2003). Up to now, various types of topographical parameters based on DEMs have been extracted to describe the terrain features. Research on the algorithms and the expression of individual TAs based on DEMs is comparatively mature. But the analysis of the relevance of these TAs is limited to aspects connected with social, economic, or hydrological parameters (Keck *et al.* 1993, Garner 1998, Giles and Franklin 1998, Yue 1998, Western *et al.* 1999, Li 2000, Da Silva and Alexandre 2005, Shary *et al.* 2002, Shary and Sharaya in press). Systematic research on the correlation among numerous TAs is absent. Therefore, it is important to conduct research on the correlation and the laws governing the relation between attributes and their spatial locations, in particular on the quantitative model within TAs based on DEMs of high accuracy and a spacing of 5 m. However, little work has been done on the correlation between multiple TAs and their spatial variance, and this is essential for quantitative geomorphology research. The objective of this chapter is to investigate this topic.

2 Experiment Foundation

2.1 Research area and data

The research area is located in northern Shaanxi province in the Loess Plateau (see Figure 1). The landform in the research area contains most of the representative landscapes of the Loess Plateau (Gan 1990, Luo 1988, Zou 1985) such as loess Yuan, loess Liang and loess Mao. Interestingly, landforms in the research area vary with the spatial location according to a certain law. From north to south, landforms vary from loess low-hill to loess Mao, then gradually to loess Liang, until areas around Luochuan and Huangling that contain fragmentary loess Yuan. Then coming to the south, fragmentary loess Yuan is replaced by loess Yuan in Tongchuan, and then comes the terrace of the Weihe River. This area therefore favours topographical research on the laws governing landform and spatial location. It makes full consideration of the effect and the relation among materials, and energy and time in the landform evolution process as well. Forty seven test areas were selected according to typicality, data availability, and integrity. Corresponding 5 m grid cell DEM data, produced according to the national standard of China, are selected as test data.

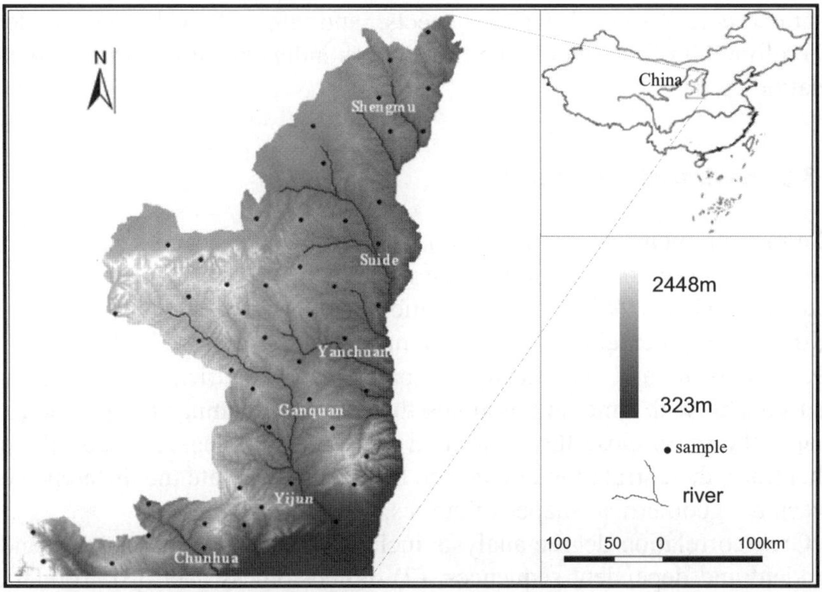

Figure 1. Distribution of the sample areas.

2.2 Selection of TAs

Three aspects were considered in choosing suitable TAs: (1) the landform shape should be fully expressed; (2) every selected TA should be independent of others; (3) the selection should be the most effective one to reveal the law governing the relation among them.

Any landform, no matter whether plain, hill or high mountain, is formed by different slope surfaces (Zou 1985). Any change of landform actually results from a change of slope surface. That is to say, every complicated area can be divided into many tiny slope surfaces. Every slope surface reflects the degree of undulation through its slope gradient. And the transition and turning between different slope surfaces can be reflected by profile curvature and plan curvature. Therefore, slope, profile curvature, and plan curvature are selected to express the undulation and change of landform surfaces.

Water erosion is a serious problem in the Loess Plateau, and the integral value of an Area-Altitude curve is a TA reflecting the stage of basin evolution. It reflects the extent to which the basin landform has been eroded and represents the stage of landform evolution. The elevation difference is the essential feature of the landform. Several TAs portray different landform characteristics from different aspects, and are relatively independent. Therefore, all of them could be regarded as independent factors in the correlation model.

2.3 Experimental Method

The measurement of the correlation between two TAs can be expressed by the grey correlation degree in grey systematic theory (Deng 1987, 1988, Liu 1999, Zhao 1997). Grey correlation analysis is a quantitative analysis of the developmental trend of a dynamic process presented by Deng (1987, 1988). This method analytically compares the geometrical shape of curves that change with time and assumes that the more similar the geometrical shape, the more close the trend, and the larger the degree of correlation. Therefore, the correlation degree can be used to evaluate the difference between the geometrical shapes of curves.

Grey correlation degree analysis includes the following steps: (1) independent and dependent sequences; (2) dimensionless data; (3) correlation coefficient; (4) correlation degree; and (5) the correlation sequence.

3 The Correlation Model for TA

3.1 Conceptual model

There is special meaning only when TAs are concretely combined with landforms. Different TAs can express different aspects of the landform, so it is very difficult to make sure that there is a clear correlation among TAs from their algorithms. In some landforms, the trend of sub-sample values for two or more TAs may be very close. But this trend may be a totally independent relationship in another landform. These phenomena are inherent representations of the fact that correlation of TAs is not same as data correlation in mathematics. The variation of this correlation can reflect the variant characteristics of landforms to a certain extent. Hence, we can conclude that when one or several TAs are evaluated in some landform, the value of another TA corresponding to them varies around a definite value over a short space according to a certain law. This kind of relationship among TAs is defined as TA correlation with uncertainty in this concrete landform. The correlation among TAs changes according to the landform.

The functional relationship and relevancy relationship among TAs in a concrete landform can exchange in certain situations. The functional relationship among TAs may change into a relevancy relationship when there is an observation error. And correlation among TAs may also be transformed into function relationship by applying all the related factors in the equation as variables. By this method, the correlation could be described as a type of functional form.

Correlation does not mean causality. Causality can be established from professional knowledge or previous considerations or theories instead of by statistics. Correlation makes preparation for causality. It may be direct causality, indirect causality, or only a relationship in statistics among the variables for remarkable correlation (Wei 2005). Only after understanding the correlation in the systems or among the factors can there be a thorough understanding of the system, allowing the determination of the leading factor or potential factor, and knowing which is superior or inferior. It is the first step of system analysis to find correlation in the data and calculate the value of the correlation degree. This is the basis of factor discrimination, predominance analysis, validation of prediction precision and system decision, etc.

According to different classified symbols, the correlation among TAs can be shown in different forms. It can be divided into significant correlation, partially significant correlation, and insignificant correlation according to the degree of correlation; it also can be divided into positive correlation and negative correlation according to relevant directions; linear correlation and non-linear correlation according to the relevant form; simply correlation, compound correlation, and partial correlation according to the number of variables concerned.

3.2 Applicable condition

There is a most suitable circumstance for every model. In this situation, the model runs steadily and its result is credible. This condition of applicability includes sampling and modelling.

The propinquity among all factors of the system is qualitatively estimated by correlation analysis so that the characteristic of correlation values sequence is discovered by comparing the independent factors with the dependent factors of the system. As the sequence of correlation degree is even more important than the value itself in correlation analysis, whether the correlation value can keep in the sequence gained by different models is a basic prerequisite and an important part of correlation analysis. Correlation coefficients of every sub-sample are the foundation of factors correlation calculation, whereas the distribution of the correlation coefficient is affected by many factors.

Different formulae for the correlation coefficient take on different characteristics in keeping the sequence of correlation values. A comparison is drawn between these formulae and it is known that Equation (1) does better in keeping the sequence of correlation values (Zhao 1997). Therefore, it is adopted in the following study.

$$\xi_i(k) = \frac{\Delta_i(k) + \rho b}{a + \rho b} \tag{1}$$

where
$$a = \min_i \min_k \left| X_0(k) - X_i(k) \right|,$$
$$b = \max_i \max_k \left| X_0(k) - X_i(k) \right|$$
$$\Delta_i(k) = \left| X_0(k) - X_i(k) \right|,$$
$$\rho \in [0,1].$$

ρ is the recognition differential and $\xi_i(k)$is the relative difference between X_i (comparing series, i=1, 2, …, n) and X_0 (reference array), which is also called the correlation coefficient at the k-th point.

$\min_k \left| X_0(k) - X_i(k) \right|$ represents the minimum absolute difference between X_i and X_0, whereas $\min_i \min_k \left| x_0(k) - x_i(k) \right|$ refers to the minimum among all the $\min_k \left| x_0(k) - x_i(k) \right|$, i.e. minimum absolute differences between reference array (X_0) and comparing series $v(X_i$, i=1, 2, …, n).

In many cases, because of the huge extent of the population, it is not possible to examine all of the population. Accordingly, sampling is necessary and then the characteristic of the population can be acquired by generalizing the result of sampling. That is to say, the sampling methods, area, and number of samples have a great influence on the credibility and stability of the model. Every sample area will be randomly sampled 20 times; 70 subsamples are sampled each time and the size of each sub-sample is 3 × 3 km. The results of the 20 samples were analysed. If the results are stable, the average value of these results will be accepted as the mean trend of this area.

Due to the algorithm, the analysis windows of slope, plan curvature, and profile curvature are focal widows, while the analysis windows of elevation difference and integral value of the Area-Altitude curve are extended to the whole area. That is to say, the analysis windows of the selected five TAs are inconsistent. Therefore two statistic variables, i.e. mean and standard deviation, are selected to express the overall characteristics of these three factors (slope, plan curvature, and profile curvature) so that the analysis windows of all factors can be unified over the entire extent of the sample area. Eight TAs are selected for inclusion in the correlation model.

If there are two factors, one of which shows close correlation with another one in different areas, i.e. this factor can be substituted by another one, it needs to be eliminated from the eight selected factors in order to reduce the workload of the operations that follow. Accordingly, each two factors are analysed and compared by simple correlation analysis in the 47 sample areas. The results are: (1) a constant, close correlation does not exist between every pair of factors from the eight selected in various sample areas; (2) a significant degree of correlation varies with different topographical areas. Based on all the above, the eight selected TAs could be included in the following calculations of the correlation model.

3.3 Correlation Analysis

3.3.1 The dependent variable for correlation

There is a need to select a dependent variable as a reference one in multivariable analysis, to express the characteristic of landform evolution in the sample area with the help of correlation between dependent variables and other variables. The dependent variable should reflect the compound character of the landform for the sample area and express the current characteristic of the landform and the potential trend of the evolution process to a certain extent. According to the above, gully density was selected as the dependent variable for the following correlation analysis.

Gully density, also named valley density or channel density, means the total length of gullies in a unit area. It is a factor to reflect the extent of ground fragmentation, split by channels. The higher the value of gully density, the more fragmented is the ground. The more fragmented the ground, the more significantly undulating are the slopes. The higher the value of gully density, the faster is the surface runoff, soil erosion, and the development of gully erosion. Therefore, gully density is an important character of landform evolution, precipitation, the ability of soil infiltration, and the resistance to erosion of surfaces, and is also a compound reflection of climate, topography, lithology, vegetation, and some other ingredients.

3.3.2 Correlation Analysis of TA

3.3.2.1 Correlation Analysis in single sample area

The sample area of Yijun was taken to analyse the correlation among slope, plan curvature, profile curvature, elevation difference, integral value of Area-Altitude curve, and gully density. Figure 2 is the average correlation chart of 20 stochastical sampling results from the Yijun sample area.

As is shown in Figure 2, the mean profile curvature shows the most significant correlation with gully density, followed by mean plan curvature and mean slope in sequence. The slope standard deviation and integral value of Area-Altitude curve are the two most highly correlated factors.

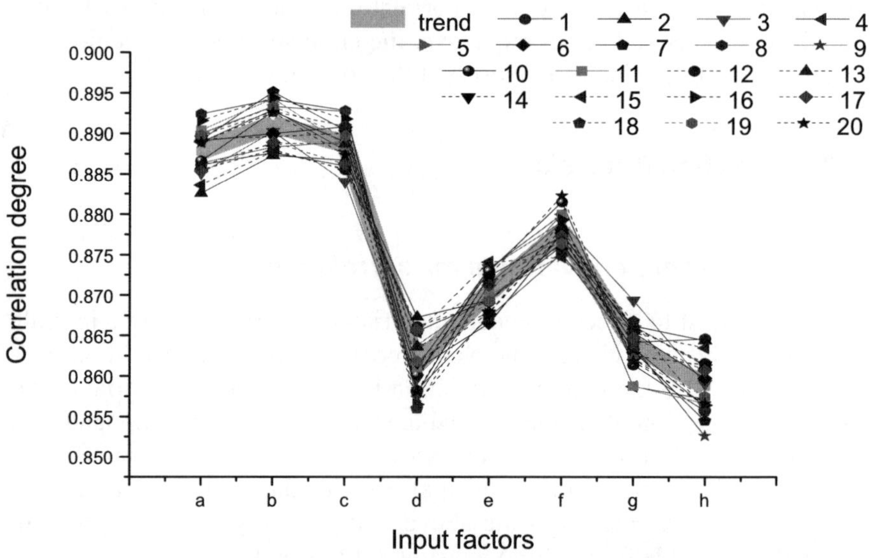

a: mean slope
b: mean profile curvature
c: mean plan curvature
d: slope deviation
e: profile curvature deviation
f: plan curvature deviation
g: elevation difference
h: integral value of Area-Altitude Curve

Figure 2. Average correlation chart of 20 stochastic sampling results in the sample area of Yijun.

The sample distributions of these three TAs are extraordinarily close to the one of gully density in the Yijun area, whereas the other five TAs go further. Figure 3 is the sample distribution and scatter diagram of mean slope and integral value of the Area-Altitude curve corresponding to gully density.

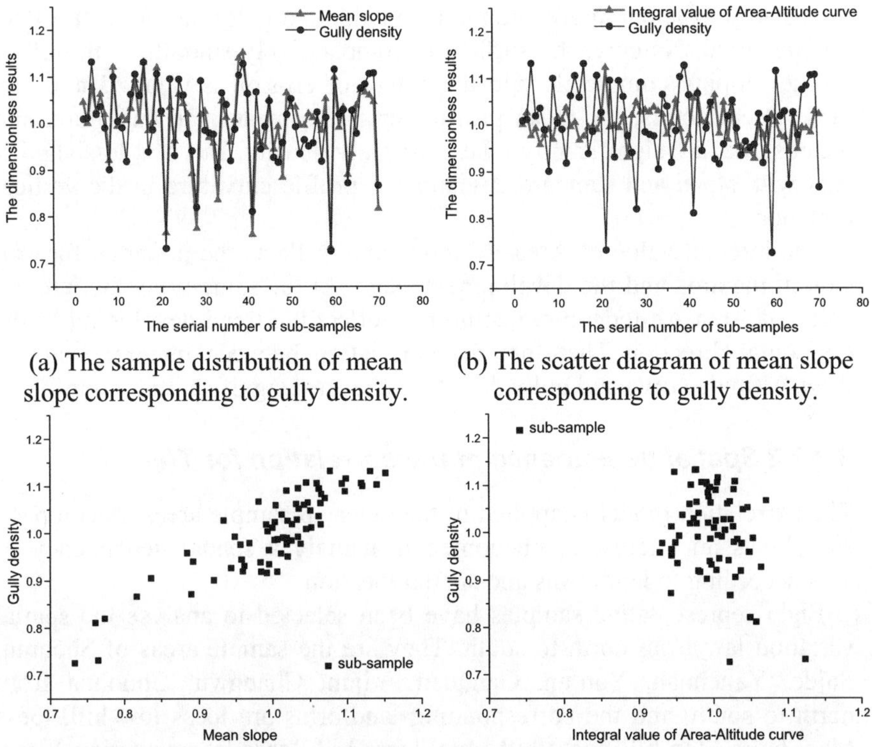

(a) The sample distribution of mean slope corresponding to gully density.

(b) The scatter diagram of mean slope corresponding to gully density.

(c) The sample distribution of integral value of Area-Altitude curve corresponding to gully density.

(d) The scatter diagram of integral value of Area-Altitude curve corresponding to gully density.

Figure 3. The sample distribution and scatter diagram of compared factors corresponding to gully density.

With the comparison in Figure 3, the mean slope shows a significant correlation with gully density. Namely, when one increases, the other one shows the same tendency, and the distribution characteristics of their samples are the same. However, there are no correlations between the integral value of the Area-Altitude curve and gully density, and the distribution characteristic of their samples is inconsistent, namely when one of them increases, the another might not vary with it. The same is true for the other TAs.

As for the whole process of loess landform evolution, the sample area of Yijun is in the early stage of development. Its erosion relies mainly on ditch erosion; the depth of ground incision is shallow, and ground erosion is in the horizontal direction. The TAs reflecting landform changes in the horizontal direction are related closely to gully density. There is also great potential to cut down to the erosion datum in the Yijun area. Correlation of the mean profile curvature and the mean slope to gully density just reflect this potential. Namely, the surface is comparatively smooth, with gullies not developing completely and the extent of erosion is weak. Therefore, these three factors, i.e. mean profile curvature, mean plan curvature, and mean slope, are significantly related to gully density. But it is less significant with slope and standard deviation of profile curvature in the vertical direction.

The integral value of Area-Altitude curve reflects the potential for erosion in the area and the development stage. At the same time, the integral value of Area-Altitude curve could not reflect the trend developing in the horizontal direction Therefore the correlation between integral value of Area-Altitude curve and gully density is the least significant in Yijun.

3.3.2.2 Spatial dependence of the correlation for TAs

The correlation model is applied in the other 46 sample areas. A comparison shows an interesting phenomenon, namely a spatial dependency of TAs according to landforms and spatial location.

Eight representative samples have been selected to analyse the spatial variation law from north to south. They are the sample areas of Shenmu, Suide, Yanchuan, Yan'an, Ganquan, Yijun, Changwu, Chunhua from north to south, and the corresponding landforms are loess low-hill, loess Mao, loess Mao-hill, loess hill-Mao, loess hill, loess fragmentation Yuan, loess Yuan, and loess table Yuan, respectively.

The tendency charts of correlation are calculated from the eight sample areas (Figure 4). The curve formed by the correlation is called the correlation curve or correlation spectrum of TAs. Figure 4 shows:

(1) The correlation among TAs varies according to the landform and spatial location. The correlation will be different when obtained from different landforms or different spatial locations. For example, the correlation results in sample areas Suide and Yanchuan are different; although the landform is similar, their locations are different.

The undulation of the correlation curve increases gradually when the landform varies from loess low-hill (such as Shenmu) to loess Mao (such as Suide) then to loess Mao-hill (such as Yanchuan), whereas it becomes smoother in loess hill-Mao (such as Yan-an, Ganquan). Subsequently, it becomes steeper when the landform transits to loess hill and fragmentary loess Yuan (such as Yijun, Changwu).

There are still some differences between the two correlation curves in Suide and Yanchuan, where the landforms are similar. This phenomenon is due to the characteristics of the landforms. The landform of sample area Suide is loess Mao, whereas the one in Yanchuan is loess Mao-hill. The landform in Suide is more seriously eroded, more fragmented on the ground, more dispersed on the curvature of the contour, more uniform in gully distribution, and less dispersed on slope distribution than the one in Yanchuan. There is the common characteristic, which determines the similarity of their correlation curves. However, the unique characteristics of their own landforms are the cause of the difference in their correlation curves.

(2) The definition, algorithm, and the geographical meanings of TAs fully depict the landform in different aspects and the correlation and their sequence among these TAs also depict the landform in different aspects. Moreover, correlation values and their sequences vary depending on landform and the spatial location.

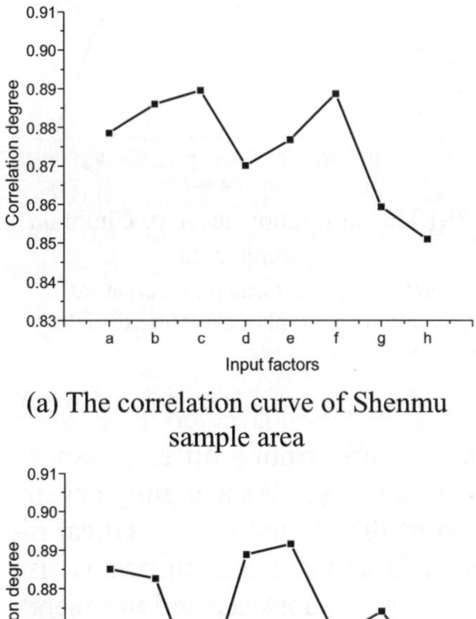

(a) The correlation curve of Shenmu sample area

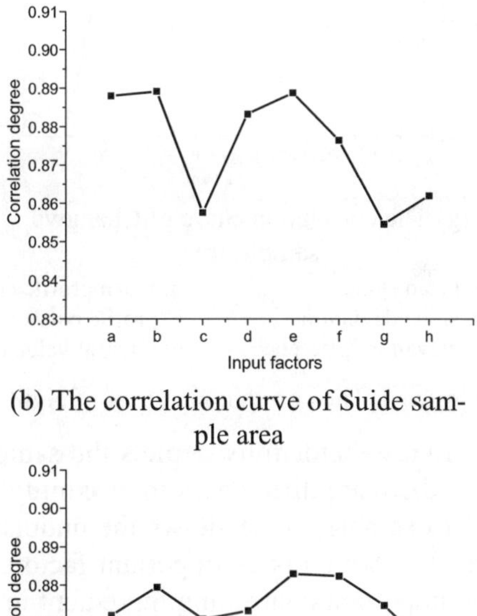

(b) The correlation curve of Suide sample area

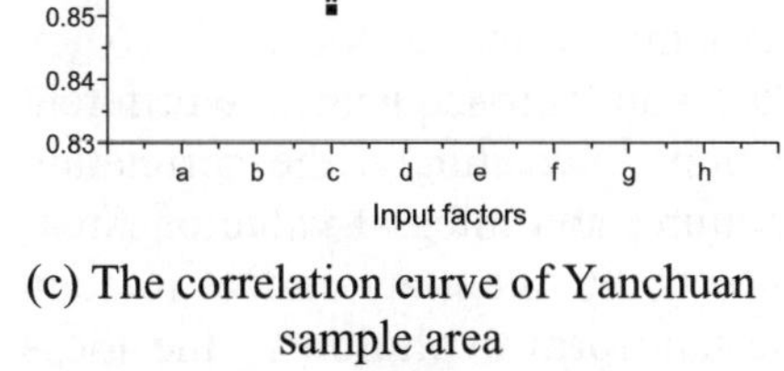

(c) The correlation curve of Yanchuan sample area

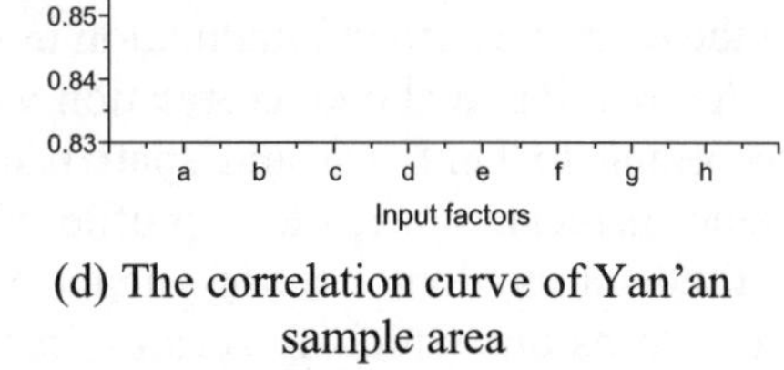

(d) The correlation curve of Yan'an sample area

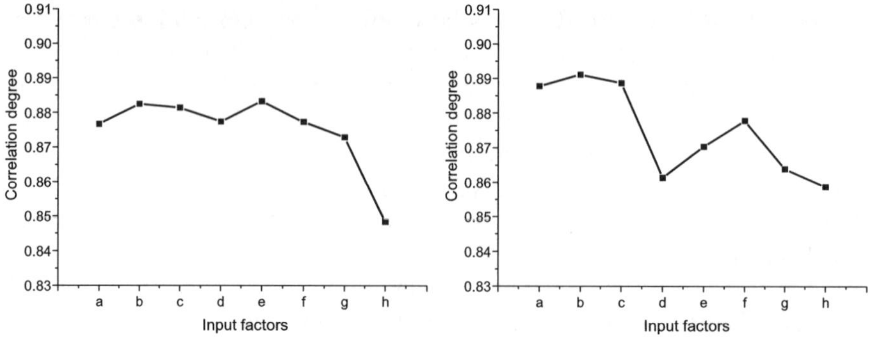

(e) The correlation curve of Ganquan sample area (f) The correlation curve of Yijun sample area

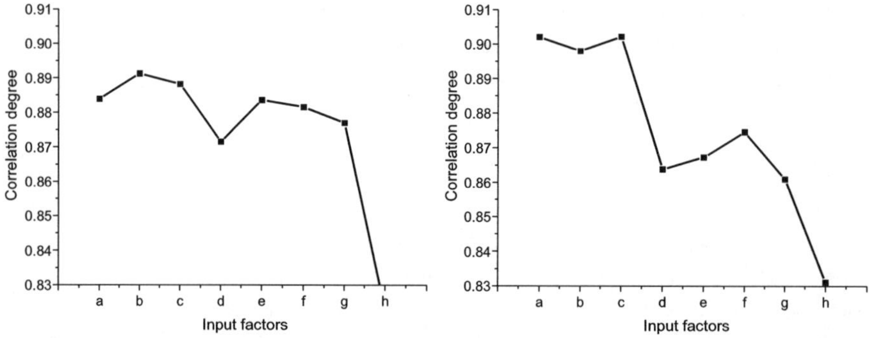

(g) The correlation curve of Changwu sample area (h) The correlation curve of Chunhua sample area

a: mean slope b: mean profile curvature c: mean plan curvature
d: slope deviation e: profile curvature deviation f: plan curvature deviation
g: elevation difference h: integral value of Area-Altitude Curve

Figure 4. The correlation curves of eight typical sample areas from north to south.

Every factor fully depicts the sample landform from a different aspect, but there are differences in proximity between these TAs and gully density For example, slope shows the undulation of the ground in the vertical direction, which is an important factor in water and soil erosion, and is also an important controlling ingredient in the flow of substance and the energy slope surface. Its average value shows the intermediate level of ground undulation in the sample area, and its standard deviation reflects the variation in the degree of ground undulation in the sample area.

As mentioned above, correlation values and their sequence are different according to landform and spatial location. Interestingly, the correlation value of mean slope, mean profile curvature, and integral value of Area-Altitude curve do not change much. Slope is the important motive force in water erosion, which is serious in the landform evolution in the loess plateau of northern Shaanxi Province. And the correlation between slope and gully density is relatively high in every case. It is particularly special for the mean slope in Yan'an and Ganquan. In these two areas, the difference between the two correlation values is not obvious, but the correlation of standard deviation of profile curvature, which could reflect the dispersion extent of changes in the vertical direction, increases suddenly so that the correlation orders of other TAs decline to various extents and tend to balance and the correlation order of mean slope drops slightly. Mean profile curvature, having significant correlation to gully density, also expresses the importance of water and soil erosion. The integral value of the Area-Altitude curve can only reflect the potential of the ground to be eroded, and is not good at expressing the current landform. Therefore, its correlation order is always located in the position with minimum influence. In the sample areas of Suide and Yanchuan, correlation of mean plan curvature reduces greatly. Therefore, the correlation order of integral value of Area-Altitude curve rises a little.

The correlation between mean plan curvature and gully density is relatively high at the primary stage of landform evolution when the development of gullies is incomplete, the deepness of surface incisions are relatively shallow, and the surface erosion is comparatively high in the horizontal direction, such as in Chunhua and Changwu, while at the mature landform evolution stage, there is violent water and soil erosion, a sufficiently developed gully network and deep incision of the surface, and the erosion on the surface is comparatively high in the vertical direction. Consequently, the distribution of plan curvature and gully density in subsamples is inconsistent, and the correlation value and its order reduce greatly.

In the primary stage of landform evolution, the gully network does not develop completely; the riverbed undercuts, side erosion is violent, and land between ditches is relatively smooth and wide. Accordingly, there are few transitional places from flat places to gullies, and the distribution of slope is relatively dispersed, which makes the subsample distribution of slope standard deviation and profile curvature standard deviation, apart from those of gully density and the correlation, relatively low (such as Chunhua and Yijun), while at the mature stage of landform evolution, the correlation is relatively significant.

The plan curvature standard deviation mainly reflects the distribution of the turnings of contours and its expression on landform, especially on the status of erosion, and is not nearly so good as mean slope and mean profile curvature. In Shenmu especially, erosion is not only by water but also by wind. There are pieces of loess Mao-hill distributed in the area, covered with dune or a thin layer of sand, with 317 m of absolute difference in elevation. Therefore, the surface erosion is comparatively high in the horizontal direction and the correlation order of the plan curvature standard deviation increases accordingly. Whereas in Yan'an, with loess hill landforms and a sufficient gully network with 368 m of absolute elevation difference, the dispersion of surface transitions is not obvious in either the vertical or horizontal directions. So the correlation of its plan curvature standard deviation and profile curvature standard deviation are relatively significant.

3.3.2.3 Analysis of mixed geomorphology

Based on the analysis of variations in the trends from the eight representative samples from north to south, the sample area of Ansai with mixed landforms has been chosen to validate the law. This area extends from 109°00'00"E to 109°07'30" E, from 37°10'00" N to 37°15'00" N, and lies near to the Yanhe River in Ansai county. Figure 5 is the surface shaded relief map.

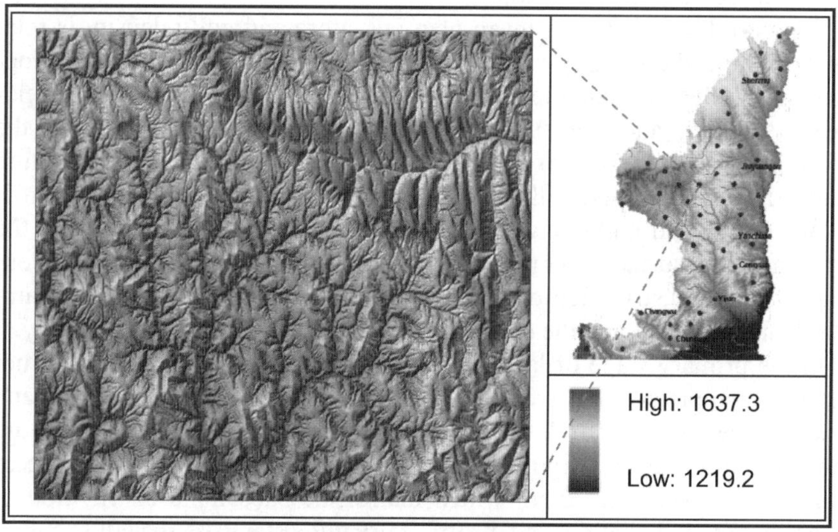

Figure 5. The hill shaded map of the sample area in Ansai.

As is shown in the chart, the mostly asymmetric erosion happens in this area and the surface is fragmented. Landforms in the southwest corner are mainly loess Mao, with loess hill in the northeast corner, i.e. landform in this area is heterogeneous. Accordingly, the correlation curve might be different from the one of simply loess Mao landform or loess hill landform.

Figure 6 depicts the correlation percentage and sequence after the sorting of all factors. Comparing Figures 6 and 4, the following observations can be summarised:

(1) The correlation curve of the Ansai area possesses the same characteristics as not only the one of loess Mao but also of loess hill. The form of the curve corresponds to its landform.

The reason is that this sample area is located in the continuing part of Baiyu Mountain and close to the Yanhe River and its surface erosion is accelerating. Consequently, landforms in this area are heterogeneous, i.e. there is a wide extent of loess Mao as well as beam shaped loess hills. The landforms determine the distribution of TAs, the ultimate characteristics of correlation among TAs, and then the form of the correlation curve.

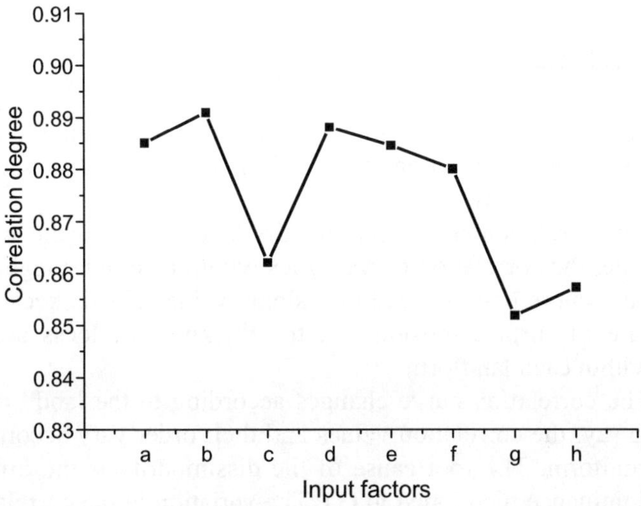

a: mean slope b: mean profile curvature c: mean plan curvature
d: slope deviation e: profile curvature deviation f: plan curvature deviation
g: elevation difference h: integral value of Area-Altitude Curve

Figure 6. The correlation curve of TAs in the Ansai sample area.

(2) In this area, the correlation of the TAs expressing the change on the surface in the vertical direction is relatively significant, while the one in the horizontal direction is further correlated. Among these factors, mean profile curvature shows the most significant correlation to gully density, followed by slope standard deviation and mean slope in sequence, while elevation difference shows the least correlation.

The landform of the sample area is the ultimate reason for this kind of correlation curve. The heavy erosion, especially in the vertical direction, leads to the significant correlation of these TAs, which can reflect the changes in the vertical direction (such as mean profile curvature, slope standard curvature, etc.), while the relatively insignificant correlation of those TAs reflects the small changes in the horizontal direction (such as

mean plan curvature and plan curvature standard deviation). The correlations of elevation difference and elevation curve integral are the two farthest for the heterogeneity of the landform because of the heterogeneity of the landform which leads to the asymmetric distribution of their value in sub-samples, are the two of the least significant correlated.
(3) Figure 6 has validated the results in representative sample areas, and it also confirmed that the correlation curve changes along with the landform, different forms of curve indicating different landforms, and similar correlation curves indicating similar landforms.

4 Conclusions

In this chapter, we proposed a novel correlation model, and the correlation characteristic and its variation laws for TAs have been discussed. Several conclusions can be drawn.

1. TA correlation curves can identify landforms. For the same terrain type, the correlation curve varies within a certain range, and takes on a curve band instead of a single value. This is because the nature of surface erosion and the thickness of loess are different within each landform.
2. The correlation curve changes according to the landform. That is to say, the correlation values and their order vary according to the landform. The root cause of the dissimilarity is the difference of dominance mentioned in (1). The variation in the correlation spectrum reflects the essence of the dissimilarity.
3. The topographical area and evolution pattern of the test areas can be expressed by the TA correlation spectrum perfectly. A new idea and method are proposed for geomorphology research on the Loess Plateau.

This work could be helpful for landform recognition research in the Loess Plateau, because correlation curves are sensitive to the landform and evolution pattern. Future work may focus on the mechanism of the correlation characteristics of TAs, combining with parameters such as manner of erosion on the surface and thickness of loess, etc., and the relationship between these parameters and the correlation value of TAs is also an interesting research topic.

References

Zou, B. (1985), *Principle of small geomorphology,* Beijing: Commercial Press (in Chinese).

Da Silva, J.R.M. and Alexandre, C. (2005), Spatial variability of irrigated corn yield in relation to field topography and soil chemical characteristics, *Precision Agriculture*, 6: 453–466.

Deng, J. (ed.) (1988), *Grey System Book*, Windsor: Sci-Tech Information Services.

Garner, M.E. (1998), Effects of anthropogenic activities upon land cover change in Johnson County, Arkansas, PhD thesis, University of Arkansas.

Giles, P.T. and Franklin, S.E. (1998), An automated approach to the classification of the slope units using digital data, *Geomorphology*, 21: 251–264.

Deng, J. (1987), *Basical method of gray system,* HuaZhong Technology College Press (in Chinese).

Keck, T.J., Quimby, W.F. and Nielsen, G.A. (1993), Spatial distribution of soil attributes on reconstructed mine soils, *Soil Sci. Soc. Am. J.*, **57**: 782–786.

Li, B. and Tao, S. (2000), Correlation between AVHRR NDVI and climate factors, *Acta Ecologica Sinica*, **5**: 898–902.

Shary, P.A. and Sharaya, L.S. (in press), Criteria of TAs selection in predictive soil and vegetation mapping, *International Journal of Ecology & Development.*

Shary, P.A., Sharaya, L.S. and Mitusov, A.V. (2002), Fundamental quantitative methods of land surface analysis, *Geoderma*, 107: 1–32.

Luo, S., Sun, X. and Chen, Y. (1988), *Research of natural conditions in Loess Plateau,* Xi'an: Shaanxi People's Press (in Chinese).

Liu, S., Guo, T. and Dang, Y. (1999), *Theory and application of gray system,* Science Press (in Chinese).

Tang, G. and Zhao, M. (2003), Modeling slope uncertainty derived from DEMs in Loess Plateau, *Acta Geographica Sinica*, 58: 824–830.

Tang, G. (1991), Foundation and application of digital slope model, *Journal of Northwest University*, 21: 33–41.

Western, A.W., Grayson, R.B. and Bloschl, G. (1999), Observed spatial organization of soil moisture and its relation to terrain indices, *Water Research*, 35: 797–810.

Wei, Y. (2005), Principle and method of geographical modelling, Science Press (in Chinese).

Yue M. (1998), The relation among ecological factors and diameter growth of platycladus orientalis in the south of Northern ShaanXi Province, *Journal of WuHan Botanical Research*, 16: 47–53.

Zhao, Y., Long, Y., Zhao, Q. and Luo, Z. (1997), *Application of gray system theory in geography,* HuaZhong Polytechnic University Press (in Chinese).

Gan, Z. (1990), *Research of geomorphology of Loess Plateau and soil erosion,* Xi'an: Shaanxi People's Press (in Chinese).

Terrain-based Revision of an Air Temperature Model in Mountain Areas

YANG Xin and XIAO Chenchao

Abstract

The estimation of air temperature in mountainous areas at a local scale is of great significance in both agriculture and local climate investigation. However, because of the lack of sufficient field observed data, the temperature at each spatial site is always difficult to estimate. Based on digital elevation models (DEMs) of 25 metre resolution, a terrain reversed temperature model is constructed, with which spatial estimation of temperature is realized in detail by simulating insolation in each grid unit. Then two comparisons between the new model and the traditional model, the estimated temperature and the radiation temperature from a Landsat Thematic Mapper thermal infrared image (TM6), are conducted to validate the new model. An experiment in Yaoxian County, part of the Loess Plateau of China, shows that the new model estimated a refined and accurate temperature in detail at local scale under the conditions of very little observed data. The model is effective and available for use in those mountainous areas where there is a lack of observed data or where it is difficult to obtain enough observed data.

Keywords: DEM, solar radiation, temperature, simulation.

1 Introduction

Mountain terrain is characterized by irregular and rapid changes in elevation over short distance. The air temperature in mountainous areas varies widely in spatial distribution. As it is well known, air temperature in mountainous areas is influenced by many factors such as altitude, latitude, distance to large water bodies and mountain structure. However in certain local regions the impact of terrain on air temperature (called local factors including slope gradient, aspect, terrain shading and character of land cover) is rather obvious (Fu *et al.* 1996, Weng *et al.* 1990). In practical terms, due to insufficient climate observation data, the temperature could not be estimated accurately in detail at local scale. The global and comprehensive data cannot meet the needs of precision agriculture and local climate exploration.

Spatial interpolation is the traditional method for estimating air temperature in rugged areas (Fu *et al.* 1996, Weng *et al.* 1990, Pan *et al.* 2004). With a small number of observed data, those precise interpolation methods such as the Inverse Distance Weighting approach (IDW), the Spline method, and the Kriging method, etc, are usually not available and suitable for obtaining a correct and refined result. Statistical regression is usually employed to estimate the air temperature (Ouyang *et al.* 1996, Collins and Bolstad 1996, Oleg 2001, Pan *et al.* 2004). Even so, the results can only reflect the global trend of temperature in spatial distribution that varies with altitude, latitude and longitude. It is still hard to obtain the detailed and refined temperature in each grid unit at higher resolution.

In recent years, with the development of digital terrain analysis methods and multi-scale national DEM databases, more research in applying DEMs to simulate the surface temperature of rugged areas has become possible (Li *et al.* 1999). Zhang (2002) introduced topographic factors into a regress model to simulate the temperature in the area of Qinghai Lake, Qinghai province of China. The estimated result is coarse and it cannot reflect the topographical impact on land surface temperature. An even more complicated model was brought forward in which more factors, such as field measured temperatures and remote sensing images are required in the simulation (Wilson and Gallant 2000). As a matter of fact, there are quite a few weather stations located in China, but sometimes there has been only one station for a county, which is insufficient to acquire enough field information to estimate a refined result. Because of these limitations, some complicated temperature models that call for much factual data are not applicable in a large area with high resolution simulation. Therefore, a new method is put forward to estimate temperature with some higher accuracy under the condition of limited field measurement of temperature data.

2 Test area and data

2.1 Test Area and Data

Yaoxian County, located in the centre of Shaanxi Province (108°34′–109°6′E and 34°50′–35°20′N), in the southern part of the Loess Plateau (see Figure 1), was chosen as the test site because of its ideal terrain conditions and research basis. Yaoxian County has three landform regions: low hill, a gully-hill region, and the loess tableland, with a total area of 1,613 km^2. Along the N-S direction, elevation reduces gradually, ranging from 1,713 to 543 metres above sea level, with a mean elevation of 1,114 metres. The relief is complex and with many steep gradients. The gradient slope ranges from 0 to 68 degrees with the mean value of 16.8 degrees. Yaoxian County lies in the warm temperate zone with a continental semi-humid, semi-dry monsoon climate, with low rainfall and rich solar radiation. From north to south, the temperature decreases gradually. Owing to the lower relative height, the regional wind advection has less influence on the local temperature. Thus, this chapter focuses mainly on the impact of slope gradient, aspect, and the terrain shadings on the local surface temperature distribution.

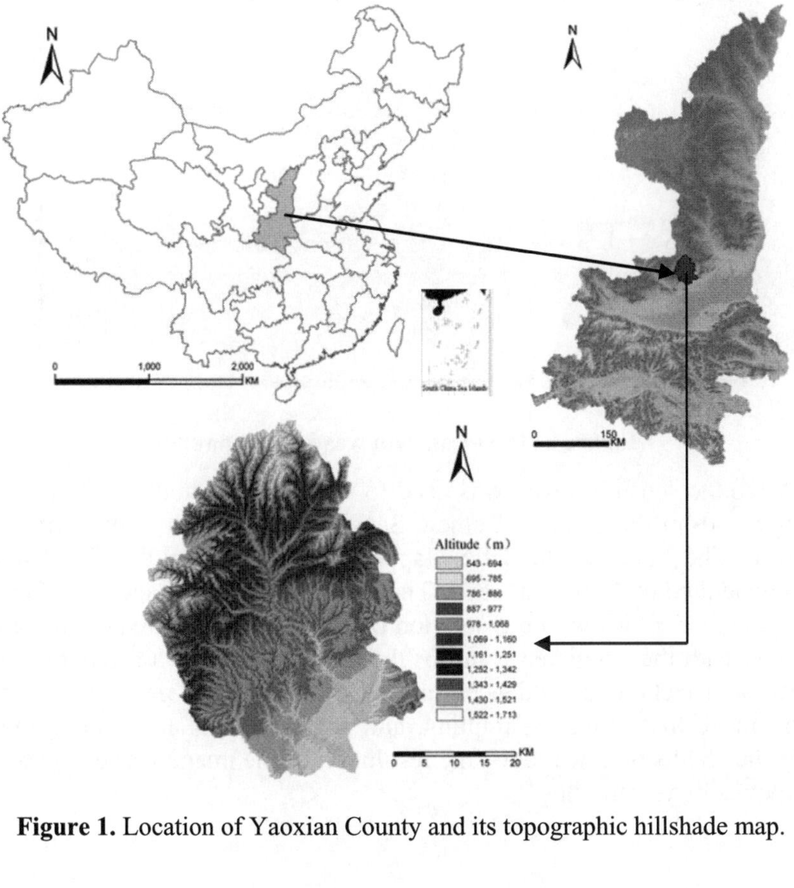

Figure 1. Location of Yaoxian County and its topographic hillshade map.

2.2 Experimental Data

DEMs, observed climate data, and remote sensing images are used in this study. DEMs were generated by digitizing contour lines from 10 sheets of 1:50,000 scale topographic maps and interpolating in a regular grid mesh of 25 metres.

Climate data was provided by the Shaanxi Meteorological Bureau. The average temperature per month is recorded from 1970 to 2000 at 15 weather stations around Yaoxian County. Figure 2 shows the location of weather station sites, which are Yijun, Baishui, Yongshou, Pucheng, Dali, Weinan, Gaoling, Jingyang, Qianxian, Binxian, Xunyi, Tongchuan, Chuhua, Fuping, and Yaoxian.

Figure 2. The location of weather station sites

A remote sensing image was used to validate the simulated result. The image is from the Shaanxi Remote Sensing Information Centre for Agriculture. The Landsat Thematic Mapper (TM) image of Yaoxian County was generated on May 28th 1997. The thermal infrared image of TM (band 6, also called TM6) with a resolution of 120 m was prepared for the comparison with the simulated result of the new model after a series of image processing techniques had been carried out, including geometric correction, image matching, re-sampling, and normalization. In order to match with the results of new model, the resolution of the image was converted to 25 metres by re-sampling.

3 Models

3.1 Traditional models

Various interpolation methods are used to estimate the spatial distribution of temperature. Many of them, such as Inverse Distance Weighting (IDW), Spline and Kriging interpolation, etc., need enough sampling points to acquire an exact and refined result. Therefore, statistical regression of temperature against altitude, latitude and longitude (see Equation 1) is usually applied.

$$T = a_0 + a_1\varphi + a_2\lambda + a_3 H \quad (1)$$

where, φ, λ, H denote temperature, latitude, longitude, and altitude of the observation point, respectively, and parameters a_0, a_1, a_2, a_3 are the regression coefficients. In this model, the latitude, longitude, and altitude reflect the change of temperature from the equator to the two poles, from ocean to inland, and from sea level to high mountains, respectively. Although the model could give a comprehensive figure in revealing temperature distribution at the regional scale, it still fails to reveal the difference of temperature at the local scale.

Although the former temperature model had contained 'elevation', one factor of topography, the estimated result for temperature is still coarse. In order to improve the simulation result, some models have also integrated other topographic factors, such as slope and aspect, directly into the regression model (Zhang *et al.* 2002). However, it is known that many weather stations are located in flat areas where they suffer little terrain influence. Thus, it is hard to import topographical factors into models directly when the statistical data of the regression model come from such weather stations.

As we all know, solar radiation is the greatest thermal energy source. The more insolation, the higher is the temperature (Weng 1997). Some studies introduce insolation and vegetation index (e.g. NDVI index) into the temperature model to achieve a more accurate result. However, it is not applicable in areas where there is a lack of many field measurements, e.g. temperature, the amount of cloud cover, etc. and RS data that provides the required dataset for the temperature model.

Therefore, a simple and effective terrain based revision of the temperature model is put forward in this chapter. In this model, a refined result of temperature distribution is estimated by only a few observed field temperature data.

3.2 Revised temperature model based on terrain

The idea of this improved temperature model is to introduce insolation as a coefficient, which is the ratio between insolation from an inclined surface and insolation from a plane surface. The improved model is put forward as follow:

$$T' = T + \Delta T \tag{2}$$

where, *T'* is the revised air temperature. *T* is the temperature calculated by formula (1), ΔT is the temperature difference between the inclined surface (named 'slope unit') and the correlated plane surface (named 'flat unit'). ΔT presents the local temperature difference with the scale of each grid unit.

When the astronomical parameters, atmospheric circulation and geographical factors are similar in a certain area at a large scale, the influence of slope gradient, aspect, and terrain shadings are relatively obvious at the local scale. However, it is very difficult to introduce these topographic factors directly into the temperature model. As mentioned above, weather stations are usually located in flat areas, or on the tops of hills where the terrain is very similar in these sites. So the impact of terrain on temperature is counteracted when statistical regression is built (Ouyang *et al.* 1996). In order to better reflect the influence of terrain on land surface air temperature, total insolation is introduced to the traditional temperature model. As is well known, solar radiation is the main source of land surface temperature. Total radiation received by the land surface can reflect the influence of terrain on temperature synthetically. Many researches have shown that insolation and temperature have a close relationship (Tan *et al.* 2001). Weng (1997) proved that a linear relationship exists between air temperature and total insolation, and indicated that the discrepancy of surface temperature results from the diversity of solar radiation received. Fu and Rich (2002) also proved that there are linear relationships between the maximum and minimum soil temperatures and solar insolation. Therefore, the relationship between temperature in the slope unit and flat unit, as well as total insolation in slope unit and flat unit, can be expressed as follows:

$$\frac{Q_{slope}}{Q_{flat}} = \frac{T'}{T} \tag{3}$$

where, Q_{slope} and Q_{flat} stand for global radiation amounts (MJ/m^2) at each slope unit and flat unit, respectively. *T'* is the temperature in the inclined surface (the temperature after revision by Equation (2)). Hence, according to Equation (2), the temperature difference ΔT between each slope unit and flat unit is:

$$\Delta T = \frac{Q_{slope} - Q_{flat}}{Q_{flat}} T \tag{4}$$

where *T* is the mean temperature per month, calculated by Equation (1). Q_{flat} denotes total insolation in each flat unit. Q_{flat} reflects the impact of those global factors on temperature. Q_{slope} is the total insolation in each slope unit, which can reflect the impact of local terrain factors on temperature.

In rugged areas, total insolation is the sum of direct insolation, sky diffuse insolation, and reflected insolation, which is part of the solar short wave radiation from terrain reflection. The reflected radiation is less than 5% of the others, which is small enough to be ignored in mountains without snow (Li *et al.* 1988). Under the same climate conditions in the loess hill area, the proportion of the direct insolation is relatively larger than the diffuse insolation. The spatial distribution character of total insolation is dominated by direct insolation. Furthermore, the spatial distribution of direct solar radiation is dominated by astronomical radiation.

As a result, the ratio of total insolation from each slope unit to each plane unit is approximately equal to the ratio of astronomical radiation from each slope unit to each plane unit under the same climatic conditions.

$$\frac{Q_{slope}}{Q_{flat}} \approx \frac{S_{0slope}}{S_{0flat}} \tag{5}$$

where, S_{0slope} is the astronomical radiation in each slope unit, S_{0flat} is the astronomical radiation in each flat unit (projected plane of the real surface in each grid unit) so that the terrain based revised model for estimating air temperature can be expressed as follows:

$$T' = \frac{S_{0slope}}{S_{0flat}} \left(a_0 + a_1\varphi + a_2\lambda + a_3 H\right) \tag{6}$$

$$S_{0slope} = \frac{I_0 T E_0}{2\pi} \sum_{i=1}^{n} g_i \begin{bmatrix} u \sin\delta(\omega_{s,i} - \omega_{r,i}) + \\ + v\cos\delta(\sin\omega_{s,i} - \sin\omega_{r,i}) - \\ - w\cos\delta(\cos\omega_{s,i} - \cos\omega_{r,i}) \end{bmatrix} \tag{7}$$

$$\begin{aligned} u &= \sin\varphi\cos\alpha - \cos\varphi\sin\alpha\cos\beta \\ v &= \sin\varphi\sin\alpha\cos\beta + \cos\varphi\cos\alpha \\ w &= \sin\alpha\sin\beta \end{aligned} \tag{8}$$

$$S_{0\,flat} = \frac{24}{\pi} I_0 E_0 (\omega_s \sin\varphi\sin\delta + \cos\varphi\cos\delta\sin\omega_s) \tag{9}$$

where E_0 is the revised factor of Earth orbit; I_0 is the solar constant (4.9212 MJ/m^2·h); T is the time period of a day (24 hours); δ is the solar declination angle; n is the dispersed number of insolation-duration; $\omega_{r,i}$ and $\omega_{s,i}$ are sunrise hour angle and sunset hour angle on the slope incline in ith differential time units, (it is zero at noon and positive before noon); g_i is the terrain shading, which reflects the impact of the terrain on the insolation-duration; ω_s is the sunset hour angle in plane. E_0 and I_0 are the periodic function of t. Their Fourier progression expression is as follows (Zuo *et al.* 1991):

$$\begin{aligned} E_0 &= 1.000109 + 0.033494\cos\theta + 0.001472\sin\theta + \\ &+ 0.000768\cos 2\theta + 0.000079\sin 2\theta \end{aligned} \tag{10}$$

$$\begin{aligned} \delta &= 0.006894 - 0.399494\cos\theta + 0.07207\sin\theta - \\ &+ 0.006799\cos 2\theta + 0.00089\sin 2\theta \end{aligned} \tag{11}$$

where θ is called the day angle.

$$\theta = \frac{2\pi}{365.2422} t \tag{12}$$

$$t = N - 1 \tag{13}$$

where N is accumulation date. It is the arranged number of a date in a year.

The above equations show that only several field observed temperatures are needed in the new temperature model. All of the other factors could be generated from DEM data.

4 Methods

There are two major processes to realize the temperature estimation. One is statistical regression of climate observed data, which are needed to fix the parameters of the models. Equation (14) is the regression equation of the annual mean temperature using the monthly mean temperature information collected from 15 weather stations.

$$T = -2.456 - 1.0 \times 10^{-6} X + 4.7 \times 10^{-6} Y - 4.88 \times 10^{-3} H \quad (14)$$

In this equation, X is the distance to east (metres), Y is the distance to the north (metres), the role of X, Y is same as φ, λ of Equation (1). H stands for the elevation (metres). The multiple correlation index (R) is 0.97, and it passed the significance test when F is 0.01.

The other process is the simulation of solar radiation on both flat and slope areas based on the DEM. The technology for this has already come to maturity (Dozier and Frew 1990, Corripio 2003, Wilson and Gallant 2000, Yang *et al.* 2004, Zeng *et al.* 2003).

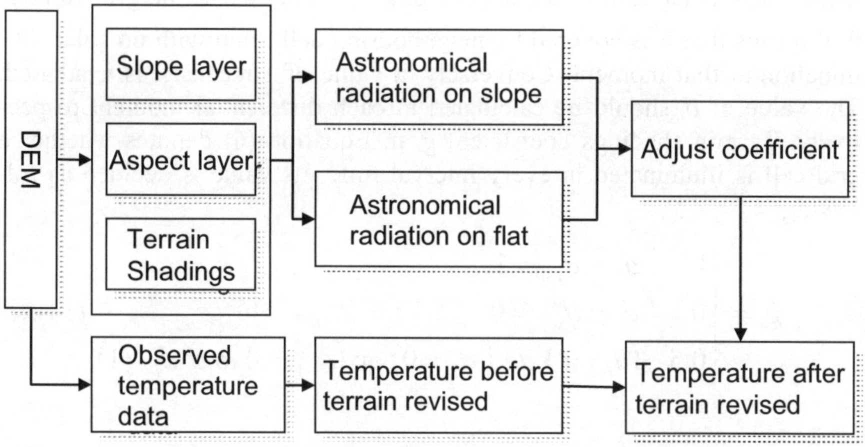

Figure 3. Procedure for the temperature simulation.

Figure 3 shows the procedure of the temperature simulation. Key problems in the realization of the estimation of air temperature are the simulation of astronomical radiation on both slope and flat surfaces. In rugged areas, the exact sunshine hours on each grid unit are most important for the correct simulation. For the convenience of integral computing, the duration of insolation on each DEM grid cell is dispersed. Along with the shift of solar position, the reachable sunlight is deduced for every differential time unit by considering the terrain factors. The effective sunshine hours in a day in every single grid cell is calculated by summing up every differential time unit. Solar simulation is realized with the support of Arc/Info AML (Arc Macro Language) language. The simulation process is as follows:

1. Extracting slope gradient and aspect based on DEM in ArcGIS software.
2. Calculating the sunrise hour angle and sunset hour angle on each specific day and giving the dispersed number of durations of insolation for 36 differential times.
3. Calculating the sunrise angle hour ω_{ri} and sunset angle hour ω_{si} for each differential time, and calculating the corresponding solar altitude angle and solar azimuth angle in each solar position.
4. Simulating the terrain shadings in each differential time.

An illumination tracking algorithm is employed to calculate terrain shading (see Figure 4). Searching all grids along the projected line of solar position, which is the solar azimuth angle at ith time, 0 is set as grid cell-value, namely $d_i = 0$, under the condition that the angle (α_2), being the elevation angle between current grid position (A), and others grid (B) is greater than solar altitude angle (α_1); otherwise, $d_i = 1$. If the grid value is 0, it means that it is covered by neighbouring cells, and with no solar illumination at that moment. Conversely, a value of 1 means it is irradiated. The value of d_i should be calculated in each differential moment respectively. Terrain shadings coefficient g_i in Equation (6) denotes whether a grid cell is illuminated in every interval time. Its value is decided as following:

$$g_i = \begin{cases} 1 & d_i = d_{i-1} = 1 \\ 0 & d_i = d_{i-1} = 0 \\ 0.5 & (d_{i-1} = 1 \ and \ d_i = 0) \ or \ (d_{i-1} = 0 \ and \ d_i = 1) \end{cases} \tag{14}$$

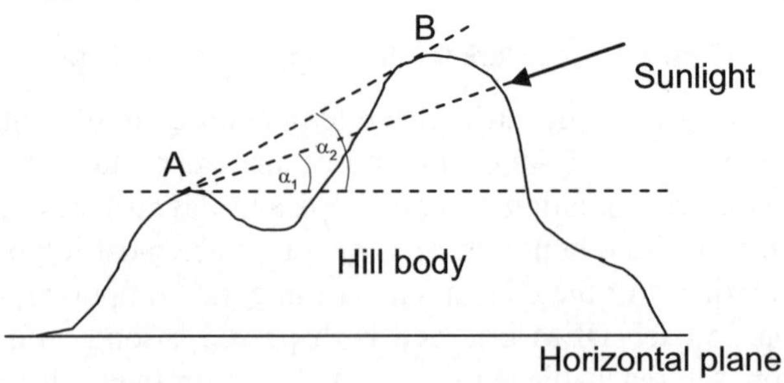

Figure 4. Concept of terrain shadings d_i.

The above equations indicate that the illumination of a cell is controlled by the illumination of two ends of a interval time from the hour angle of ω_{i-1} to ω_i. If both ω_{i-1} and ω_i are illuminated, the whole time of the interval is illuminated. If one of them is illuminated, then half of the whole interval time is illuminated.

The daily astronomical solar radiation is available by accumulating the solar radiation in each differential time. A month's radiation is the sum of the daily radiation.

After the simulation of plane astronomical radiation and slope astronomical radiation, the revised coefficients could be decided and a cell by cell revised surface temperature matrix could be achieved.

5 Results

5.1 Simulation result and comparison with traditional model

The annual mean temperature of Yaoxian County is mapped and compared with that from the traditional model (statistical regression). Experimental results (see Figure 5) show that the new model (terrain based revision) is more reasonable in expressing the characters of land surface air temperature. It shows that the spatial distribution of air temperature has a strong relationship with relief. The traditional model can present the general trend of temperature along with latitude and altitude. From North to South the grey colour of the map changes from dark to light, which shows the air temperature increases gradually. Even more details could be found in the map of temperature after terrain revision. Table 1 shows that the range and standard deviation of annual mean temperature are becoming larger after terrain revision. This reflects the greater diversity and more differences are appearing in the map after terrain revision. The temperature in northern low hill areas falls greatly and the minimum value changes from 7.32°C to 1.41°C. This is mainly because in some shady areas, which hardly receive direct insolation, the insolation is relatively small and causes the decrease in the revised coefficient of the new model. Therefore the temperature in this area is correspondingly low.

A comparison of annual mean temperature histograms (see Figure 6) between the two temperature methods shows that the temperature ranges from 6 divisions to 13 divisions. The diversity of the temperature distribution is enhanced in the model of terrain based revision.

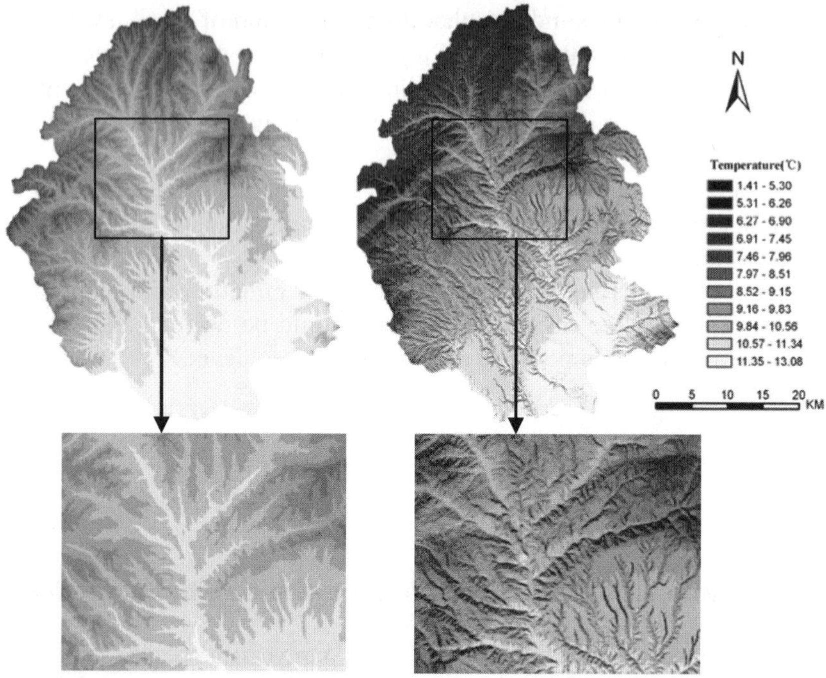

(a) Temperature before terrain-based revision (traditional model)

(b) Temperature after terrain-based revision

Figure 5. Comparison of temperature map between traditional model (a) and terrain based revised temperature model (b).

Table 1. Comparison of simulated temperature.

Items	*Max*	*Min*	*Rang*	*Mean*	*Std dev.*
Before terrain revised	12.86	7.32	5.54	10.22	1.17
After terrain revised	13.08	1.41	11.67	9.25	1.52

Figure 7 shows that temperatures fall quickly with the increase of slope gradient in the model of terrain based revision; a trend cannot be found in the traditional model.

Figure 8 shows that the trend of simulated annual mean temperature after terrain revision varies with slope aspect in a rather regular pattern. The temperature is higher in sunny slopes than in shady slopes, which seems quite balanced in eastern and western oriented slopes. The statistical result shows that the mean temperature in sunny exposed slopes (South, Southeast and Southwest slope) is 9.83°C, 1.44°C higher than that in shady slopes (North, northwest and northeast slope).

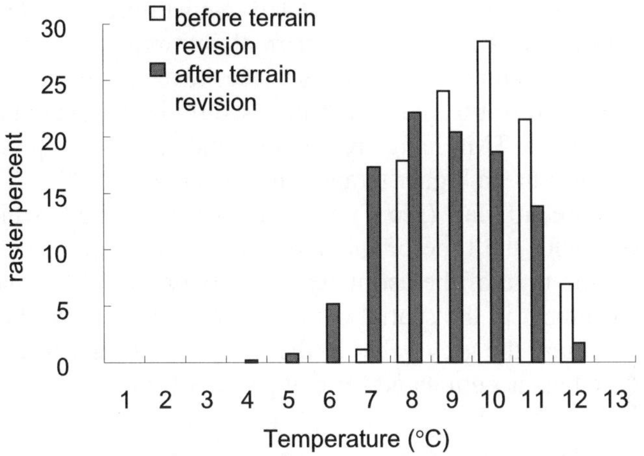

Figure 6. Comparison of simulated annual mean temperature.

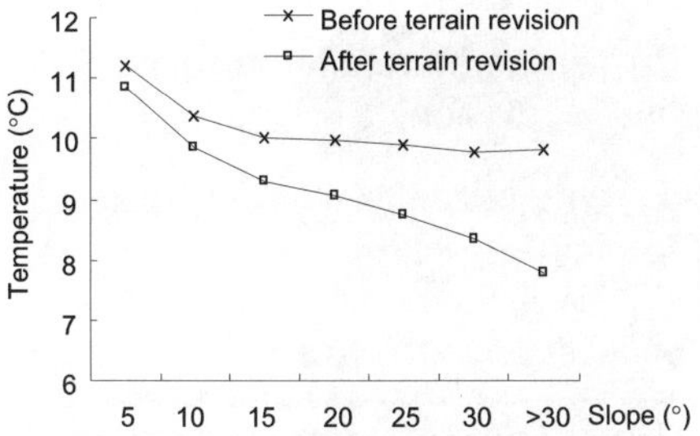

Figure 7. Temperature against slope before and after terrain revision.

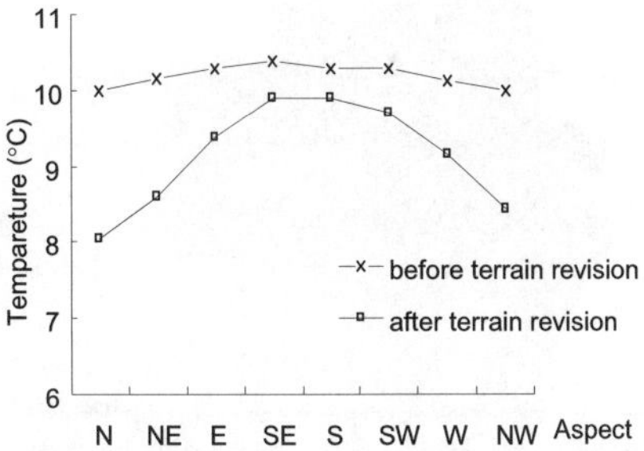

Figure 8. Temperature against aspect before and after terrain revision.

After terrain based revision, temperature becomes lower (see Table 1, Figure 7 and Figure 8). It is mainly due to the revised coefficient. The ratio of astronomical insolation between plane and slope units is different from the ratio of total insolation between plane and slope units. The gird unit shadowed by terrain usually has a smaller astronomic insolation, causing the decrease in the revised coefficient.

5.2 Comparison and validation with TM6 image

In order to validate the accuracy of the simulated result, this chapter also reports a comparison between the simulated temperatures and the radiation temperature from the Landsat TM6 thermal infrared image. This image (see Figure 9b) contains the ground surface heat information, which is compared with the result from the terrain based revised temperature model. The grey tones of the TM6 image reflect the land surface temperature. The brighter grey tones mean higher temperatures (Tan *et al.* 2001). The annual mean temperature in May (see Figure 9a) was simulated, which corresponds closely with the time of the TM image. Although there are some differences in the time of the estimated temperature and TM image, some similarities still exist in the spatial distribution of temperature because the temperature through the grey value of the image can be accepted as a consequence of the heat accumulated for that period of time.

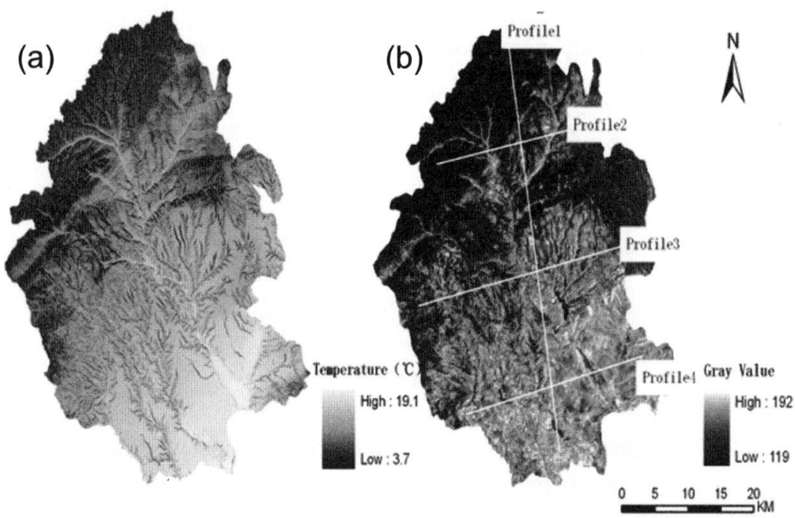

Figure 9. A comparison between average temperature simulated on May (a) and TM6 thermal infrared image and location of test section plane (b).

A few pre-processes, i.e. geometry adjustment, re-sampling, and standardization are carried out to make the TM6 image match the simulated temperature. Both the simulated temperature data and grey value of the image are standardized by normalization. The normalization equation is:

$$Z_i' = \frac{Z_i - Z_{mean}}{Z_{\max} - Z_{\min}} \tag{15}$$

where Z_i' is the new value after standard, Z_i is the old value, Z_{max}, Z_{min} and Z_{mean} are the maximum, minimum and mean value of the simulated temperature and TM6 image grid layer.

Figure 9 shows a majority of similarity between the spatial distribution of land surface heat from estimated temperature and the temperature from the TM6 image. It is clear that the relief plays a very important role in the land surface heat redistribution process on a local scale. There still are some differences of temperature on the sites of different land cover, as shown in the following profiles.

Several typical profiles are selected in the test area, located in different landforms of Yaoxian County (see Figure 9b).

Figure 10 makes a comparison between the simulated temperature and the TM measured one along four profiles. Profile 1 goes from north to south in Yanxian County and across three different landforms. Profile 2, 3 and 4 are located in low hill region, gully-hill region and loess tableland, respectively. A general similarity could be found in their distribution pattern, except for some specific land cover areas, i.e. a reservoir and river can lead to the falling of temperature (see A in profile 1, C to G in profile 4), and an urban heat island effect will increase temperature (see B in Profile 2). Owning to the instantaneous nature of satellite data, some errors also appear caused by the time differences.

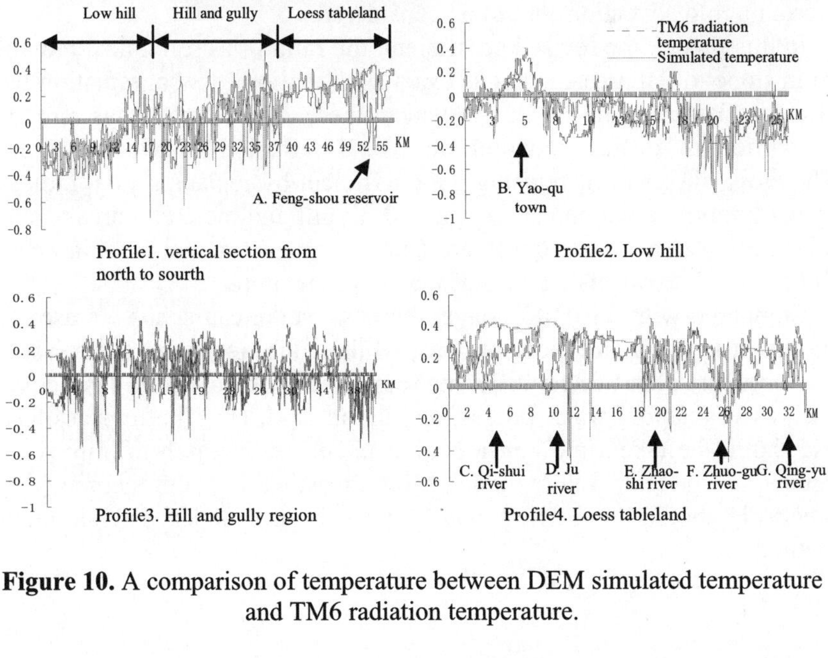

Profile1. vertical section from north to sourth

Profile2. Low hill

Profile3. Hill and gully region

Profile4. Loess tableland

Figure 10. A comparison of temperature between DEM simulated temperature and TM6 radiation temperature.

The result proves that the terrain based revision of the temperature model has more reasonable results than the traditional one in low hill, gully-hilled loess regions. In the loess tableland, the errors are relative large because of the complexity of land use and land cover in this area.

6 Conclusions and Discussions

Terrain based revision of the temperature model can present the spatial distribution of temperature in detail. Taking insolation as an adjusted coefficient of temperature can overcome the difficulty of integrating terrain factors into a statistical regression model and it can also improve the precision of temperature estimation. Although the model is simpler than some complex temperature models, the simulated result for temperature is valid and refined. It can be widely and effectively used in the rugged area without much observed temperature data.

The improved models involve local terrain factors such as slope gradient, aspect and terrain shadings. The obstructing effect of the terrain to atmosphere advection is not within the scope of this research. In some high mountains, wind is blocked off by terrain and causes great differences in temperature between windward slopes and leeward slopes (Fu *et al.* 1996). Therefore, in high mountain areas, the influence of terrain obstructing warm or cold advection should be considered.

In this model, the revised coefficient, the ratio of astronomical radiation from slope to flat areas, is not as exact as the ratio of global radiation from slope to flat areas. In the Loess Plateau area, clouds are always rare. The proportion of diffuse radiation in global radiation is relatively small. Therefore, the error of revising the coefficient by replacing total radiation by astronomic radiation can be ignored. In hill and mountain areas, generally lacking in adequate observed data, the revised model is valid and refined enough to estimate land surface air temperature.

Compared with the TM6 image, the new model can achieve reasonable results in low hill and gully regions, while exhibiting large differences in the loess table land. Most differences are due to the different land covers, such as vegetation, water bodies, and urban land, etc. The impact of land use should be taken into consideration in further research to improve the veracity of the temperature model. More investigation, both theoretical and practical, should be carried out to perfect the model and simulation method.

Acknowledgments

Thanks for financial support from the National Natural Science Foundation of China (No.40571120, No.40671148), Specialized Research Fund for the Doctoral Program of Higher Education (No.20050319006) and the National High Technology Research and Development Program of China (2006AA12Z212).

References

Collins, F.C. and Bolstad, P.V., (1996), A comparison of spatial interpolation techniques in temperature estimation, *Proceedings of the Third International Conference/Workshop on Integrating GIS and Environmental Modeling*, Santa Fe, New Mexico, January 21-25, Santa Barbara, California: National Center for Geographic Information Analysis.

Corripio, J.G., (2003), Vectorial algebra algorithms for calculating terrain parameters from DEMs and solar radiation modeling in mountainous terrain, *Int. J. Geographic Information Science,* **17**(1): 1-23.

Dozier, J. and Frew, J., (1990), Rapid calculation of terrain parameters for radiation modeling from digital elevation data, *IEEE Transaction on Geoscience and Remote Sensing*, **28**(5): 963–969

Fu, B., Yu, J. and Lu, Q., (1996), *Mountains Climatic Resources and its Application,* Nanjing: Nanjing University Press.

Fu, P. and Rich, P.M., (2002), A geometric solar radiation model with applications in agriculture and forestry, *Computers and electronics in agriculture*, **37**: 25-35.

Li, X., Cheng, G. and Chen, X., (1999), Optimization of solar radiation model in any situation, *Chinese Science Bulletin*, **44**(9): 993-998.

Li, Z. and Weng, D., (1988), Model of global solar radiation in hill and mountain, *Acta Meteorologiga Sinca,* **46**(4): 461-468.

Oleg, A., (2001), Prediction patterns of near-surface temperature using empirical data, *Climatic Change,* **50**: 297-315.

Ouyang, Z., Zhao, X. and Zhao, Y., (1996), A small net-method for local climate calculation in mountain area, *Transaction of the CASE*, **12**(3): 144-148.

Pan, Y., Gong, D. and Deng, L., (2004), Smart distance searching-based and DEM-informed interpolation of surface air temperature in China, A*cta Geographic Sinica,* **59**(3): 366-374.

Tan, Z., Zhang, M. and Karnieli, A., (2001), Mono-window algorithm for retrieving land surface temperature from Landsat TM6 data, *Acta Geographica Sinica*, **56**(4): 456-466.

Weng, D. and Luo, Z., (1990), *Mountains Topographical Climate*, Beijing: China Meteorological Press.

Weng, D., (1997), *Radiation Climate in China,* Beijing: China Meteorological Press.

Wilson, J.P. and Gallant, J.C., (eds.) (2000), *Terrain Analysis: Principles and Applications*, New York: John Wiley & Sons, Inc.

Yang, X., Tang, G. and Wang, L., (2004), Modeling of global radiation over rugged areas based on DEM, *Geography and Geo-information Science,* **20**(5): 41-44.

Zeng, Y., Qiu, X. and Liu, C., (2003), Study on astronomical solar radiation distribution over Yellow River basin based on DEM data, *Acta Geographica Sinica,* **58**(6): 810-816.

Zhang, H., Ni, S. and Deng, Z., (2002), A method of spatial simulating of temperature based Digital Elevation Model (DEM) in mountain area, *Journal of Mountain Science,* **20**(3): 360-364.

Zuo, D. and Zhou, Y., (1991), *Research on Earth Surface Radiation,* Beijing: Science Press.

National Mapping of Landform Elements in Support of S-Map, A New Zealand Soils Database

James R.F. BARRINGER, Allan E. HEWITT, Ian H. LYNN and Jochen SCHMIDT

Abstract

This paper describes an application of terrain analysis where a national dataset of landform elements derived from a 25 m resolution national digital terrain model has been used to underpin soil-landscape modelling in the complex steepland terrain of New Zealand. This research supports S-map, a new initiative to improve New Zealand's soil databases. Landform elements are derived from a fuzzy classification based on local geometry (curvature and slope) and landscape context, and provide a primary foundation for mapping soil distribution in steepland land systems, which significantly improves knowledge of soil distribution on hillslopes in complex landscapes, and over a wide range of land systems. So far, land elements and soil-landscape models have been used to model soil distribution in more than 10 distinct land systems (covering over 10% of New Zealand's total steeplands area).

Keywords: Digital Elevation Model (DEM), landform elements, steeplands, soil mapping.

1 Introduction

Soil survey/mapping in New Zealand has a history, in common with many countries, of polygon-based mapping at multiple scales with publication of paper maps and associated reports dating from the 1930s to 1990s. Recent demand for more quantitative soil data to support environmental simulation and production models with comprehensive national coverage has highlighted the need for national consistency in scale, soil correlation, and data standards. For example, coverage of current spatial data is patchy in its distribution, scale, and quality. Better data exist for arable lowlands, but lack of data for non-arable hill and mountain lands is limiting sustainable development plans, ecosystem studies (Leathwick *et al.* 2003), and hydrological modelling.

The need for new soils data to better meet this demand has been recognized, but to generate new and improved soils data using traditional mapping techniques based on aerial photograph interpretation and extensive field survey is constrained by limited resources. In addition, traditional methods involve elements of subjectivity that can make results inconsistent and hard to reproduce. Over the last decade studies, both internationally (Gessler *et al.* 1995, Wilson and Gallant 2000, Grunwald 2006), and in New Zealand (Dymond *et al.* 1995, McLeod *et al.* 1995, Harmsworth *et al.* 1995), have shown the utility of digital terrain analysis to complement manual soil survey. This applies in particular to steepland and hill soils, where soil hydrology and soil physical properties are controlled by topographic variation such as changes in slope, aspect, and elevation. With medium to higher resolution digital elevation models (DEMs) now readily available over large areas, terrain analysis methods are being used extensively in digital soil mapping (e.g. McBratney *et al.* 2003).

2 S-Map

New Zealand's response to the demand for improved soil data is called S-Map (Lilburne *et al.* 2004). Completion of S-Map will for the first time provide consistent and comprehensive national soil data layers to support applications at local, regional, and national scales. It builds on previous soil mapping by filling gaps with new mapping and upgrading the information content and associated database to meet a new national standard. The S-Map database will have national coverage and contain predominantly new digital data at a scale that will resolve hill slopes (1:50,000 scale).

The vision for S-Map is to provide excellent access to relevant soil information through the Internet, with consistent soil map scale, key soil attributes derived for all soils, centralized curation and quality assurance, ease of integration of soil with other data sets/models, and application from field to national scales.

The primary map layer of S-Map consists of soil classes, i.e. polygons that are labelled with the taxonomic family of the soil. Each soil family is defined as a unique combination of attributes at the New Zealand Soil Classification (NZSC) soil form level (Hewitt 1998, Clayden and Webb 1994). Attributes included are the NZSC Order, Group and Subgroup, parent material, rock class, dominant texture, and permeability class. Soil classes are further characterized as siblings according to their depth to rock, stoniness, drainage, texture, and functional horizons (Webb 2003). The uncertainty of each of these family and sibling attribute classes is specified.

Associated with this soil class layer are additional map layers comprising base properties developed from expert knowledge and derived soil properties based on a model or pedotransfer function. The base soil property layers are depth to contrasting material (diggability), depth to slowly permeable layer, rooting depth, rooting barrier, horizon thickness, stoniness, clay and sand content, and a set of up to five functional horizons that best describe soil morphology vertically (Webb 2003). The derived soil layers are generated from a variety of models from simple lookup tables that depend only on the soil class to more complex mathematical models that combine soil, land use, vegetation, climate, or topographic attributes. Derived layers will include available water (mm), macroporosity, water retention, bulk density, total carbon, total nitrogen, phosphorus, calcium, cation exchange capacity, pH and phosphorus retention.

The S-Map methodology differs depending on which of two major landscape types are being mapped.

1. Lowlands, dominantly flat to rolling land. Landforms are of such low relief that DEMs (based on current 20 m contour data) cannot be used for soil-landscape modelling. Soil mapping uses conventional methods, based on air photo interpretation and free survey techniques.
2. Uplands, dominantly hill and mountain terrain. Relief allows application of soil-landscape modelling based on DEMs and other spatial information. The actual modelling used will depend on the land system and the sampling cost and availability of data. The predominant technique will be to derive soil distribution rules from available data, literature, and new sampling from representative windows, and apply these to modelled landform elements.

Our chapter focuses on progress in mapping and presenting soil classes and soil properties in upland environments based on DEM analysis of landform elements and soil-landscape modelling of soils and soil properties.

3 Derivation of Landform Elements

The mapping method in upland landscapes also has two components (Schmidt *et al.* 2005). The first component involves analysing the DEM to delineate landform elements that occur within the land system being modelled and the second places these landform elements in terrain context (Schmidt and Hewitt 2004).

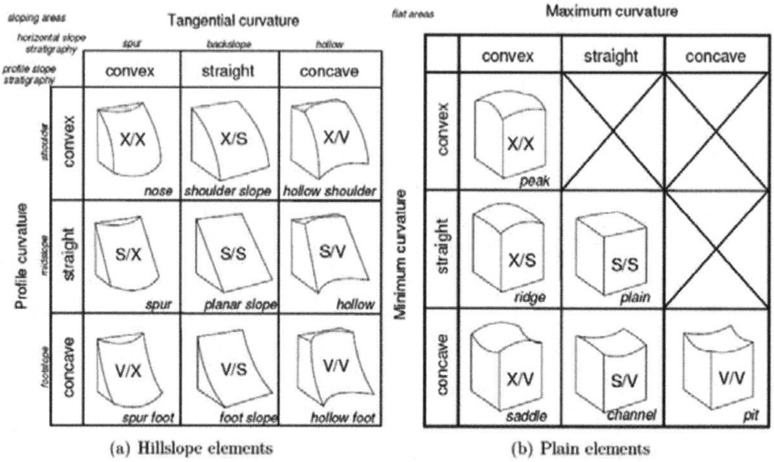

Figure 1. Schema defining 15 local landform elements based on classification of local slope and curvature (after Schmidt and Hewitt 2004).

For classifying the land surface, we used a combination of the approaches described by Dikau (1989) and Wood (1996), which describe curvature for sloping and for low-gradient/flat areas (Figure 1). For sloping areas, profile curvature and tangential curvature are used to classify nine form elements (Figure 1a). Curvature is classified according to three classes: concave, straight, and convex. For flat areas, maximum and minimum curvatures (Shary *et al.* 2002) are used to describe six fundamental form elements (Figure 1b). The final classification results in 15 elements, which can be defined by geomorphographic terms as shown in Figure 1 (after Schmidt *et al.* 2003).

Landform elements are strongly scale-dependent, and the calculation of curvature derivatives at varying window sizes is available in our model, based on Wood (1996) and Schmidt *et al.* (2003). However, attempts to utilize multi-scale analysis (Schmidt and Andrew 2005) have tended to show that significant landform elements at within-hillslope scale will be ignored in favour of larger-scale forms. We found swale and hollow terrain at a scale of 10s to 100s of metres was often ignored in favour of the ridges and valleys in which they are set, which have spatial scales in the order of 100s to 1,000s of metres. For our initial national land element analysis, we therefore chose to use a fixed kernel size for curvature calculations of 7 × 7, 25 m DEM cells, measuring curvature over a spatial scale of 175 m.

Because the "form elements" defined by this analysis are subject to considerable uncertainty (fuzziness), particularly in terms of the DEM derivatives we use to describe them, we have applied a fuzzy classification system. The rules defining a form element (Figure 1) can then be implemented by applying a fuzzy classification model with a minimum/maximum classifier and AND/OR connections. These algorithms generate membership-value maps for each of the 15 form elements, which can be used to derive a map of form elements from the maximum membership value for each cell. Additionally, uncertainty in the classification can be quantified by the derived maximum membership value, the "confusion index" (Burrough *et al.* 1997), or entropy values (Wood 1996). These maps express the uncertainty of the classification for each classified unit (e.g. raster cell).

The classification of form elements, based on fuzzy local curvature and gradient characteristics, must also be placed within a terrain context. To do this, a model for terrain position was implemented based on a second, higher-scale landform classification. The BLACK TOP HAT approach (Rodriguez *et al.* 2002) was extended to produce fuzzy membership values for grid cells in a DEM, which are a measure for landscape position (Schmidt *et al.* 2003). The algorithm requires three global (spatially homogeneous) parameters: two window sizes identifying the horizontal spatial scales of valleys and hills and an elevation threshold defining the minimum magnitude for valleys and hills (i.e. how much hills/valleys are exposed-above/submerged-below the surrounding area). These measures can be estimated from the DEM. In similar fashion to the form elements analysis, the membership functions are mapped (i.e. hills, hillslopes, and valleys) and a combined surface based on the maximum membership of each of these three landscape contexts derived.

This approach worked very well in experimental areas, but presents issues of variable landform scale, as well as being a significant performance bottleneck when mapping larger areas. The TOPHAT measure is quite spatially coarse, and certainly does not require a particularly high resolution DEM to acquire satisfactory results. Experiments with a variety of resampled DEMs indicated that a resolution of 100 m, as opposed to 25 m, gave results that were effectively indistinguishable, while the performance gain achieved by reducing the number of cells in the TOPHAT neighbourhood analysis by a factor of 16 was substantial. During processing the 25 m DEM is resampled to 100 m, TOPHAT is run, and then the result is interpolated back to a 25 m grid. For our initial national scale analysis, we chose to fix the three global parameters to the following values (hill radius = 1.5 km, valley radius = 1.5 km and elevation threshold = 20 m) rather than attempting to implement spatially varying scale parameters.

The final step in the landform element classification is to combine the higher-scale, landscape position, TOPHAT model with the form element model. Planar, flat landform elements can occur on ridges, backslopes (as terrace treads), and in valley bottoms (both as valley bottom and as pronounced terrace treads). These significantly different landscape units are not distinguished by the form element model alone. The solution lies in the consideration of the higher-scale landscape context: a flat, planar area on a hilltop, for example, would be a ridge, or in the case of more extensive areas, a plateau. Utilizing the landscape position model and form element model provides up to 45 different combinations of position and element (3 × 15). In reality the soil scientist cannot deal with this level of differentiation in landform elements, so a rule set is used to derive a final set of landform elements simplifying this complexity down to 10 classes by combining similar landform elements. This rule set can of course be modified according to need.

For the current S-Map project, we have initially used a 25 m resolution national DEM for New Zealand (Barringer *et al.* 2002) to generate a national coverage of landform elements (Figure 2). Although a fixed spatial scale approach has limitations, the scale chosen broadly matches the scale at which soil scientists are currently most confident at large-scale delineation of soil variability in hilly and steepland terrain. This process is carried out in a series of embedded Arc Macro Language (AML) scripts using workstation Arc/Info version 9.1 (© Environmental Systems Research Institute), which divides New Zealand into tiles of 30 km by 40 km according to the map sheet layout of the 1:50,000 scale 260 series topographic map sheets (Land Information New Zealand). Once the processing for each main island is completed, the tiles are reconstructed into a single Arc/Info Grid for each of the North and South islands of New Zealand.

4 Linking Soils to Landscapes

The final phase of the soil-landscape modelling process involves linking a set of soil-landscape rules for each land system with the landform element map to derive a spatial representation of soil classification. This is achieved in ArcGIS in grid (raster) format using a simple coding/lookup classification system. But, in order to achieve compatibility with the S-Map data structure for the traditionally polygon-based mapping of lowlands, ArcGIS "focal majority" and "gridpoly" functions are applied to the soil classification grid to produce a vector (polygon) soil class map. The distribution of soil attributes is then derived from their association with soil classes or by soil pedotransfer functions. Derived soil information may include soil classification (family, sibling, New Zealand Soil Classification); physical properties such as soil depth, texture and soil water storage capacity; classification of functional horizons; chemical properties, and soil fertility.

Currently, soil-landscape models are being derived based on conceptual modelling as used in traditional soil survey. This practice relies heavily upon expert pedological knowledge. It is currently the primary methodology because in almost all cases there are insufficient measured soil data from the field to drive empirical modelling systems.

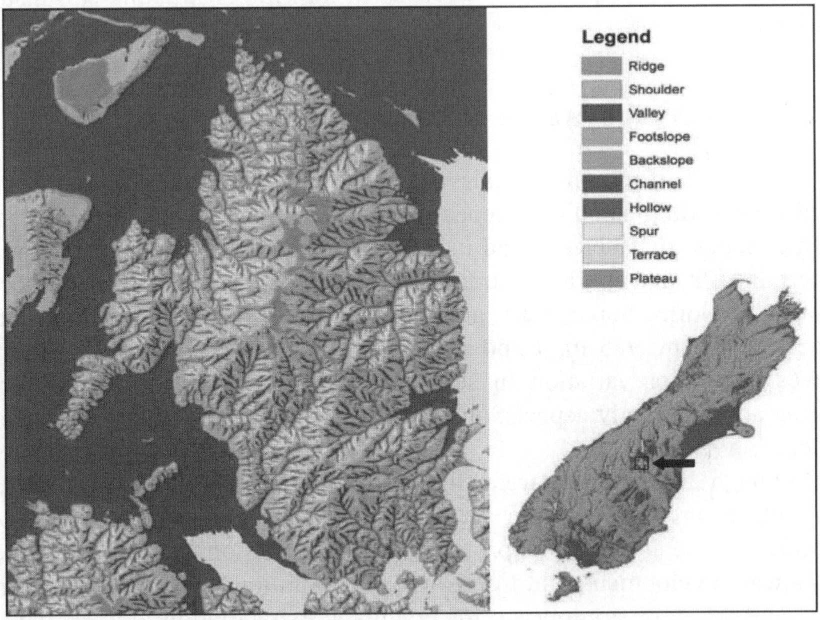

Figure 2. Left - Part of the national landform element map of New Zealand. The Benmore Range is a dry greywacke landscape, and the landform elements shown are based on the combined TOPHAT landscape position model and form element model (local slope and curvature). The final set of landform elements contains 10 classes describing the most common landform features. Right - The landform elements for the South Island of New Zealand based on the combined TOPHAT landscape position model and form element model (local slope and curvature) with the Benmore Range arrowed.

5 Results

We discuss the results of mapping soils for two land systems. One is a relatively simple steep mountain land with simple geology (dry greywacke mountains) and approximately 125 soil sample points. The other is a mixture of steepland and hill country with complex geology and significant climatic and topographic variation containing some seven land systems. Of these, one has 65 sample points, another 25 sample points, but the remainder have less than 10 soil sample points per land system. However, previous published and unpublished work has been used to derive soil-landscape relationships, which are used in turn in constructing soil distribution rules.

5.1 Simple dry greywacke mountainous terrain

In the dry greywacke mountains where rainfall is less than 700 mm, the landscape is dominated by a repeating pattern of angular ridges with long planar slopes, sharp crests, and narrow valley floors. The soil pattern was determined from a stratified sampling programme focusing on four aspects (magnetic North, South, East, and West), at low, medium, and high altitudes (c. 650 m, 985 m, 1,260 m). Variation in leaching and weathering drives the major variation in soil properties with more developed and deeper soils on shady aspects, and with increasing elevation on all aspects (McIntosh *et al.* 2000).

Although 20 soil series are recognized in the Benmore Range model establishment area, these were amalgamated into four clearly defined soil families. These are based on profile depth, texture, stoniness, degree of pedological development, and trends in base saturation and phosphate retention, and are more appropriate for pragmatic extrapolation to areas where data are scarce or absent. In this land system, concave profile footslopes are rare, shoulder slope components are generally of restricted extent, and the area of low-gradient slope components is also limited; hence the DEM-derived "backslope" and "footslope" landform elements were combined to distinguish "backslopes"; and all "ridge", "shoulder" and "spur" landform components were combined to distinguish "ridges" (Tables 1 and 2, and Figure 3).

Table 1. Soil-landscape model for dry greywacke steepland land system (Benmore Range). See Table 2 for soil details.

Backslopes and Footslopes on elevations less than 750 m	
South	Tg (80%) + Om (20%)
North, West, East	Om (65%) + Tg (35%)
Backslopes and Footslopes on elevations between 750 and 1,100 m	
North	Tg (70%) + Om (30%)
East	Tg (60%) + Bm (40%)
South	Bm (90%) + Tg (10%)
West	Bm (70%) + Tg (30%)
Backslopes and Footslopes at elevations above 1,100 m	
North	Br (50%) + Om (25%) + Bm (25%)
South	Bm (45%) + Tg (45%) + Br (10%)
West + East	Bm (60%) + Br (25%) + Om (15%)
Ridge sites at all elevations	
All aspects	Om (85%) + Tg (15%)
Easy rolling plateau	
All aspects	Bm (100%)

Table 2. Explanation of soil classification and soil characteristics for soils of the dry greywacke steepland land system.

Soil code	*Series name*	*Characteristics*	*NZSC classification*	*Soil taxonomy*
Om	Omarock	Shallow, high base saturation, low P retention	Typic Rocky Recent	Orthents, Haplustrepts
Tg	Tengmeyer	Deep, fertile, high base saturation, low P retention	Typic Immature Pallic	Haplustrepts
Bm	Benmore Acid	Deep, acid, low base saturation, moderate-to-high P retention, low fertility	Acid Orthic Brown	Dystrudepts

Preliminary testing of the soil-landscape relationships on the adjacent Kirkliston Range, and the <750 m elevation component on the Doctors Range, and in the Wadsworth Stream catchment (Figure 3), 200 and 300 km north of the model establishment area, respectively, show good agreement, giving confidence that the model developed near the southern extremity of the dry greywacke land system can be extrapolated to the whole land system (Lynn *et al.* 2002).

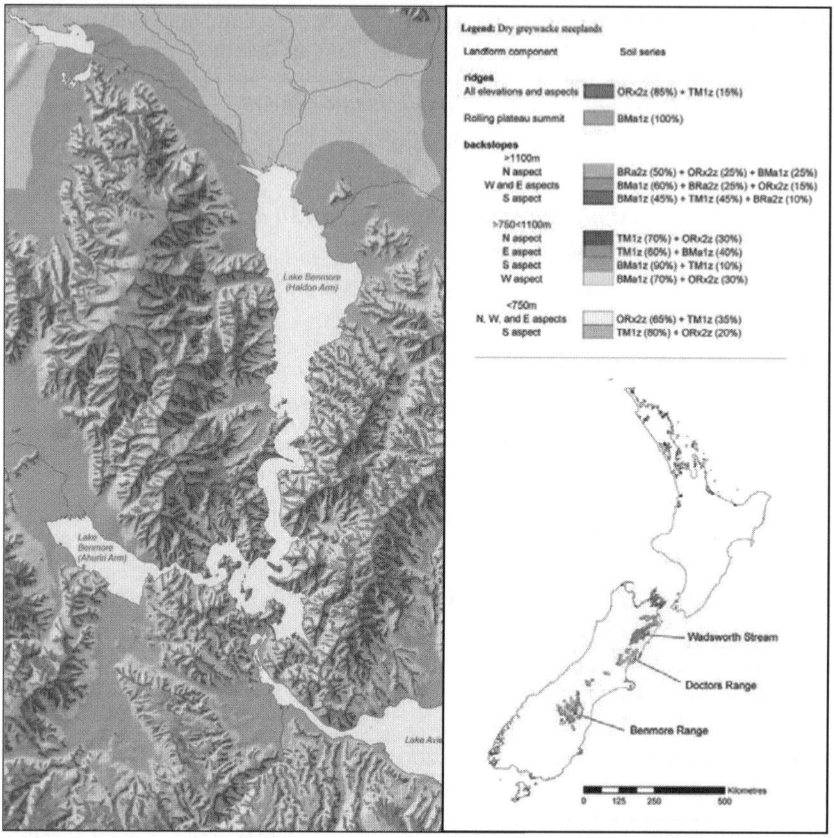

Figure 3. Left - The distribution of soils based on landform elements and a soil-landscape model for the Benmore Range (dry greywacke landscape). Right - Shows the extent of this land system (1.1 million ha), which represents approximately 4% of the total land area of New Zealand.

5.2 Mountainous terrain with multiple parent materials and variable climate

The overall landscape in the Nelson-Motueka region is much more complex, ranging from high mountains in the west and south through to hill country and rolling hills nearer the coast. There is a steep rainfall gradient from less than 950 mm to greater than 3,500 mm over a distance of about 30 km near the main ranges. Parent materials range from early Permian ultramafic and indurated sandstones, siltstones and argillites and Devonian and Cambrian marbles, limestones, greywackes, argillites and schist in the mountains, to younger Cretaceous but deeply weathered and highly erodible granites and granodiorites in the western foothills, and thick deposits of weakly indurated, deeply weathered gravels of late Pleistocene to early Pliocene age in the central Moutere Hills terrain.

Land systems are defined by grouping rock types of similar compositions, induration, and erosion pattern. Within each land system, a soil-landscape model is derived by grouping landform elements with similar soil associations. The models take the form of lookup tables that relate grouped landform elements to soil associations. The soils in these associations are described in detail in the S-map database.

Because tectonic uplift rates are generally high in the western mountains, soil distribution is strongly related to erosion patterns. This means that the soils are strongly related to land surface form, and the soil patterns that can be resolved at a scale of 1:50,000 are readily related to the land units delineated by the national map of landform elements. Grouping of landform elements differs between land systems, which reflects differences in the dominant soil-landscape processes. Within the granite land system, for example (Adams 1970, Chittenden *et al.* 1966), the soil distribution is mapped by only two groups of landform elements, but divided further by slope, lithology, and degree of weathering between Separation Point Granites and Separation Point Granodiorites (Tables 3 and 4).

Table 3. Soil-landscape model for granite steepland land system (Motueka catchment). See Table 4 for soil details.

Land element	Soils map unit label
Granite – Rolling and low hills – very strongly weathered and leached soil	
Steep slopes (>25 deg)	PoP (80%) + PoS (20%)
Crest, ridge, spur, shoulder, plateau	KaP (60%) + KaD (40%)
Side slope and hollow	KaD (70%) + KaP (30%)
Granite - Dominantly steeplands – rainfall > 1,400 mm per year	
Crest, ridge, spur, shoulder, plateau	PoP (60%) + PoS (20%) + PoD (20%)
Side slope, hollow, terrace	PoD (70%) + PoP (30%)
Granodiorite - Dominantly steeplands – rainfall less < 1,400 mm per year	
Very steep slopes (> 35 deg)	NgL (80%) + OrP (20%)
Crest, ridge, spur, plateau	OrP (60%) + OrD (40%)
Side slope, hollow, terrace	OrD (70%) + OrP (30%)

The relatively simple soil pattern in the granite landscape contrasts with a more complex pattern of the adjacent marble karst land system (Bell 1973), which uses three groupings of landform elements, slope, and three soil moisture regimes to map soil distributions, which includes associations of up to five soils occurring together within a single mapping unit.

Table 4. Explanation of soil classification and soil characteristics for soils of the granite steepland land system.

Symbol	Soil	Characteristics	NZ Soil Classification	Soil Taxonomy
KaD	Kaiteriteri deep	Deep colluvium	Typic Yellow Ultic	Hapludalt
KaP	Kaiteriteri sap	Moderately deep saprolite	Typic Yellow Ultic	Hapludalt
OrD	Oronoko deep	Deep colluvium	Typic Orthic Brown	Hapludalf
OrP	Oronoko sap	Moderately deep saprolite	Typic Orthic Brown	Hapludalf
NgL	Ngatimoti lithic	Thin colluvium on rock	Typic Orthic Brown	Dystrudept
PoD	Pokororo deep	Deep colluvium	Acidic Orthic Brown	Dystrudept
PoS	Pokororo stony	Deep stony colluvium	Acidic Orthic Brown	Dystrudept
PoP	Pokororo sap	Moderately deep saprolite	Acidic Orthic Brown	Dystrudept

6 Discussion

Our aim is to establish a structure within which to update and express spatially our national soils knowledge. This is a daunting task for a small group of pedologists and spatial modelling staff. The limited resources available mean that it will be many years before S-Map is completed. Therefore, we need to construct a database structure and modelling system that will provide a sound foundation for the future. That structure also needs to be pragmatic, so that we can make useful progress with limited knowledge now, while also being flexible and "automated" (as far as possible) so that structural improvements, new input layers, improved modelling processes or methods, or improved soil-landscape models can all be utilized to easily upgrade existing S-map data. In this context, we have recognized the (current) *ad hoc* approach to constructing soil-landscape rule sets and models. We have investigated modelling approaches such as EXPECTOR (Lilburne *et al.* 1998) and SOLIM (Zhu *et al.* 2001). In most land systems, we fall well short of having sufficient data points to drive geostatistical analysis techniques or similar data-driven analytical approaches. Approaches that utilize a more objective approach to rule-based modelling such as SOLIM are attractive since they offer a more repeatable, transparent and robust approach to deriving soil-landscape models. Such approaches are also more likely to lead to more consistent model structures, which will assist in automation of soil-landscape model implementation, both for new land systems and for improved/updated existing models. It is also important to recognize that the need for a flexible approach to soil-landscape modelling will not change over time, since for more remote mountainous areas, particularly in the South Island, it is unlikely that there will, in the foreseeable future, be sufficient justification to collect the sort of raw data required to implement rigorous geostatistical analyses.

With respect to DEM terrain analysis, we have recognized some potential problems. Terrain parameters are sensitive to the algorithm used in DEM analysis and the quality of DEM data (Schmidt *et al.* 2003). There is also a strong dependency of terrain features on the scale of observation (Schmidt and Hewitt 2004). For example, a particular point in the landscape can be part of a backslope, a small hill, and a larger valley all at the same time, depending on the scale of observation. This means in order to model landform features fully we need to consider the multi-scale character of landscapes (Schmidt and Andrew 2005). However, up to the present, none of the potential innovations that we have investigated have resulted in an improved solution over the basic land element analysis described in this chapter. Attempts to determine a dominant landscape scale have so far tended to recognize the larger features as dominant, and resulted in oversimplification of the landform elements on hillslopes (spurs, hollows, etc.) that have been recognized as significant features at the nominal 1:50,000 scale of S-map soil-landscape modelling.

Notwithstanding the comments above, we have found that mapping landform elements at a fixed scale has been a useful advance in depicting soil variation across complex landforms. Even where there appear, at least qualitatively, to be clear changes in overall landscape scale (e.g. rolling hills to steep hills to mountains) the scale-invariant landform elements seem to provide a useful and consistent view of landform elements at all of these scales.

Landform elements have thus proven useful for grouping similar areas of landscape with similar soils into aggregated mapping units (e.g. ridges, shoulders, and spurs) for mapping soil distribution within land systems. At the same time, landform elements permit us to map soil distribution at a scale that effectively disaggregates what would have been homogeneous soil mapping units (polygons) in conventional soil surveys. This means that using landform elements as a primary carrier of soils information provides significantly improved knowledge of soil distribution on hillslopes in complex landscapes, and over a wide range of land systems in New Zealand.

Acknowledgments

The authors would like to thank Linda Lilburne, Trevor Webb and David Pairman for the comments on drafts of this chapter, and Christine Bezar for editorial checking and corrections. The research described here is supported by the New Zealand Government through the Foundation for Research, Science and Technology under contract C09X0306.

References

Adams, J.A., (1970), *A study of soil sequences in relation to the growth of Pinus radiata in Nelson*, unpublished PhD thesis, Lincoln University, Canterbury, New Zealand.

Barringer, J.R.F., McNeill, S. and Pairman, D., (2002), Progress on assessing the accuracy of a high resolution digital elevation model for New Zealand, *Proceedings 5th International Symposium on Spatial Accuracy Assessment in Natural Resources and Environmental Data*, 10–12 July, (2002), Melbourne: 187–195.

Bell, C.J.E., (1973), Mountain soils and vegetation in the Owen Range, Nelson. 1. The soils, *New Zealand Journal of Botany,* 11: 49–72.

Burrough, P.A., van Gaans, P. and Hootsman, R.J., (1997), Continuous classification in soil survey: spatial correlation, confusion and boundaries, *Geoderma,* 77, 115–135.

Chittenden, E.T., Hodgson, L. and Dodson, K.J. (1966), Soils and agriculture of Waimea County, New Zealand, Soil Bureau Bulletin 30, Wellington: DSIR.

Clayden, B. and Webb, T.H., (1994), Criteria for Defining the Soilform - the Fourth Category of the New Zealand Soil Classification, Landcare Research Science Series No. 3.

Dikau, R., (1989), The application of a digital relief model to landform analysis in geomorphology, In Raper, J. (ed.): *Three Dimensional Applications in Geographical Information Systems*, London: Taylor and Francis.

Dymond, J.R., De Rose, R.C. and Harmsworth, G.R. (1995), Automated mapping of land components from digital elevation data, *Earth Surfaces Processes and Landforms,* **(20)**2: 131–137.

Gessler, P.E., Moore, I.D., McKenzie, N.J. and Ryan, P.J., (1995), Soil-landscape modelling and spatial prediction of soil attributes, *International Journal of Geographical Information Systems*, **(9)**4: 421–432.

Grunwald, S. (ed.) (2006), Environmental Soil–Landscape Modelling: Geographic Information Technologies and Pedometrics, London: Taylor & Francis.

Harmsworth, G.R., Dymond, J.R. and McLeod, M., (1995), Automated mapping of soils in hilly terrain using DTMs: a New Zealand example, *ITC Journal,* 2: 87–94.

Hewitt, A.E., (1998), New Zealand Soil Classification, Landcare Research Science Series No. 1, Lincoln: Manaaki Whenua Press.

Leathwick, J., Wilson, G., Rutledge, D., Wardle, P., Morgan, F., Johnston, K., McLeod, M. and Kirkpatrick, R., (2003), *Land Environments of New Zealand*, Ministry for the Environment, Auckland: David Bateman.

Lilburne, L., Hewitt, A.E., McIntosh, P.D. and Lynn, I.H., (1998), GIS-driven models of soil properties in the high country of the South Island, *Proceedings of the 10th Colloquium of the Spatial Information Research Centre*, University of Otago, New Zealand: 173–180.

Lilburne, L., Hewitt, A.E., Webb, T.H. and Carrick, S., (2004), S-map : a new soil database for New Zealand, *Proceedings of SuperSoil (2004): 3rd Australian New Zealand Soils Conference*, Sydney, Australia.

Lynn, I.H., Lilburne, L.R. and McIntosh, P.D. (2002), Testing a soil–landscape model for dry greywacke steeplands on three mountain ranges in the South Island, New Zealand, *Australian Journal of Soil Research,* 40: 243–255.

McBratney, A.B., Mendonça Santos, M.L. and Minasny, B., (2003), On digital soil mapping, *Geoderma,* 117: 3–52.

McIntosh, P.D., Lynn, I.H. and Johnstone, P.D., (2000), Creating and testing a geometric soil–landscape model in dry steeplands using a very low sampling density, *Australian Journal of Soil Research*, 38: 101–112.

McLeod, M., Rijkse, W.C. and Dymond, J.R., (1995), A soil-landscape model for close-jointed mudstone, East Cape, North Island, New Zealand, *Australian Journal Soil Science,* **(33)**3: 381–396.

Rodriguez, F., Maire, E., Courjalt–Rade, P. and Darrozes, J., (2002), The Black Top Hat function applied to a DEM: a tool to estimate recent incision in a mountainous watershed (Estibere Watershed, Central Pyrenees), *Geophysical Research Letters,* **(29)**6: 91–94.

Schmidt, J., Evans, I.S. and Brinkmann, J., (2003), Comparison of polynomial models for land surface curvature calculation, *International Journal Geographical Information Science*, **(17)**8, 797–814.

Schmidt, J. and Hewitt, A.E., (2004), Fuzzy land element classification from DTMs based on geometry and terrain position, *Geoderma,* 121: 243–256.

Schmidt, J., Tonkin, P. and Hewitt, A.E., (2005), Quantitative soil-landscape models for the Haldon and Hurunui soil sets, New Zealand, *Australian Journal Soil Research*, 43: 127–137.

Schmidt, J. and Andrew, R., (2005), Multi-scale landform characterization, *Area,* **(37)**3: 341–350.

Shary, P., Sharaya, L.S. and Mitusov, A.V., (2002), Fundamental quantitative methods of land surface analysis, *Geoderma,* 107: 1–32.

Webb, T.H., (2003), Identification of functional horizons to predict physical properties for soils from alluvium in Canterbury, New Zealand, *Australian Journal Soil Research*: 1005–1019.

Wilson, J.P. and Gallant, J.C. (eds.) (2000): *Terrain Analysis: Principles and Applications*, New York: John Wiley.

Wood, J., (1996), *The geomorphological characterisation of digital elevation models*, Ph.D. Thesis, Department of Geography, University of Lancaster, United Kingdom.

Zhu, A.X., Hudson, B., Burt, J., Lubich, K. and Simonson, D., (2001), Soil mapping using GIS, expert knowledge, and fuzzy logic, *Soil Science Society America Journal,* 65: 1463–1472.

Concluding Remarks

Progress in Digital Terrain Analysis

Brian G. LEES

We have been very fortunate to have been able to put together a volume with a wide range of international expertise. I note with interest that the TADTM initiative was the first of a number of similar projects that have revitalized interest in digital terrain analysis. Digital elevation modelling was one of the earliest application areas of GIS and it is still one of the most important. The growth of digital terrain analysis, as distinct from digital elevation modelling, highlights the difference, so long ignored, between a DEM and a DTM.

George Miliaresis' generous linking of his geomorphometry group to the TADTM community (geomorphometry@yahoogroups.com) means that, in parallel with this publication, there is an ongoing, and very vital, exchange of ideas in this field. This means that my concluding remarks here are not a static milepost along the road to the next physical meeting of researchers interested in digital terrain analysis, but one item in a continuing discussion. Whilst discussing that, it is worth mentioning Peter Shary's (http://www.giseco.info/) research homepage with its wealth of resources. We are extremely fortunate to have the introductory chapters in the book from these two eminent researchers.

So, what has preparing this volume shown us? Importantly, it demonstrates the massive movement into the international research community of our Chinese colleagues. The very significant commitment to research and to higher education in China is becoming evident in the size and quality of the postgraduate research population. This, coupled with the contribution from more senior researchers who originated in China, but now help staff research centres and universities world wide, gave TADTM a distinct flavour.

Not only that, but a keen interest in environmental problems and the management of the environment, has brought a strong practical focus to the volume. Most of the theory discussed is related to practical applications. This makes the volume an extremely valuable resource.

It is appropriate at this point in the volume to consider what lies ahead; at least to see what the current state-of-the-art as presented here suggests might lie ahead. Several chapters dealt with the SRTM-3 globally available data and the modifications necessary to make it useable at the scales we believe necessary for vegetation and erosion modelling. Other papers discussed other approaches to global data sets. A clear message is that many of the problems we face, both environmental and social, need higher resolution data over more extensive areas. Why is this? With an increasing population to feed, we need to manage our environments carefully. I also suspect that the next series of reports by the Intergovernmental Panel on Climate Change will firm their views on sea level rise, rainfall intensity, and cyclone/typhoon intensification. Let's consider these in some detail.

It is self-evident that we need much better elevation data to predict the impacts of sea level rise. As the various feed-back mechanisms start to cut in, and become measurable and predictable, estimates of sea level rise will almost certainly increase from the modest IPCC (2007) predictions of up to 0.59 m by 2099, which policymakers clearly believe not to be a significant threat, to possibly an order of magnitude greater over time. Once IPCC forecasts start to include future dynamical changes in ice flow, policymakers will be knocking on our doors wanting to know if we can show them where the impacts will fall. At present, with current data sets, we are unable to do this with any precision.

Similarly, modelling the impacts of more frequent, and more intensive, cyclones/typhoons really requires better digital elevation data and better models of how inundation works as a process. Simply raising sea level, digitally, and colouring it blue is not an adequate solution when millions of people are at risk.

The other phenomenon I mentioned above, erosion, also changes with changes in climate. Most of the models of climate change predict an intensification of rainfall events. More rainfall, falling in a shorter period of time, is going to have significantly more effective erosional and transporting power. Hillslopes will be affected, as will river channels. A steepening and shortening of the flood hydrograph will, in many places, make occupation of river floodplains untenable. Increased levee rupturing and crevasse splays will sterilize valuable cropland.

All of these processes are processes which we, as a group, model with a greater or lesser degree of success. Social expectations of our ability to provide useful predictions for planning and policymakers are certain to rise. There is a lot to do to meet these expectations. Crude modelling, such as the USLE or RUSLE, will not suffice beyond the initial reconnaissance level. Neither will crude estimates of inundation. These things are not beyond us. We saw, at the meeting, colleagues discussing global data sets of various sorts. A decade or so ago such widespread access to DEM data was only a dream. Now that it is becoming a reality we can move forward to apply our craft to giving those who fund us the support and information they expect in dealing with the serious problems which confront us all.

Intergovernmental Panel on Climate Change (2007), *Climate Change 2007: The Physical Science Basis; Summary for Policymakers,* IPCC Secretariat, c/o WMO, 7bis, Avenue de la Paix, C.P. N° 2300, 1211 Geneva 2, SWITZERLAND. 18p.